LAKE CHILWA

MONOGRAPHIAE BIOLOGICAE

Editor

J. ILLIES
Schlitz

Volume 35

Dr W. Junk bv Publishers The Hague–Boston–London 1979

LAKE CHILWA

Studies of change in a tropical ecosystem

Edited by

MARGARET KALK,
A. J. McLACHLAN
& C. HOWARD-WILLIAMS

Dr W. Junk bv Publishers The Hague–Boston–London 1979

The distribution of this book is handled by the following team of publishers:

for the United States and Canada

Kluwer Boston, Inc.
160 Old Derby Street
Hingham, MA 02043
USA

for all other countries

Kluwer Academic Publishers Group
Distribution Center
P.O. Box 322
3300 AH Dordrecht
The Netherlands

Library of Congress Cataloging in Publication Data   CIP

Lake Chilwa: studies of change in a tropical ecosystem.

(Monographiae biologicae; v. 35)
Papers written by participants in the Lake Chilwa Co-ordinated Research Project, University of Malawi, 1966–1976.
Bibliography: p.
Includes index.
1. Lake ecology – Chilwa, Lake, Malawi and Mozambique. 2. Chilwa, Lake, Malawi and Mozambique. I. Kalk, Margaret. II. McLachlan, A.J. III. Howard-Williams, C. IV. University of Malawi. Lake Chilwa Co-ordinated Research Project. V. Series.

QP1.P37 vol. 35 [QH195.M47]   574'.08s [574.5'2632]   79-16288
ISBN 90-6193-087-1

Computer enhanced colour composite photograph of the Chilwa Basin, taken on 8th October 1973 during a minor recession of the lake. Lake Chilwa is in the centre with the lake bed exposed on north, west and south. The red colour indicates actively growing plants such as algae in Lake Chilwa and vegetation around the mouths of the Domasi and Likangala rivers. Where the water has receded from the swamp, the vegetation above ground is dead and appears yellow. A comparison with Fig. 2.1 will enable geographical features to be identified. (By courtesy of the Earth Satellite Corporation, sponsored by the World Bank for the Malawi Government.)

Contents

Part 1. The environmental setting

Part 2. The response of plants and animals to changes

Part 3. The people of the Chilwa area Page

Part 4. Conclusions

Participants in the Lake Chilwa Co-ordinated Research Project
University of Malawi, 1966-1976

Authors in this monograph are indicated by an asterisk

*Swanzie Agnew, M.A., F.R.G.S.,
c/o African Studies Department, University of Edinburgh, Scotland.

J. M. Berreen, B.Sc.,
Zoology Department, University of Birmingham, England.

D. M. Bourn, M.Sc.,
Animal Ecology Research Group, University of Oxford, England.

*H. D. Brown, D.Sc.,
Head, Red Locust Control Service, Pretoria, South Africa.

*M. A. Cantrell, Ph.D.,
Biology Department, University of Malawi, Zomba, Malawi.

A. Chilivumbo, Ph.D.,
Department of Sociology, University of Zambia, Lusaka, Zambia.

*C. Chipeta, Ph.D.,
Department of Economics, University of Malawi, Zomba, Malawi.

A. Cockson, A.I.M.L.T.
Zoology Department, University of Western Australia, Perth, Australia.

*C. O. Dudley, Ph.D.,
Biology Department, University of Malawi, Zomba, Malawi.

*H. R. Feijen, Ph.D., Biology Department, Eduardo Mondlane University, Maputo, Mozambique.

Cobi Feijen,
Biology Department, Eduardo Mondlane University, Maputo, Mozambique.

*M. T. Furse, B.Sc.,
Freshwater Biological Association, River Laboratory, East Stoke, Wareham, Dorset, England.

*C. Howard-Williams, Ph.D.,
Institute of Freshwater Studies, Rhodes University, Grahamstown, South Africa.

Wendy Howard-Williams, M.Phil.,
Institute of Freshwater Studies, Rhodes University, Grahamstown, South Africa.

McG. Hutcheson, Ph.D.,
Department of Geography, University of Aberdeen, Scotland.

*Margaret Kalk, Ph.D.,
Professor Emeritus of Biology, University of Malawi, c/o Zoology Department, University of the Witwatersrand, Johannesburg, South Africa.

*R. G. Kirk, Ph.D.,
Fisheries Division, Department of Agriculture, European Economic Community, Brussels.

*N. Lancaster, Ph.D.,
Department of Geography, University of the Witwatersrand, Johannesburg, South Africa.

Margaret Magendantz, Ph.D.,
School of Tropical Medicine and Hygiene, London.

*A. J. McLachlan, Ph.D.,
Department of Zoology, University of Newcastle-upon-Tyne, Newcastle, England.

Sandra M. McLachlan, M.Sc.,
Department of Medicine, Wellcome Research Laboratory, University of Newcastle-upon-Tyne, Newcastle, England.

A. Morgan, M.Sc., F.R.I.C., F.R.S.H.,
S. African Bureau of Standards, Pretoria, South Africa.

*P. R. Morgan, Ph.D.,
Blair Research Laboratory of Parasitology, Salisbury, Rhodesia.

*B. Moss, Ph.D.,
School of Environmental Studies, University of East Anglia, Norwich, England.

Joyce Moss, B.Sc.,
c/o School of Environmental Studies, University of East Anglia, Norwich, England.

N. P. Mwanza, Ph.D.,
Project Co-ordinator, Programme Activity Centre for Environmental Education and Training (PACEET) in Africa, Nairobi, Kenya.

F. Nicholson, Ph.D.,
Director, McGill Sub-Arctic Research Laboratory, Schefferville, Canada.

*B. Pachai, Ph.D.,
Director of International Education Centre, St. Mary's University, Halifax, Canada.

Pauline Phipps, M.Sc.,
Postgraduate School of Social Sciences, Yale University, Boston, U.S.A.

*G. G. M. Schulten, Ph.D.,
Royal Tropical Institute, Amsterdam, The Netherlands.

Toos Schulten-Senden,
c/o Royal Tropical Institute, Amsterdam, The Netherlands.

J. Shroder, Ph.D.,
Department of Geology, University of Omaha, Nebraska, U.S.A.

*D. E. Stead, Ph.D.,
Department of Biology, University of Malawi, Zomba, Malawi.

*D. Tweddle, B.Sc.,
Fisheries Department, Ministry of Natural Resources, Lilongwe, Malawi.

T. D. Williams, Ph.D.,
Department of Economics, University of Edinburgh, Scotland.

Acknowledgements

The multi-disciplinary Lake Chilwa Co-ordinated Research Project, as a whole, was supported by the Leverhulme Trust, London for six years. Besides financing research needs of the staff members of the University of Malawi, these grants subvented three successive Research Fellows and their field/ laboratory assistants. The Centre for Overseas Pest Control, London supplied a locust ecologist and his equipment for three years.

Most field and laboratory equipment was obtained through grants from the University of Malawi and from the Overseas Development Administration, England. All the transport requirements: boats, engines, a research cabin cruiser and Land Rovers were donated by WENELA African Interest Fund, South Africa (in recognition of the contribution by Malawians in the South African gold mines); it also assisted in the expenses of the preparation of this monograph.

Membership of the International Biological Programme (Productivity Freshwater and Human Adaptability) enabled several members of the Chilwa Project to participate in international conferences through the generosity of the IBP(PF) U.K.

The Fisheries Research Officers in Malawi shared much of their equipment and field staff with the Chilwa Project and thanks are particularly due to D. Eccles, R. G. Kirk, A. Mtotho, C. Ratcliffe and J. Stoneman, to whom we are grateful for advice and assistance. We wish to acknowledge the considerable kindness of members of the Meteorology Department, Water Resources Division, Geological Survey, Department of Surveys, Ministry of Agriculture and Natural Resources (particularly the soil analysis laboratory at Bvumbwe), the National Statistical Office and the Government Printer. The South African Atomic Energy Agency kindly analysed samples for radio-activity. In each of the Chief's Areas where we worked, the administrators and officials were helpful and the villagers co-operative, which considerably facilitated our research.

The participants of the Chilwa Project are very grateful for the competent work of the field and laboratory assistants in many disciplines: C. A. L. Bai, H. Bafuta, G. H. Chikwapula, M. Chirwa, W. J. Jinazali, J. E. A. Lupoka, L. J. Kalilombe, M. S. Kamanja, S. M. Kuntambila, B. W. Makwiti, A. Mpote, S. F. Nchema and T. D. Thawale and several others who assisted for short periods.

On behalf of the Lake Chilwa Co-ordinated Research Project, the editors wish to express their thanks for the unofficial invaluable expert laboratory and field work contributed by the graduates, Patricia Bradley, Cobi Feijen, Debbie Furse, Wendy Howard-Williams, Sandra McLachlan, Joyce Moss and Toos Schulten-Senden, several of whom were co-authors of Chilwa publications. Dr. T. G. Dyer kindly subjected our data on lake levels and rainfall to his mathematical model of spectral analysis for prediction, at the University of the Witwatersrand. Mr. E. A. K. Banda of the University of Malawi Herbarium helped considerably with plant identifications. We are particularly grateful to Judith Lancaster for redrawing and standardizing all maps and figures in this monograph and we wish to thank Zubaiola Jeewa for typing the manuscript.

The participants in the Project and their addresses are fully listed.

The Chancellor of the University of Malawi, Ngwazi, His Excellency Dr. H. Kamuzu Banda, Life President of Malawi gave the Project his encouragement at its inception. Our thanks must also be recorded for the many tangible expressions of support by Chancellor College of the University of Malawi, in addition to research grants.

Through the Royal Society, London we have been fortunate in having the advice of a limnologist with long experience in Central Africa, Professor Leonard Beadle. He spent some weeks at the University of Malawi and Lake Chilwa in 1967, soon after the Research Project had started, and helped to define our objectives more clearly. The editors wish to thank Professor Beadle for his guidance and encouragement and for writing a Foreword.

List of donors

*Centre of Overseas Pest Research, London.
Fisheries Research Department, Zomba, Malawi.
International Biological Programme (PF), U.K., London, England.
International Red Locust Control Centre, Lusaka, Zambia.
*Leverhulme Trust, London, England.
*Overseas Development Administration, London, England.
Royal Society, London, England.
S.A. Atomic Energy Agency, South Africa.
The Employment Bureau of Africa Ltd. (WENELA Division).
U.K. Atomic Energy Agency, London, England.
*University of Malawi, Zomba, Malawi.
*WENELA African Interest Fund.

*Major Financial Support

Foreword

Leonard C. Beadle

In contrast to the more stable oceans, inland waters are, on the geological time scale, short-lived and are subject to great fluctuations in chemical composition and physical features. Very few lakes and rivers have existed continuously for more than a million years, and the life of the majority is to be measured in thousands or less. Earth movements, erosion and long-term climatic changes in the past have caused many of them to appear and disappear. No wonder then that most freshwater organism are especially adapted to great changes and many even to temporary extinction of their environment.

Recent studies of residual sediments from existing and extinct lakes in tropical Africa have told us much about their age and the past history of their faunas and floras, from which we may deduce something about the climate and the conditions in the water in the past. The forces that have formed and moulded the African Great Lakes have been catastrophic in their violence and effects. They are not yet finished, but the present rate of change is, in human terms, too slow for direct observation of the ecological effects. The large man-made lakes are providing very good opportunities for studying the chemical and biological consequences of the initial filling but, once filled, they are artificially protected against major fluctuations. At the other extreme are small patches of water, especially in arid regions, which appear at irregular intervals and disappear within a week or two or even a few days. Their fauna and flora are very restricted.

Lake Chilwa is so situated that the annual fluctuations in level and water composition, with recessions of the lake every six years or so and two periods of complete desiccation during the past century, have had exceptionally great ecological consequences. The number of species that have survived this ordeal in drought resistant forms or by taking refuge in peripheral swamps and streams is greater than might be expected. Events are thus proceeding at a pace admirably suited to a research project of a few years' duration. The entire cycle of changes can be studied more than once without undue haste. It was therefore fortunate, though at first sight depressing, that by the time plans were completed and the Lake Chilwa Co-ordinated Research Project properly launched the lake had all but disappeared. It was, in fact, a lucky chance that a major recession was included in the programme, giving the opportunity for studying the process of recovery.

Dr. Margaret Kalk, at the time Professor of Biology in the University of Malawi, must be congratulated for recognizing the Chilwa basin as an object for concerted study likely to produce results of scientific and practical value, and for directing the energy and enthusiasm of the young scientists engaged in the project. The entire basin with its inflowing streams and periodically flooded lands has very properly been treated as one ecosystem, in which *Homo sapiens* is clearly the most influential of all the species. There are several small closed lake basins in Africa and in other continents that are subject to similar drastic changes, but none, so far as I know, involves to such an extent the economy of a country. Chilwa in its 'good' years is one of Malawi's productive regions for agriculture, fisheries and stock-raising. Its 'bad' years are correspondingly catastrophic. The practical importance of basic research is thus unusually obvious.

Few readers will fail to be interested in the main themes developed in this book – the geological, climatic and human history of the basin, the origins of the lake's fauna (especially the fish), the fluctuations in the fauna and flora in response to the great changes in the environment, and the effects of these on the welfare of the people of the region.

Which of the many simultaneous chemical and physical changes associated with a swing in climate is the direct cause of an observed change in the number or behaviour of a species is always a difficult question for ecologists and has led to much speculation both reasonable and unreasonable. In principle, it can be solved by experiments, preferably based on some knowledge of the relevant physiology of the organism. But to plan, perform and interpret an experiment that is relevant to the natural situation is a formidable task and may require sophisticated equipment, though much can be done by simple means in a field laboratory or even in the open. With a few exceptions this approach has rarely been adopted in investigations on the biology of tropical African Lakes. It is therefore gratifying to see that some of the many problems of Lake Chilwa have been tackled experimentally.

Beside giving us an interesting description of the cycle of changes which the fluctuations of the climate impose on the ecology of the basin, this work has made it clear that Chilwa offers a grand opportunity for investigating some fundamental problems whose bearing on important practical matters needs no special emphasis. We hope that others will follow the lead.

Leonard C. Beadle
Honorary Senior Research Fellow
University of Newcastle-upon-Tyne
Emeritus Professor of Zoology and
Wellcome Research Professor,
Makerere University, Uganda

Part 1. The environmental setting

1 Introduction: Perspectives of research at Lake Chilwa

Margaret Kalk

1 Introduction: Perspectives of research at Lake Chilwa

Lake Chilwa is in the Southern Region of Malawi, a small land-locked country in the east of Central Africa (Fig 1.1). The country is called Malawi today, because it had been inhabited by the Maravi People since the 13th Century. Their name has been adapted by transliteration of the consonants to those used in Chichewa, one of the official languages, to spell 'Malawi'. From 1891 to 1964 it was called Nyasaland, after its very large and deep Lake Nyasa, now Lake Malawi; Nyasa was the Yao word for 'lake'.

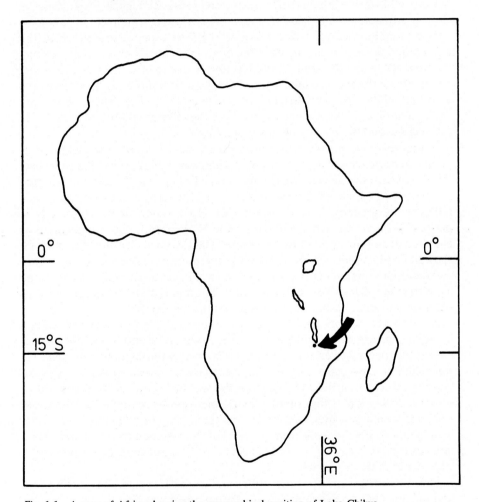

Fig. 1.1 A map of Africa showing the geographical position of Lake Chilwa.

The southern end of the western Rift Valley of Africa holds lakes Malawi, Malombe and Shire River (a tributary of the Zambesi river), but Lake Chilwa lies 50 km to the east of it and is 100 km southeast of the tip of Lake Malawi. Lake Chilwa is at an altitude of about 622 m above sea level and has no outlet to the sea (see Frontispiece and Fig. 2.1).

Lake Chilwa can be seen from any vantage point on the Shire Highlands of Southern Malawi, occupying the centre of hot low-lying plains about four times the area of the lake at its high level. The lake is greyish, turbid, often turbulent, less than 5 m deep and almost 2000 km² in area at the end of the wet season when in flood. The plains around the lake are almost treeless, with meandering streams and a few conical hills (inselbergs); they extend towards the highland in all directions. A broad belt of dense swamp surrounds the lake, in which the dominant plant is *Typha*, a plant known as reedmace, cattail or bulrush. A floodplain around the swamp becomes inundated in the wet season. These three areas: open water, swamp and floodplain have roughly equal proportions.

Lake Chilwa is classified by the International Biological Programme in Project Aqua as an A II lake, that is, an important research site in an almost natural condition (Luther & Rzóska 1971). The ecosystem with which this monograph is concerned, includes the open lake, the swamp and marshland, the floodplains and the affluent streams. It is bounded by a beach terrace about 4–5 m above the average lake level of today and adjoins drier plains, where a dense population lives by farming, fishing and fish trading. These people are dependent on the lake and the areas where they live are therefore included in the socio-economic studies of this monograph.

Explorers of the nineteenth century gave a vivid account of the lake and its people in those days. The water of the lake in April 1859, at its highest level at the end of the rainy season, was higher than it has been in the last one hundred and twenty years, since Livingstone and his party viewed it from Mt. Myupyu, a hill near the western shore (Wallis 1956). Open water then surrounded the peripheral islands of Nchisi in the west and Njalo in the south, which are now engulfed in swamp. A painting by Moore (1903) who visited the lake in 1888, showed Chidiamphiri (Python) Island in the northwest surrounded by water, whereas today it is enclosed in swamp. The water was described by Livingstone as 'bitter, but drinkable'. Later in 1888, Drummond (1903) described it as brackish and undrinkable, but 'relished by wild animals'.

A different description of the lake was given by O'Neill (1884) who walked along the eastern shore, in October 1883, towards the end of the dry season, having approached overland from the Indian ocean through Mozambique. He was told that a few years before it had been possible to walk over the dry lake bed during a dry period (1879), from the western shore to Nchisi Island (a distance of 3–5 km at that time). Elderly men spoke of a time in their youth when the swamp was much narrower than then. They recalled that at one time men had been able to walk from one end of the lake bed to the other. This is indeed credible because the eastern lake bed is firm sand and a recession of the lake would make it possible.

These are the first written indications that the lake level of Lake Chilwa rose and fell and that occasionally part of the lake bed was dry.

There was a tendency in some early reports to overestimate the size of the lake, since its length and breadth could not properly be judged from a nearby hilltop. Several recent visits to the same hilltops, confirm that the distant shores could not be seen clearly and the low hills visible on the eastern horizon and the massive Mt. Mulanje in the south are many kilometres from the water's edge. Livingstone seemed to have assumed that Lake Malombe, which is about 50 km distant to the northwest and an expansion of the Shire river at its exit from Lake Malawi, was a continuation of Lake Chilwa (Wallis 1956). In 1883 O'Neill walked around the northern shore of Lake Chilwa across the sandbar which separates it from Lake Chiuta, and was the first to report that Lake Chilwa had no outlet to the sea (O'Neill 1884) (Frontispiece)

Even so, the open water in the mid-nineteenth century was both longer and wider than today. As the water level of the receding lake fell during the last hundred years (probably not gradually but intermittently after spells of low rainfall), the periphery of open water was replaced by the dense growth of *Typha* more especially on the north and west. Nevertheless the boundary of the floodplain, marked by the 4–5 metre terrace has not changed. During the intervening years periodic recessions of the water exposed various amounts of the lake bed at different times (Chapter 3). At the end of the wet season in 1978 the lake had again risen and spread over the floodplain, to the 4–5 m beach terrace almost to the high level of 1859.

In the nineteenth century, the plains were teeming with elephant, buffalo, zebra, antelope and buck of several kinds. The scene was described enthusiastically in 1888 by Drummond (1903). 'Nowhere else in Africa did I see such splendid herds of the larger animals. The zebra was especially abundant . . .' During his visit to Lake Chilwa, Drummond had been' . . . much surprised to discover near the lakeshore . . . a path so beaten by multitudes of human feet that it could represent only some trunk route through Africa. . . . Due to the ravages of the slaver, the people of Shilwa [Chilwa] are few, scattered and poor. . . . The densest population is to be found on a small island [Nchisi Island], heavily timbered with baobabs, which form a picturesque feature. . . .' Until the end of the nineteenth century, the surrounding hills and plains, including the sandbar, which separates Lake Chilwa from the freshwaters of Lake Chiuta were well-wooded (O'Neill 1884). Now the larger mammals have all but disappeared, the hills and plains have become deforested, the people have increased in number, settlement on the land is stable, crops are grown, cattle find pastures on the plains and there is now a thriving fishing industry in the lake and swamp. Some aspects of this transition are described in the various chapters of this book.

1. The Chilwa scene in modern times

The plains extending down from the hills to Lake Chilwa are now heavily cultivated and only isolated areas of the original *Brachystegia* savanna remains on the highest parts. As the soils become progressively more sandy towards the upper beach terraces (indications of a larger, deeper prehistoric lake), savanna type vegetation with the silver-leaved *Terminalia sericea* trees and scattered *Combretum* shrubs occur with tracts of open grassland. As one travels down towards the lake across the old lake terraces on to the poorly drained soils, a few huge trees of *Acacia albida*, *Combretum imberbe* and the peculiar 'sausage tree' *Kigelia africana* appear. In the northern and southern regions of the lake, groves of the yellow-barked 'fever tree' *Acacia xanthophloea* can be seen and on the edges of water courses, large figs (*Ficus* spp.) are found. The remaining trees become shorter and shorter as one approaches the lake, and the grasslands become more extensive. The trees disappear at about the level of the

4–5 m terrace, the nominal boundary of the ecosystem, and now the floodplain takes on the appearance of flat grassland, interspersed with many small clumps of bushes. Closer inspection reveals that each clump is associated with a termitarium where the soils are better drained. Eventually even the 'anthills' and their bushes disappear and finally the floodplain becomes a wide flat expanse of short grasses and sedges over which large herds of cattle roam in the dry season.

Near the water's edge the Lake Chilwa *Typha* swamp begins as a dark, high green wall of vegetation demarcating the end of the floodplains. On the lake side, the swamp ends abruptly, but small clumps or 'islands' of vegetation can be seen extending out into the lake itself. These are composed of the tall sedge *Scirpus littoralis*, and the grass, *Paspalidium geminatum*. Only a few perennial rivers penetrate directly through the swamp to the lake. Many others lose themselves in the extensive swamps. Nevertheless, the swamps are penetrated from the open water by a large number of man-made channels, which traverse the swamp from the lake to the marshes and to villages on the edge of the floodplain. Some of these lead to large lagoons or man-made clearings where fishes congregate and subsistence fishing can continue during the dry season.

Due east across the lake is Nchisi Island, the remnants of a long extinct volcano, of which half has been eroded away, exposing a semicircular 'arena' below the hilltops. The lower slopes are speckled with baobab trees (*Adansonia digitata*), and the hillsides are densely clothed in limestone vegetation and there is no *Brachystegia*. The picturesque 'candelabra tree', *Euphorbia ingens* stands out from the rest.

Beyond Nchisi Island, looking east, the eye travels blankly across the featureless water, missing out the distant low-lying opposite shoreline, and the ends of the lake to the north and south are invisible from the shores or a boat on the water. Looking north, the twin islands of Tongwe interrupt the water surface, but the peripheral islands of Njalo to the south and Chidiamphiri (Python) on the northwest are engulfed in swamp.

At night the lake offshore is dotted with fishermen in dug-out canoes, and by day the few beaches are crowded with fishermen, drying mats, smoking kilns and traders with their baskets and bicycles (Fig. 1.2). These small pockets of activity on the lake fringe interrupt the smooth green of the swamp. At the largest fish distribution centre at Kachulu (opposite Nchisi Island) where the road from Zomba ends, an embankment was constructed in 1961 to enhance the use of a natural harbour in the swamp fringe. Fish catches are weighed there by the Fisheries Department and sold to traders. Beyond it a wooden jetty slopes down to the seasonally changing water level.

Fishing is a thriving industry, but it is still almost entirely carried out by traditional methods, using dug-out canoes, which are propelled by a single man with a long pole. Fish-traps, long-lines and nets of various kinds are used, and yet in 1965, almost half the total catch of fish in the whole country was landed at Lake Chilwa (Eccles 1970). In 1976, 20 000 tonnes was caught in this way, and fish trading is a major occupation. Supplies are taken by traders with baskets and bicycles on lorries and buses to the growing urban population in the newly industrialised areas on the Shire Highlands and to the workers in the tea plantations.

Since the nineteen sixties, the population of the Chilwa area has become one

8

of the densest in rural Africa, south of the Sahara. The people still use hoes to cultivate the maize, cassava, rice, cotton, tobacco and pulses; and they keep chickens, goats, pigs and some cattle, although very, very few are used as oxen to prepare the ground or for transport.

Fig. 1.2   The fishing beach at Kachulu in 1965 (photo: W. Plumbe).

2. The problems of the Chilwa area and the research aims

In 1965, when the new University of Malawi was opened, a closer look showed that the whole Chilwa area was beset with problems. Oral history recalled periodic and sometimes devastating recessions of the lake waters when fishing became impossible. Fish, which had usually been plentiful were becoming less so and seemed stunted. It had been suggested that the swamp was 'rapidly advancing and choking up the lake' (Garson 1960). A fisheries cold storage development scheme had been closed down for lack of freshwater for the ice plant.

Rainfall over the plains had been uneven and unreliable from year to year. Frequent plagues of red locust had destroyed the crops. The lake itself was saline and the very few boreholes were brackish. Water supplies were inadequate for domestic use. Malaria, enteritis, bilharzia, hookworm and leprosy were common. Natural radio-activity on Nchisi Island had been considered a health hazard (Weir 1962).

Communications and transport were very difficult. The area was isolated: it had been a peaceful backwater for eighty years, since slavery had ended, because since the beginning of the century the main axis of communication through the country had been developed by water, road and rail from south to north. The Chilwa basin lay apart, bordering on Mozambique. It was served by one poor third-class road to Zomba, then the capital of Malawi, and by a seasonally usable track to Mulanje, the centre of the tea plantations. There was no electricity, nor telephone and few buses. Travel was by foot or bicycle. Hospitals, schools, markets and shops were some distance away, 30 km to Zomba, 40 km to Phalombe on the southern plain and 60 km to Mulanje up the southern escarpment, yet the population was growing rapidly by immigration as well as by natural increase.

There were rumours of conflict between rice-farmers and cattle keepers, since both used the floodplain. It was hinted that the deflection of water from rivers to rice farms might lower the lake level. At that time the Chilwa area did not figure in any national development plans.

Against such a backdrop in 1966, the small staff at the new University of Malawi formulated a research policy which, for the sake of economy of time, effort and funds, required convergence on a single focus in a multi-disciplinary project, so that members of staff from many disciplines might gain from co-operation with one another. The academic staff, mainly foreign but including some Malawians, would thus achieve some insight into their totally new environment. It was the policy of the University of Malawi to relate the teaching and expertise of the staff to local problems, since it was hoped that in this way graduates might better contribute to national development. Advice from the representatives of various Government Departments on the University Consultative Committee for Natural Resources encouraged the University to undertake a predevelopment survey of the Chilwa basin. It was believed that the lake and plains offered some potential for the future and that problems needed study. A multi-disciplinary investigation was formally launched as the Lake Chilwa Co-ordinated Research Project under the directorship of Professor Margaret Kalk. This monograph reports its first ten years work, 1966–76.

In this setting, the first programme of the Research Project, drawn up in 1966, proposed the study of four broad aspects of the ecosystem: 1. the biology of the lake and swamp; 2. the morphology and aging of the lake; 3. the ecology of the plains with regard to land use, and the nature of the potential 'outbreak' area of the red locust; 4. the social organisation, the economy and the health problems of the people on the plains who were engaged in fishing, trading, farming and stock-keeping (Kalk 1970).

The initial broad objectives of the Research Project involved seven departments. Then, as will be seen, events overtook us when the lake dried up in 1968 and refilled in 1969. These have given the studies a particular emphasis on changes in the environment and their effects on the biota, including man. Catastrophic changes often occur in the tropics where fishing and farming are important, and our findings are of wider interest than at first envisaged. Smaller changes in depth, salinity and alkalinity are features of many tropical lakes, more especially in the Rift Valley of Africa (Beadle 1974) and the inland lakes of Australia (Bayly & Williams 1966), as well as in man-made lakes. There are also some features in common with temperate endorheic lakes such as the Neusiedlersee in Austria (Lottler 1974).

3. Natural and social changes

It soon became apparent that every facet of life in the ecosystem depended in some way upon the fluctuating lake level (Kalk 1970). The responses of the biota to the environmental fluctuations were often clearcut but immensely varied. Therefore the research reported in this monograph reflects responses to natural environmental changes of four kinds: catastrophic, periodic, seasonal and long-term geological change; and in Part IV, social changes are considered.

3.1 *Catastrophic change*

The most dramatic change was the evaporation of this large lake to complete dryness in the dry season of 1968, followed by refilling in one wet season. The fall in level took three years: in the dry season of 1966 canoes had to be pushed through liquid mud to the fishing grounds offshore, but after the wet season the level rose. In the dry season of 1967, soft mud had been exposed in a belt several kilometres wide on the west and northwest around the inner swamp edge (Fig. 1.3). Villagers could walk to and from Nchisi Island; and it was recalled that this had not been possible for over thirty years. In 1968 all that remained of the lake was one main body of water about 150 km² in area in the southeast sector, where the water had been almost two metres deeper than the rest. In the rainy season, pools were scattered over the vast expanse of mud. Finally, there were ten weeks in the 1968 dry season when the lake was completely dry over the whole lake bed. This was confirmed by aerial reconnaissance. Cattle wandered over the lake bed on the eastern foreshore, and terrestrial isopods were found beneath their dung. Water remained in many lagoons in the swamps, however, and the few perennial rivers trickled slowly, sometimes damned up in their deltas. Then, in five months of rainfall, not

Fig. 1.3 Kachulu Bay with dry lake bed in 1968 (photo: M. Kalk).

unusually heavy, but early, the lake refilled to a 'normal' level of 2–3 m (Fig. 1.4), with a much lower salinity than had been recorded in 1966 and 1967 and the recovery of the fauna and flora commenced.

The changes in the ecosystem at all trophic levels have been investigated over the years 1966–72, during the 'decline and recovery' of the lake: the environmental features were monitored and the effects on the fauna, the flora and the people have been studied. Sometimes clear relationships were demonstrable and some responses have been investigated in the laboratory. There is evidence that complete drying had happened before in Lake Chilwa in both historic and prehistoric times.

Fig. 1.4 Kachulu Bay at high lake level after refilling in 1969 (photo: M. Kalk).

3.2 *Periodic change*

The Lake Chilwa hydrograph exhibits a pattern of a few years' average annual high levels followed by average annual low levels accompanied by increasing salinity and ending in a minor recession before the cycle starts again. Various degrees of lake recession at Lake Chilwa have been recorded in various ways (by explorers, hunters, duck-shooters, birdwatchers, administrators and by oral testimony of elders and chiefs) about ten times in the last hundred years. Lake levels have been recorded by gauges only since 1950 (Fig. 1.5). In this time the periodic low levels were seen in 1954, 1960, 1966 and 1973. Relatively high maxima were a feature of the 1950s and later 1970s, while the maxima of the nineteen sixties were lower and culminated in the drying of 1968. In 1977 and 1978 higher maxima still have been recorded, almost reaching the level of the 4–5 m terrace around the lake. The possibility of a

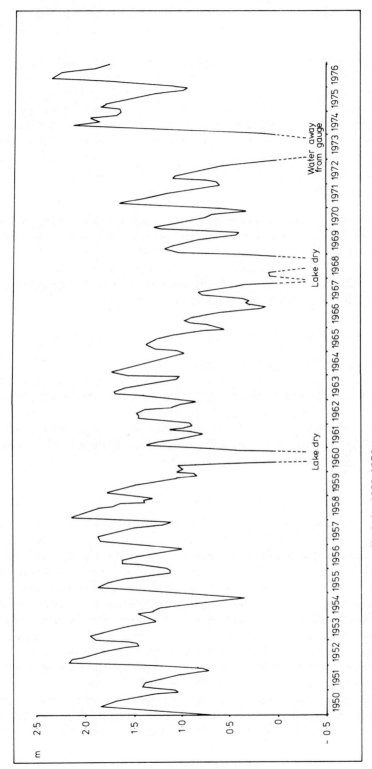

Fig. 1.5 Lake level at Ngangala gauge near Kachulu, 1950–1976.

13

periodicity inherent in the factors which control the lake level has been investigated, and it is hoped that it will be possible to make predictions of recessions (Chapter 3).

The instability of the lake has wide implications for the fundamental features of the ecosystem, such as the food-web and the encroaching vegetation. It affects the fisheries which are of economic importance to Malawi, since they are the main source of dietary protein for the major industrial area as well as for the people on the Chilwa plains. The rainfall pattern, which induces the changing lake, imposes further constaints on the economy, since it places farming in jeopardy at times.

3.3 *Seasonal fluctuations*

Annual variations in depths of Lake Chilwa result in a rise and fall of around one metre (Fig. 1.5). The main direct effect of the fluctuation is on the gradient from the lake's edge through the swamp to the marshes and the flooded grasslands. There are many other seasonal changes: in rainfall, solar radiation and evaporation, temperature, salinity and in turbidity with the influx of silt and detritus, all of which may affect the flora and fauna.

3.4 *Changes in geological time*

Geological evidence points to long-term changes in Lake Chilwa in the Quaternary period. The lake was formed from a river; it dried up and refilled and became endorheic (without outlet). It has shrunk considerably, but with the climate of today it may remain in the present stage of instability for a long time. It may become deeper as have many East African lakes in the past, or it may dry up. The long term changes will have influenced the natural selection of the surviving species of plants and animals.

3.5 *Social changes*

The historic background of this study was the first ten years of the Republic of Malawi. It was a time of slow transition from subsistence economy to partial cash economy, with greater incentives to work for profit. The denudation of the woodlands and the extermination of the mammals, the spread of permanent cultivation and the increase in fish trading had created new problems, but technological changes were beginning in several spheres, such as locust control, rice farming, fishing, use of insecticides, road-building and health measures, all of which received small amounts of international aid. The old-established well-organized matrilineal way of life was giving place to that of a modern nation without tribal rivalries. Migrations, integration, development, and the handling of cash were making contributions to social change.

In the Chilwa area, the major lake recession was superimposed on the national social changes and its effects interacted with these. Socio-economic studies have given a picture of this development.

4. Responses to changes

The authors of chapters in this book have attempted to relate the knowledge gained by the participants in the Chilwa Research Project to some of these five kinds of change in the Chilwa basin, both major and minor. Part I, The Environmental Setting, is introductory. Chapters 2 and 3 describe the background to the problems in the physical environment and the fluctuations in lake level, in both geological and modern times, and their climatic causes. An attempt is made to predict future recessions of the lake from a mathematical model. Chapters 4 and 5 describe the major chemical and physical features of the lake and swamp respectively, in 'normal' times and during the decline and recovery of the lake.

The greater part of the book, Part II, The Responses of Plants and Animals to Changes, concerns the distribution of selected organisms in the lake and swamp and on the floodplain and their varied responses to changes of several kinds. Succession of species, reproductive adaptations, aestivation, stunting, migrations, invasions and eradication are viewed against the permanence of the swamp community. Questions of low diversity of species, variations in populations with time, mechanisms for survival, the immediate causes of mortality and the conditions for recovery, and various evolutionary implications are examined in Chapters 6 to 11.

Chapter 12 examines fisheries techniques, the present state of the fishery industry and the fisheries potential of the lake and swamp in the face of the fluctuations in lake level. It puts forward the proposal of dispensing with any conservation measures since the lake (with the swamp) has its own built-in mechanism for recovery. Chapter 13 highlights the interactions of the swamp and the lake and the dependence of life in the lake on the swamp and its decay. The relationship between the vertebrates (other than fish) and the climatic instability is described in Chapter 14, followed by the effect of social change on the larger mammals. The interplay of some insects and man's activities is illustrated in Chapter 15.

Part III, The People, is concerned with socio-economic studies on the plains and the problems of the area in a time of post-Independence development. A chapter on the historical background introduces the people, so as to reach a better understanding of the human resources of the heterogenous population of intermingled ethnic groups. It traces the story of the welding together of peoples through the centuries, during the successive invasions and absorption of food-gatherers and hunters, fishermen, cultivators and later pastoralists. Their way of life today are described in the next two chapters and a case is made for an 'integrated small-scale economy' of farming, fishing and cattle-rearing, during the gradual transition from quasi-subsistence to a partial cash economy. Developments are reviewed and attention is drawn to some problems including health and education.

Finally in Part IV, the Conclusions have been divided into limnological and social. The threads have been drawn together to see what can be learnt about the biology of this ecosystem and its changes. The factors underlying the success of the impoverished lake fauna and flora and their powers of recuperation from the vigorous selection pressure of the environment are evaluated in the context of the whole lake basin, and a detritus food web, maintained by the permanent *Typha* swamp. Finally, we focus on the problems involved in the future well-being of the people dependent on the lake, and suggestions are put forward on the basis of the realistic planning which has already commenced.

A provisional checklist of animals and plants encountered is given in Appendix A.

References

Bayly, I. A. E. & Williams, W. D. 1966. Chemical and biological studies on some saline lakes of Southeast Australia. Austral. J. Mar. Freshwater Res. 17:177–228.

Beadle, L. C. 1974. The Inland Waters of Tropical Africa. Longman, London. 365 pp.

Drummond, H. 1903. Tropical Africa, 11th edit. Hodder & Stoughton, London. 228 pp.

Eccles, D. 1970. Lake Chilwa Fisheries. In: M. Kalk (ed.) Decline and Recovery of a Lake, Government Printer, Zomba, Malawi. 56–59.

Garson, M. S. 1960. The Geology of the Lake Chilwa Area. Geol. Surv. Nyasaland Bull. 12. Government Press, Zomba, Nyasaland. 69 pp.

Kalk, M. 1970. Decline and Recovery of a Lake. Government Printer, Zomba, Malawi. 60 pp.

Löffler, H. 1974. Der Neusiedlersee. Naturgesichte eines Steppensees. Verlag Fritz Molden, Wien.

Luther, H. & Rzóśka J. 1971. Project Aqua. A sourcebook of inland waters proposed for conservation. I.B.P. Handbook 21. Blackwell Scientific Publications, Oxford.

Moore, J. E. S. 1903. The Tanganyika Problem. Hurst and Blackett, London. 371 pp.

O'Neill, H. E. 1884. Journey from Mozambique to Lake Shirwa and Amaramba. Parts 1, 2 and 3. Proc. Roy. Geogr. Soc. 6:632–655; 713–740.

Wallis, J. R. P. 1956. The Zambesi Expedition of David Livingstone. II. Journals, Letters and Dispatches. Central African Archives, 9. Chatto & Windus, London. 215–462.

Weir, R. N. 1962. High background radiation in a small island. Nature, 194:265–267.

2 The physical environment of Lake Chilwa

N. Lancaster

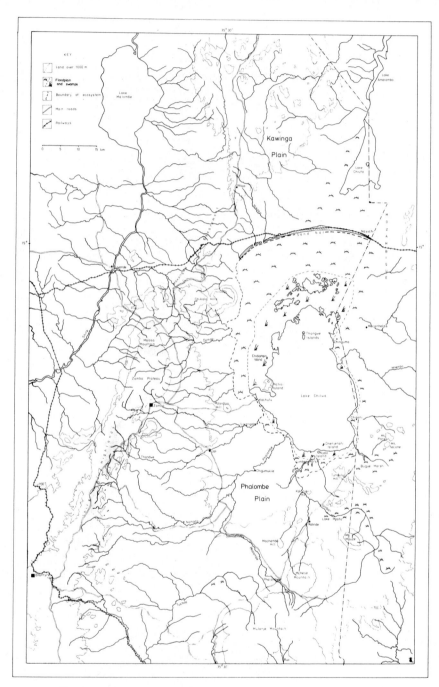

Fig. 2.1 The Chilwa basin showing the features mentioned in the text.

The physical environment of Lake Chilwa

1. General features of the lake

Today, Lake Chilwa is a shallow, enclosed saline lake situated in the southern part of a NE–SW trending tectonic depression, to the east of the main Rift Valley in southern Malawi. It is separated from Lake Chiuta, which is linked to Lake Amaramba and Lugenda River in Mozambique, by a sand bar some 25 m high (Fig. 2.1). The lake lies at a mean altitude (1950–1976) of 622 m above sea level and had an open water area of 678 km² in 1972. This was surrounded by a further 578 km² of swamps and marshes and 580 km² of seasonally inundated grasslands; the areas vary with the level of the lake in any year. The total area, including islands, of the ecosystem with which this monograph is concerned is 1850 km².

The level of the lake fluctuates annually by 0.8–1 m, but larger fluctuations in level, of the order of 2–3 m, occur over periods of 6 and 12 years. When these oscillations coincide, as in 1967–68, the water level falls drastically and the lake may actually dry up (see Chapter 3). Deposits and landforms marking former shorelines of a much larger lake in the Chilwa basin occur at altitudes of 627, 631, 637 and 652 m above sea level. These are described in detail in Section 2.6 and indicate that changes in lake level of a much greater extent have taken place during the late Quaternary, when the former outlet to the Indian Ocean became closed.

The purpose of this chapter is to describe the main features of the physical environment of the Chilwa basin with particular emphasis on the lake and the areas immediately adjacent to it, and to provide a background against which the subsequent discussion of the effects of lake level fluctuations may be seen.

Lake Chilwa lies 100 km southeast of Lake Malawi and is centred on latitude 15°30′S and longitude 35°30′E, making it the most southerly of the major African lakes. Its location relative to the other lakes of East and Central Africa is shown in Fig. 2.2. Unlike many of the lakes in East Africa, Chilwa does not lie in the Rift Valley, and is separated from Lake Malawi and the Shire River drainages by a narrow watershed. Although at present a closed lake, there is good geological and biological evidence that during the late Pleistocene, Lake Chilwa was connected with Lake Chiuta and the Lugenda River, which drains to the Indian Ocean. Chilwa is thus unusual in that it is one of two East African lakes to have had direct connection with the Indian Ocean (the other being Lake Ziway in Ethiopia), a fact of some importance to its biology.

In common with many other lakes in East Africa, Chilwa has become a closed lake in relatively recent times, certainly within the last 15 000 years. Street & Grove (1976) have demonstrated that the majority of lakes between the Tropics in Africa reached very high levels during the early Holocene period, some 8 000–9 000 years ago. This phase of high lake levels was preceded by a period 13 000–15 000 B.P. when lake levels throughout Africa were very low. How far Chilwa has followed the paleoclimatic patterns of these lakes, with high levels in the early Holocene and generally low levels subsequently, is

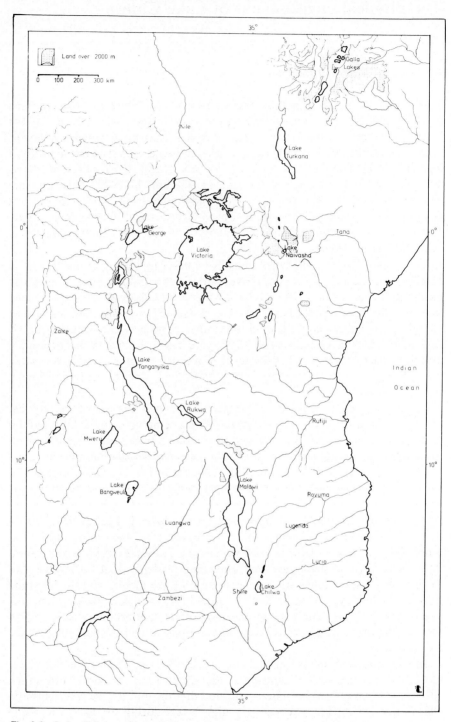

Fig. 2.2 Lake Chilwa and its relationships with the inland waters of East and Central Africa.

not as yet clear. Interestingly, the modern fluctuations of Chilwa discussed in Chapter 3 have not followed the pattern of those of other lakes in East Africa. This is probably the result of both local catchment factors and of regional variations in climate.

2. Physiography of the Chilwa basin

The Chilwa basin (Fig. 2.3) is approximately rectangular in shape and covers an area of some 7500 km². It measures 100 km across at the widest point and 160 km along its north–south axis. The western margins of the basin are the Shire Highlands which trend NE–SW along the edge of the Rift Valley. The Shire Highlands are at an altitude of 700–1200 m above sea level, rising westwards from the Phalombe Plain to the crest of the Rift Valley escarpment. A number of steep sided isolated mountain masses rise out of this area of dissected uplands: Chiradzulu in the southwest (1770 m) and Zomba–Malosa Plateaux in the west (2000 m). The Chikala Hills run due east from the northern part of the Malosa massif and extend close to the edge of the lake. The Shire Highlands is one of the areas of highest population density in Malawi and is extensively cultivated. Most of their former cover of *Brachystegia–Julbernardia* (miombo) woodland has been cleared and remnants survive only in forest reserves. Isolated patches of montane forest and grassland can be found on Zomba and Malosa Plateaux and in the Chikala Hills. The line of the Shire Highlands is continued north of the Chikala Hills by the dissected Makongwa Escarpment zone, which forms a low and narrow watershed between the Chilwa–Chiuta and Shire drainages.

To the east of Lake Chilwa in Mozambique, is an area of low hills at an altitude of 800–900 m. These are broken by large isolated hills, such as Mount Pera and Mount Tecone rising to 1000 m, and the more extensive mountains east of Mecanhelas, which reach altitudes of 1300 m.

The southern margins of the Chilwa basin are rather flat and marshy, with areas of indeterminate drainage between the Tuchila and Phalombe Rivers. The Mulanje massif (2000–2998 m) is a prominent relief feature, but most of its drainage is directed southwards to the Ruo River.

Between the Shire Highlands and the lake lies an area of extensive gently undulating plains, with isolated low rocky hills, called by Stobbs (1973) the Chilwa–Phalombe Plain. The main subdivision of this area is the Phalombe pediplain, an eroded platform at the foot of the Shire Highlands, at an altitude of 670–940 m. This is an area of broad convex divides separating marshy valleys, which merges locally with the area of colluvium eroded from the mountain slopes, to the north of Mulanje Mountain. In the southeast of the Chilwa basin it is continued as the Mozambique Plain, an eastward sloping depositional area with frequent marshes. Soils of the Phalombe pediplain are generally loams, sandy loams and sandy clays, which support good crops of maize, tobacco and cotton.

Towards the lake, at an altitude of 640–660 m, lies the Chilwa–Phalombe terrace, largely formed on old alluvial and lacustrine deposits, frequently overlain by sandy colluvium. In many parts of the Chilwa basin it lies between the 652 and 637 m fossil shorelines (Fig. 2.6) and forms a zone of flat clay plains with marshy alluvial deposits in the lower areas. Cultivation is con-

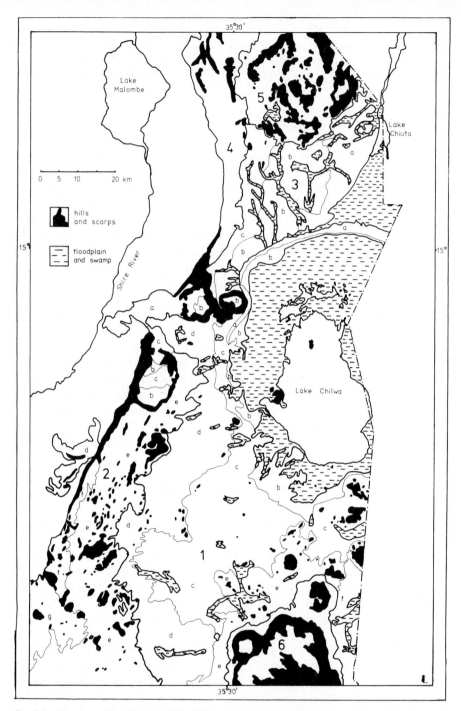

Fig. 2.3 Physiographic divisions of the Chilwa basin (redrawn from Malawi Natural Regions and Areas, sheet 3, Southern Malawi); Key: 1. **Chilwa Phalombe Plain**: (a) Sand Bar; (b) Outwash zone; (c) Chilwa–Phalombe Terrace; (d) Phalombe pediplain; (e) Mulanje Pediment; (f) Mozambique Plain; 2. **Shire Highlands**: (a) Chikala Hills; (b) Dissected high plateaux; (c) Zomba plateau; (d) Zomba–Blantyre escarpment zone; (e) Dissected uplands: (f) Dissected escarpment; (g) Blantyre area; 3. **Kawinga Plain**: (a) Chiuta terrace; (b) Level plain; (c) Undulating plains; 4. **Namizimu escarpment**; 5. **Namwera Plain**: Lungwe Hills; 6. **Mulanje Mountains**.

22

centrated on the areas of lighter, better drained, soils where maize, cassava and groundnuts are extensively grown.

Bordering the Chilwa grasslands and including the area associated with 637, 631 and 627 m shorelines is a zone of mixed lacustrine and fluvial deposits. These consist of the sandy beaches and dunes associated with the 631 m shoreline and the deltaic areas adjacent to the Domasi, Likangala and Phalombe Rivers. Soils vary considerably over short distances, from sandy alluvium or beach deposits offering good settlement sites, to heavy water-logged clays in low lying areas, suitable for rice growing. Between the 627 m shoreline and the fringing swamps and marshes of Lake Chilwa is a zone of seasonally flooded grasslands. These are most extensive north and northwest of the lake, where they support a large cattle population. North of the Chikala Hills, the Chilwa–Phalombe Plain is continued as the Kawinga Plain, a gently undulating area of colluvial soils with broad marshy valleys draining southeast to Lake Chilwa or east to Lake Chiuta. Soils here are generally sandy, and this is a good area for crops of groundnuts, cassava and, more recently, tobacco.

3. Drainage systems in the Chilwa basin

Five major rivers drain into Lake Chilwa from the Shire Highlands and Zomba Mountain (Fig. 2.1). They are, from north to south, the Domasi, Likangala, Thondwe, Namadzi and Phalombe. Between them they contribute 70 per cent of the total inflow to the lake. As they drain an area of intensive traditional agriculture, these streams, particularly the Likangala, Namadzi and Thondwe, carry a high suspended load during the rainy season. Other important influent rivers include the Sumulu and Lingoni from the Chikala Hills; the Sombani, which drains the eastern flanks of Mulanje and Mchese Mountains and enters Chilwa via Lake Mpoto (or Sombani lagoon) and its associated extensive marshes in Mozambique; and the Mnembo and Nalaua rivers which drain the hills east of Mecanhelas. These ten streams are perennial, although the flow of some of them, for example the Phalombe and Sombani, may be very small during October and November, at the end of the dry season. In addition, there are many small seasonal streams draining to the lake, especially on the wetter Malawi side.

The middle courses of all the influent rivers are incised by up to 10 m in the Chilwa–Phalombe Plain, the degree of incision decreasing towards the lake. Many of the smaller, seasonal streams cease to have well defined channels below the 637 m shoreline, and their flow is dispersed by the permeable lacustrine deposits. In their lower reaches the Phalombe and Domasi rivers have well marked terraces, up to 4 and 10 m above the present river level. These are associated with the deposition of alluvium in the river valleys during the 631 and 637 m stages of the lake, which have since been cut into by the rivers as lake levels have fallen.

The Phalombe River has created an extensive delta downstream of Chigumukile, a village 15 km from the lake edge. Abandoned meandering channels, clearly visible on aerial photographs, show that, in the past, the river's course has been up to 10 km north and south of its present position. Smaller deltas have been built by other major influent rivers, particularly the Domasi and the Mnembo.

There are a number of interesting cases of river capture of the upper reaches of some of the rivers draining to Lake Chilwa, which may be of importance in determining the distribution of fish species. In the northwest of the basin the Lingoni River, which once drained Malosa Mountain, has been beheaded by the northwesterly flowing Likwenu River, which enters the Shire River north of Liwonde. In the south of the Chilwa basin the Tuchila River, a tributary of the Ruo, has diverted a number of streams draining the northern slopes of Mulanje Mountain from the Phalombe River (see Fig. 11.1). This appears to be a relatively recent event, as drainage in this area is still indeterminate, and there are extensive areas of marsh. The inflow of water into Lake Chilwa may thus have been somewhat reduced.

4. Geology

Most of the Chilwa basin is underlain by ancient metamorphic and igneous rocks of the Malawi Basement Complex (Fig. 2.4). These are exposed in the Shire Highlands and along the Makongwa Escarpment, and dip below the lacustrine and alluvial deposits of the Chilwa–Phalombe Plain. Rocks of an essentially similar nature also underlie the eastern parts of the basin in Mozambique. The Basement Complex is represented in the Shire Highlands by a group of high grade metamorphic rocks, mostly charnokitic granulites of quartz and feldspar, with a NE–SW trend. North of the Chikala Hills in the Makongwa Escarpment, and around Mulanje Mountain these are replaced by biotite

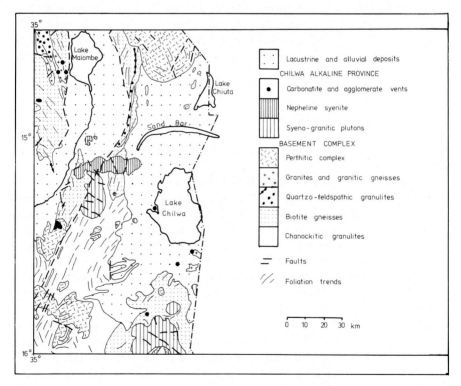

Fig. 2.4 The geology of the Chilwa basin (redrawn from the Provisional Geological Map of Malawi 1973).

gneisses, coarse grained crystalline rocks derived from granite, which may weather to give rather sandy soils. Intruded into the gneisses and granites are a series of Pre-Cambrian granitic and perthitic rocks, which frequently form prominent hills, such as Chiradzulu Mountain and the Lungwe Hills north of Lake Chiuta.

Lake Chilwa has given its name to a group of Upper Jurassic to Lower Cretaceous (140–100 m.y.) alkaline igneous intrusions, volcanic vents and dykes of a distinctive composition, described by Dixey *et al.* (1955) and Garson (1965). The intrusions are of two types: granites and syenites (which contain less quartz), which form the prominent mountain masses of Zomba–Malosa and Mulanje, as well as smaller features such as Mpyupyu Hill; and nepheline syenites (with still less silica and richer in sodium and potassium) which form the Chikala Hills.

The volcanic vents are very distinctive in composition. At Tundulu, south of Lake Chilwa, there is a complex of alkaline silicate rocks, carbonatites, rocks rich in sodium and calcium carbonates, and agglomerates, which once formed the centre of a large composite volcano, similar to Mount Elgon in Kenya, Garson (1965). One igneous centre is associated with apatite, a potentially important source of raw material for phosphate fertilizer manufacture. Nchisi Island is the largest carbonatite centre. Here, ring like intrusions of sovite, feldspathic breccia and agglomerates enclose an area of carbonatites. Pyrochlore rich dykes cut the complex, and are a potential source of the metal Niobium. Alkaline dykes occur widely throughout the area, particularly northeast of Mulanje and Mchese Mountains, and in the Zomba area, but their effects on relief are of local importance only. Hot springs at Mpyupyu and off Nchisi Island represent the final remnants of this period of igneous activity.

A large part of the Chilwa basin is underlain by Quaternary alluvial and lacustrine deposits, which increase in depth eastwards to a line extending from Nayuchi, on the northeast of the sandbar, to the Phalombe River. They were laid down in a slowly subsiding NE–SW trending downwarp, on an ancient line of crustal weakness, closely associated with the initiation and development of the Shire Rift Valley. According to Bloomfield & Young (1961), the rift was in-itiated in the Late Cretaceous or Early Tertiary times. During the late Tertiary there was further movement along the Rift Valley faults, and the area to the east of the Rift Valley, including the Shire Highlands, was tilted to the east. It was probably at this time that the Chilwa–Chiuta depression was initiated. Further fault movements, of the order of 230 m, took place during the Quaternary, and were accompanied by further uplift in the Shire Highlands and relative down-warping of the central part of the Chilwa–Phalombe Plain, leading to the devel-opment of a lake in the southern part of the Chilwa–Chiuta depression.

The sequence of deposits is variable, reflecting the changing balance be-tween fluvial and lacustrine sedimentation in the basin. Commonly, close to the present influent rivers, the deposits are mainly fluvial, with sands and silts dominating. Elsewhere lacustrine clays and sandy clays occur. Sand lenses in the deposits provide local water supplies from shallow wells. More reliable aquifers, often sub-artesian, are found in deeper and thicker sandy deposits. However, the most reliable supplies seem to come from an extensive sand layer just above the bedrock surface.

5. Morphology of the lake

Lake Chilwa is approximately oval in shape and, at average levels, measures some 40 km from north to south and 30 km from east to west. The area of open water varies with lake level. For example, in 1969 it was 594 km², but in 1972 it had increased to 678 km², as a result of a rise in lake level of only 0.4 m. At the mean lake level of 622 m, the area of the lake is approximately 600 km². In the

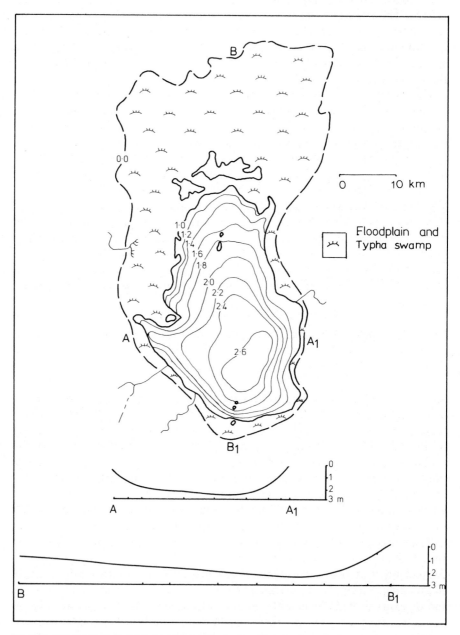

Fig. 2.5 The bathymetry of Lake Chilwa. Vertical profiles are shown below: west to east, A–A₁; north to south, B–B₁. Depths in metres (after McLachlan et al. 1972).

wet season of 1978 the lake level rose about 2 metres higher than this and flooded large areas of the plains. Lake Chilwa is a shallow lake, as the map in Fig. 2.5 shows. Its greatest depth is in its southeastern section. It shelves most rapidly on its southern and eastern margins, but very gently to the north and northwest. The lake bed is generally a thick layer of fine-grained clays, easily stirred by wave action, but on the eastern side of the lake and adjacent to the mouths of the Likangala, Phalombe and Sombani rivers it is sandy.

There are a number of islands in the lake, only two of which are inhabited permanently. The largest is Nchisi Island, off Kachulu in the west, which measures 4 km across and rises 430 m above lake level, and is large enough to support four villages. In the northern part of the lake are the Thongwe Islands rising to 147 m above lake level, and Chidiamphiri Island (Python Island). Off the southern shore are Njalo Island, an important fisheries beach, and the low Chenjerani Islands.

The open water is surrounded by an area of *Typha* swamp and marsh. This area increased lakewards temporarily, when the lake receded (1967–1968) (see Chapter 7) and the surrounding grassland flooded in years of high lake level (1976–1978). The latter provides good grazing for cattle in the dry season. In the north, certain sandy areas are breeding grounds for the red locust (see Chapter 15). The width of the swamp and marsh zone is variable. It is greatest, at 15 km, in the north and northwest, and on the western shore north of Kachulu. On the eastern side of the lake and south of Kachulu it is much narrower, (1–2 km), except in the embayments of the Limbe Marsh. The distribution of the swamp seems to be strongly influenced by the bottom contours, and its width reflects the area of water less than 2 m deep. The swamp zone restricts access to the lake. In many areas the barrier is overcome by the cutting of boat channels through the swamp by local fishermen. Beaches from which fishing can take place have a very restricted distribution, and often their sites vary over time, as the distribution of swamp varies with changes in lake level. The most important beaches are Kachulu in the west, Swang'oma and Njalo Island in the south and Chinguma on the northeastern shore.

The north–south orientation of the lake coincides with the direction of the two dominant winds affecting the area, northeast and southeast. As a result, wave action might be thought to be an important factor in shaping its shores. However the fringing swamps absorb most of the wave energy and only locally, as on the eastern shore of Nchisi Island, do wave cut beaches exist. Elsewhere, most of the higher sandy areas adjacent to the lake, which provide settlement sites, can be related to the 627 m fossil shoreline.

6. The age and evolution of the lake

The origins of the Chilwa–Chiuta basin, in the southern part of which lies Lake Chilwa, may be traced back to the close of the Cretaceous Period, some 65–70 million years ago. At this time the area of the present Upper Shire valley and the Shire Highlands was occupied by a broad up-arched highland area. Drainage from this flowed east and north to an ancestral Phalombe–Lugenda River, which drained northwards along the axis of the present depression.

According to Bloomfield (1965) the main period of rift valley formation occurred in the early Tertiary. As a result, the Upper Shire area to the west of

Zomba Mountain became lowered by at least 1100 m. This and subsequent smaller movements during the later Tertiary and Quaternary (see Section 2.4), were accompanied by uplifts along the Shire Highlands axis, and relative downwarping of the central part of the Chilwa–Phalombe Plain. The ancestral Phalombe–Lugenda River was thus unable to maintain its course across the basin, and was ponded back to form the ancestral Lake Chilwa. The formation of the ancestral Lake Chilwa thus resembles in many ways the formation of Lake Victoria, with ponding back of rivers crossing the area of the present lake by tectonic movement.

Evidence from boreholes in the area around the present lake tends to support the hypothesis that a deeply cut river system once crossed the basin. The depth of sediments at 40–80 m is greatest along the axis of the basin and in the vicinity of major influent rivers. In many areas, the character of the sediments in the lower parts of the boreholes is typically fluvial rather than lacustrine. For example, Dixey (1933) reports 'upwards fining sequences of gravels, sands and clays typical of braided rivers', in boreholes in the area north of the Phalombe River.

In the early history of the Proto–Lake Chilwa there is no evidence, geological or biological, to support Dixey's contention (Dixey 1926) that Proto–Lake Chilwa was ever linked to Lake Malawi, or that Lake Malawi ever had an outlet via the Chilwa area to the Ruo River. Equally there is no evidence to support the suggestion of Garson & Walshaw (1969) that Proto–Lake Chilwa was linked to Lake Malawi and that the Malawi–Chilwa system drained southeast to the Indian Ocean near Quelimane.

Much more detailed information about the later stages of the evolution of the Chilwa–Chiuta basin and the events leading to the formation of the closed Lake Chilwa exists in the form of high level lake shorelines and associated deposits, the existence of which was first recognized and described by Garson (1960) and continued in Garson & Walshaw (1969). They have recently been re-examined in the light of new data, particularly the precise benchmarks around the basin.

Four main shorelines may be identified in the Chilwa–Chiuta basin. They lie at altitudes of 652 m, 637 m, 631 m and 627 m above sea level, or 30, 15, 9 and 5 m above mean lake level (622 m). Their location is shown in Fig. 2.6 and their relationships by the shoreline profiles in Fig. 2.7.

The 652 m shoreline is best developed between Mpyupyu Hill and the Chikala Hills. In this area there is a well marked shoreline slope 10–20 m high, at an altitude of 645–665 m, cut into partially weathered or unweathered Basement Complex gneisses. Frequently, spreads of quartzite or laterite pebbles similar to the Dwangwa gravels of Lake Malawi, mantle the shoreline at an altitude of around 652 m. Elsewhere this shoreline is poorly defined, and often covered by colluvium. In the southern parts of the basin the 652 m lake extended around Mount Mauze and included Lake Mpoto, on the Sombani River, and its associated marshes. At this stage Chilwa and Chiuta were a single open lake with an area of some 5500 km² draining to the Indian Ocean by the Lugenda River.

The 652 m lake was succeeded by one at an altitude of 637 m. It is probable that a long interval of low lake levels intervened, during which colluvium locally covered the 652 m shoreline, and laterites were formed in lacustrine

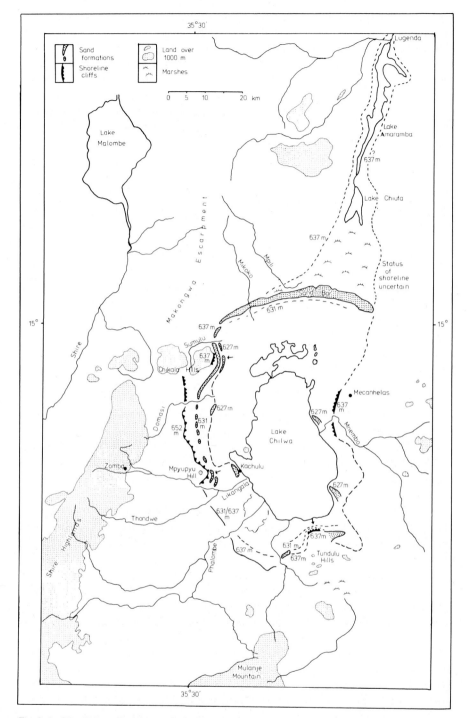

Fig. 2.6 The Chilwa–Chiuta basin showing the location of previous high level shorelines. Arrows locate profiles in Fig. 2.7.

deposits. Between Mpyupyu and Chikala extensive spreads of alluvium were laid down by rivers draining the Zomba–Malosa massif. The 637 m shoreline is visibly cut into these deposits and laterites in many localities.

When the lake was at the 637 m stage, with an area of 3200 km², it must have been similar to that at the 652 m stage, but smaller and shallower. In both cases open lake conditions prevailed and wave energy was considerable, leading to the development of wave cut cliffs in bed rock or well consolidated deposits. The dominant winds appear to have been northerly, as the best developed shorelines tend to face this direction.

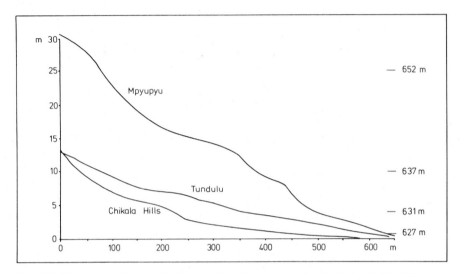

Fig. 2.7 Vertical profiles showing former shorelines at the points indicated in Fig. 2.6.

After the formation of the 637 m shoreline there was a slow fall in lake level to the 631 m stage, with a possible short halt at around 633 m. Accompanying this was an important change in the sedimentary environment of the lake, as indicated by a change in the type of sediments associated with the shoreline. On the western side of the Chilwa basin, north of the Phalombe River, the 631 m shoreline is associated with extensive sandy beaches, beach ridges and spits and bars. However, it is north of the Chikala Hills that the most extensive sand formations occur. North of the Sumulu River, the 631 m shoreline is represented by a wide sandy beach or series of beach ridges backed by low sand dunes. The dunes gradually rise in height northeastwards and merge with those of the sand bar which now separates Lake Chilwa from Lake Chiuta, and its outlet to the Indian Ocean.

The Chilwa–Chiuta sand bar (Fig. 2.8) stretches from some 30 km from the Mikoko River to a point 4–5 km east of the Mozambique border near Nayuchi. In its widest parts it may be 1–1.5 km across, and in places it reaches a height of 25 m above the north Chilwa plains. To the east, the sand bar is much narrower and lower, and only reaches a height of 12 m above the plain. Detailed studies of the composition of the sand bar (Lancaster 1979) show that it is a composite feature. On the southern side of the bar are the coarse, sub angular quartz sands of the 631 m beach. These tend to become finer and somewhat better sorted towards the east, indicating a move-

30

Fig. 2.8a View east on sand bar showing 631 m shoreline to right. (photo: N. Lancaster)

Fig. 2.8b View of 631 m shoreline north of Mpyupyu Hill where it is cut into bedrock. (photo: N. Lancaster)

ment of beach material from west to east. In its western and central sections, the sand bar is capped by hummocky sand dunes composed of fine, well-sorted quartz sand. These sand dunes were formed by deflation of sands from the adjacent 631 m beach by strong southeasterly winds. The Mpili River has infilled the area behind the sand bar with alluvium, creating an extensive marshy area 4–5 m above the Chilwa plain.

The sand bar was thus formed by an easterly movement of beach sand across the northern end of Lake Chilwa. Its formation was initially by the growth of a series of recurved spits from the western shore of the lake. As these extended eastwards, wind action on the exposed beach sands resulted in the formation of dunes on the crests of the spits, in the manner of marine coastal dunes. The formation of compound sand spits and dunes in this way is a common feature of marine environments characterised by strong onshore winds and an active supply of sandy sediments. Similar features are also common in relatively large lakes, for example on the western shores of Lake Malawi.

The source of the sands making up the sand bar was the rivers flowing into the western side of Lake Chilwa, particularly the Domasi and Sumulu Rivers. The combination of active sand supply and sediment transport, indicated by the nature of the sand, imply that both these rivers formerly carried much higher bed loads. This is borne out by the composition of the lower terrace of the Domasi River, which is associated with the 631 m stage of the lake. Deposits of this terrace are commonly coarse to medium sands and closely resemble those found in the sand bars around the Chikala Hills and in the 631 m beach along the Chilwa–Chiuta sand bar. Generally, it seems that the 631 m stage of the lake was one of strong wave action generated by southeasterly winds.

The 627 m shoreline is the best defined of a series of minor shorelines which postdate the isolation of Lake Chilwa by the formation of the sand bar. It is represented on both sides of the lake by a series of small sandy beaches and sand spits and bars, occasionally with a capping of dune sand. Locally the 627 m shoreline cuts into deposits laid down during the 631 m stage of the lake. This shoreline is the boundary of the ecosystem of the lake, swamp and floodplain.

The evidence from high level shorelines and associated deposits demonstrates that, during the late Quaternary, there have been 4 periods of high lake levels in the Chilwa basin, separated by one, or probably two, periods of lake levels lower than at present. These fluctuations have resulted from continent-wide climatic changes, rather that tectonic changes, although the effects of downcutting of the former outlet by the Lugenda River are uncertain.

Lake Chilwa is today a sensitive indicator of changes in rainfall and runoff over its catchment, as Chapter 3 describes. There is no reason to doubt that it was similarly sensitive in the past, even when it was an open lake. Evidence from large open lakes such as Victoria (Beadle 1974) and Malawi (Water Resources Division 1976) shows that comparatively small changes in rainfall amounts can lead to significant changes in lake level. The effects of such changes are much more pronounced in closed lakes, such as Chilwa.

That substantial climatic changes have occurred throughout the African continent during the last 20 000 years is now well established. The evidence for lake basins has been reviewed by Street & Grove (1976) who demonstrate that dry climates prevailed over much of intertropical Africa during the period 13 000–15 000 B.P.

This was followed by an early Holocene humid phase, with much increased lake levels, reaching its maximum extent 8000–9000 years ago. The evidence from lake level changes is supported by biogeographical studies, reviewed by Hamilton (1976), who concluded that, at the height of the last Glacial, rainforest ecosystems contracted dramatically to isolated refuges, and dry savanna ecosystems spread over wide areas of central and east Africa. The pattern in southern Africa is not as clear. However, there are indications that the subcontinent was substantially cooler and wetter during the period, 15 000–20 000 B.P., and warmer and drier during the early Holocene (Van Zinderen Bakker & Butzer 1973).

The paleoclimatic implications of high level shorelines in the Chilwa basin are clear. The water balance of Lake Chilwa is discussed fully in Chapter 3 and indicates that, under present conditions of rainfall, the balance between inflow to the lake and evaporation from the lake is sufficient to maintain a lake with an area of 600 km^2, and 3–5 m deep. Higher lake levels and larger lakes thus imply increased rainfall and inflow, or reduced evaporation, or a combination of both. Given present evaporation rates, substantial increases in inflow and thus basin precipitation would have been necessary to sustain higher lake levels.

For example, to maintain a closed lake of the size and depth of the 631 m stage of Lake Chilwa (1900 km^2 in area and about 10–15 m deep) increases in runoff and thus inflow of 2.6 times present totals would have been necessary, assuming present day temperatures and evaporation rates. Using Langbein's (1949) set of curves relating mean annual precipitation, temperatures and runoff, this implies an increase in mean basin precipitation of some 35–40 per cent. Sedimentary and geomorphic evidence, in the form of extensive sand formations, supports this hypothesis of much increased runoff and sediment production.

At the 637 and 652 m stages of the lake, Chilwa was an open lake and computations of the water balance can only give minimum values for increases in precipitation and runoff, as no estimate of outflow can be given, but increases in precipitation of 40–50 per cent are again implied. The substantial increase in lake area to 3200 km^2 is balanced by the addition of the Lake Chiuta catchment area and its runoff.

In the present absence of ^{14}C dates it is impossible to make definite correlations with other areas, and to give a definite age for Lake Chilwa. The nearest comparable lake basin to Chilwa for which ^{14}C dates for high lake levels are available is Lake Rukwa, some 800 km to the north-northwest, which reached a high level 9000 years ago (Butzer et al. 1972).

In the circumstances the simplest scheme for dating and correlation which can be put forward is as follows. It is probable that 631 m stage relates to the early Holocene humid phase, which occurred some 8000–9000 B.P. The 637 m stage is slightly older that this, and probably represents a late Glacial rise in lake levels. The form of the 652 m shore line suggests that it is of considerable antiquity and probably predates the last Glacial period. Supporting evidence for this may be sought from the period of low lake levels and lateritization of lacustrine deposits, which occurred in the interval between the 652 m and 637 m stages of the lake, when the lake level was at least as low as today.

Dating of the 627 m shoreline is possible by indirect means. Around many parts of the lake, Iron Age pottery, similar to that from sites dated to 1000–2000

B.P., is found associated with the 631 m shoreline. Therefore, at this period, the 631 m shoreline probably provided the nearest dry sites to the lake, suggesting that lake levels may well have been at or around the 627 m stage at that time. This may be close to the depth of the lake that Livingstone saw in 1859.

7. The climate of the Chilwa basin

All aspects of the climate of the Chilwa basin show a strongly seasonal pattern, resulting from the interaction of the two main systems of pressure and winds which affect this part of Africa, the Southern African anticyclone, and the Inter-Tropical Convergence Zone. These move their position seasonally with the sun, and affect the climate of southern Malawi at different times of the year. Very briefly, the pattern has one hot wet season of six months and one dry season, with four cooler months, when there is a little rain over high ground, and two hot dry months.

7.1 *Rainfall*

In the period May to September, the climate of the sub-continent is dominated by the Southern African anticyclone, situated over the eastern Transvaal. Outblowing winds from this reach Malawi from the south and southeast and bring dry, stable air, which gives rise to warm days and cool nights. At this time of year, drainage of cold air from the highlands surrounding the Chilwa basin frequently causes early morning mists over the lake and the Phalombe Plain. From time to time, particularly in the early part of this period, movements of the anticyclone towards the Natal coast bring about invasions of cool moist air from the southeast, which result in mists and light rain, *chiperone* conditions over Mulanje and the Shire Highlands, and cool cloudy weather over Lake Chilwa. Occasionally rapid pressure rises over the Mozambique coast may give rise to strong southeasterly winds, accompanied by cold clear conditions.

From September to November, the South African anticyclone moves southeast, and a weak thermal low pressure area develops over the Kalahari. Winds over southern Malawi back towards the northeast, as a further high pressure centre develops over southern Tanzania. Initially this airstream is dry, but it gradually becomes more humid and unstable as the Inter-Tropical Convergence Zone approaches from the north. In late October and November convectional storms begin to develop in the late afternoon, particularly over high ground, where orographic effects induced by the interaction of warm wet winds and the mountain slopes, trigger off the latent instability of the air.

During the period, late December to March, southern Malawi comes within the Inter-Tropical Convergence Zone. Rainfall occurs as a result of the convergence of deep moist unstable airstreams from the northwest (Congo Air) and from the east and southeast (Maritime or Monsoon Air). In this situation vertical movement of air takes place freely and convectional rainfall is often widespread and heavy. The amount of rainfall and the duration of the rainy season is strongly influenced by the position and movement of the Inter-Tropical Convergence Zone. Strong convergence and the persistence of the Zone over southern Malawi results in heavy rainfall throughout the region.

Table 2.1 Rainfall (mm) for selected stations in the Chilwa basin. (Source: Meteorological Dept., Malawi).

Area	Nov.	Dec.	Jan.	Feb.	Mar.	Apr.	May.	June	July	Aug.	Sep.	Oct.	Annual Total
Mulanje													
Tuchila Plateau	132	397	412	371	258	176	76	63	58	27	8	62	2040
Fort Lister	122	287	396	268	247	125	39	13	16	8	0	39	1560
Shire Highlands													
Changalumi	86	232	334	237	156	95	13	8	4	1	0	19	1185
Chiradzulu	120	243	227	163	153	81	22	11	6	5	5	13	1049
Makoka	87	208	283	189	173	71	22	7	3	1	1	20	1065
Mbala Estate	100	236	366	249	213	100	23	12	5	0	0	16	1320
Zomba–Malosa													
Zomba Town	151	245	246	251	195	77	17	18	37	6	9	35	1287
Zomba Plateau	155	421	377	360	344	199	76	35	26	10	3	37	2043
Domasi College	74	296	320	226	180	82	28	15	3	1	4	24	1253
Malosa Mission	96	303	377	282	250	83	26	5	4	2	2	21	1451
Chilwa–Phalombe Plain													
Mpyupyu Prison Farm	79	205	320	201	157	46	26	6	1	2	2	13	1058
Jali	71	184	229	196	138	44	15	2	4	0	1	11	895
Phalombe	74	161	290	226	170	170	40	6	7	2	1	17	1154
Ntaja	81	168	239	180	176	62	18	11	3	2	1	6	947

For location of stations see Fig. 3.5.

Conversely, periods of weak convergence and the influx of the stable north-easterly air tend to result in dry spells. The relative proportions of these conditions during the period December to March largely determine whether rainfall totals for that year will be above or below the long term average. Periodically, intense tropical depressions, or cyclones, may cross the Chilwa basin. Normally these pass through the Mozambique Channel, but occasionally their tracks may pass over southern Malawi, giving rise to concentrated heavy rainfall, often leading to flooding. The most notable occurrence of a cyclone within the area was in December 1946 (Talbot-Edwards 1948) when over 915 mm of rain fell in Zomba in 36 hours. More recently Morgan & Kalk (1970) report a cyclone affecting the area in March 1967, during which 200 mm of rain fell in 24 hours.

During March, the Inter-Tropical Convergence Zone begins to move north, and southeasterly air masses gradually extend over the region. Initially these are frequently very moist and unstable, bringing orographic rainfall to Mulanje and the Shire Highlands. In time these air masses become drier and more stable. Periods of rainfall and cloudy conditions become less frequent and the intervals of warm sunny 'dry season' weather longer.

Table 2.1 shows that rainfall over the Chilwa basin declines from a maximum in the highlands surrounding the basin, which receive 1100–1600 mm per year, rising to 2000 mm or more on Mulanje and Zomba Mountains, towards an area of low rainfall over the lake which extends towards the southern Phalombe Plain, in the rain shadow of Mulanje Mountain. Rainfall totals in this zone are 800–900 mm per year.

There is a strongly seasonal pattern at all stations, which is most marked in the drier areas. Maximum rainfall occurs in late January or early February in the rainy season which lasts from late October or November to April or early May. Stations in the Phalombe Plain receive very little dry season rainfall, but those in the southeasterly facing highland areas received up to 10 per cent of their annual totals during June and July. Annual variability is a function of rainfall totals. It is greatest, at 30 per cent in the Phalombe Plain. As a result reliability of rainfall in this area is low, and crop failures not uncommon.

Rainfall over the lake and the floodplain is best described by the record from Mpyupyu Prison Farm, some 10 km west of Kachulu. Records for this station are depicted in Fig. 2.9.

7.2 *Temperature*

The Chilwa–Phalombe area is one of the hottest areas in Malawi, with a mean annual temperature of 24°C. Once again, a seasonal pattern occurs, with maximum temperatures of 32–34°C in October, November and December, when solar radiation is at a maximum. Temperatures drop slightly with the onset of the rains, but may still reach 30°C. The coolest months are May to August, when maximum temperatures fall to 24–25°C. Clear skies at night, together with cool air draining from surrounding highlands give minimum temperatures at this time of year of 13–15°C (Fig. 2.9). The temperature range is greatest, at 13–14°C, during September and October, and least during the rains when it is 9–10°C.

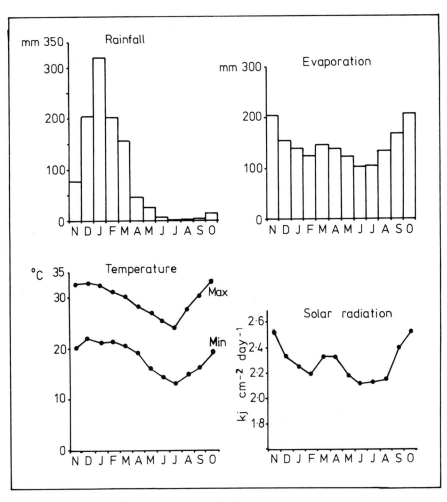

Fig. 2.9 Seasonal variation of climatic elements over Lake Chilwa: Rainfall: at Mpyupyu Prison Farm, 15 km west of Kachulu, mean monthly values (1961–1971): Mean monthly values for evaporation and temperature (1961–1971) and solar radiation (1967–1971) at Khanda, 10 km west of Kachulu.

7.3 *Solar radiation and sunshine*

Solar radiation has been measured at Khanda, near Kachulu, from 1967. The pattern shown in Fig. 2.9 indicates that it is at a maximum in October and November, with a subsidiary peak in March and April, before dropping rapidly to a minimum in June and July. There are no sunshine records for a station adjacent to Lake Chilwa, but records for the period 1969–1976 for Mangochi in the southern end of Lake Malawi, and Makoka at the edge of the Shire Highlands, may be taken as a guide with the actual figures probably closer to those at Mangochi (Table 2.2). Both stations follow a similar pattern, with maximum sunshine in September and October (9–10 hours per day), falling somewhat during the rains to 5.5–7 hours a day. At both stations minimum values are recorded during the cloudy months during the latter part of the rainy season, and rise thereafter through the dry season.

Table 2.2 Mean monthly values for sunshine (1965–76) at two stations in Southern Malawi; units are hours per day.

	J	F	M	A	M	J	J	A	S	O	N	D
Makoka	5.7	5.9	6.5	6.1	7.1	7.2	7.5	8.2	9.4	9.0	7.8	5.4
Mangochi	6.7	6.6	8.1	8.0	8.9	8.4	8.6	8.8	9.8	10.0	9.1	6.8

7.4 Winds

Wind directions were recorded at Kachulu during the period 1970–1971 by the Lake Chilwa Coordinated Research Project. The records are presented in Fig. 2.10 as a wind rose. These show that winds on the lake were predominantly from easterly directions. North and northeasterly winds were most common during the period September to November, and southerly and southwesterly winds during the period March to September. Winds during the rainy season were somewhat variable in direction. Although there are no records of wind strength, personal observations suggest that the strongest winds are from the south and southeast during the dry season. This is confirmed by records for other stations in Malawi, which record the highest percentage of moderate to high wind speeds in July.

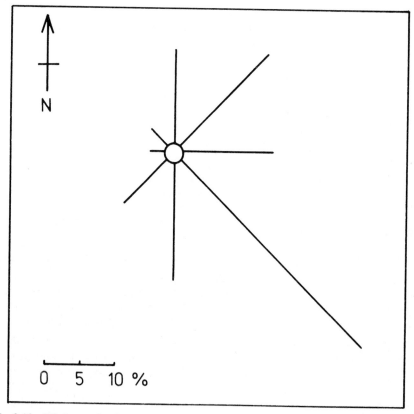

Fig. 2.10 Wind rose showing per cent frequency of wind directions recorded on Lake Chilwa (1970–1971).

7.5   *Evaporation*

Evaporation rates in the vicinity of Lake Chilwa are high, due to the high temperatures and large amounts of solar radiation received, and pan evaporation at Kachulu averages some 1757 mm per year. Fig. 2.9 indicates the relationship between the rate of evaporation and amount of solar radiation received. Evaporation is highest in October and November when solar radiation and temperatures are high. Humidity is low and winds are moderate in strength. With the onset of the rains, evaporation declines as the air becomes more humid, and increased cloudiness reduces the amount of solar radiation received.

8. Conclusions

This chapter has described the main features of the physical environment of Lake Chilwa, the most southerly of the major lakes of Africa. Comparisons with other presently closed lakes of East Africa show that Chilwa has experienced a similar paleo-environmental history, with low late Glacial lake levels being succeeded by early Holocene high levels. It was during the later stage of this period that Chilwa became a closed lake, by the formation of the sand bar separating it from its outlet via Lake Chiuta to the Lugenda River.

In many other respects Lake Chilwa contrasts with lakes elsewhere in Africa. For example, the relief of the Chilwa basin is much lower than that of most of the East African lake basins, principally because it lies outside the Rift Valley. This means that the climatic contrasts within the basin are much less, and the diversity of habitats is reduced. Geologically, the Chilwa basin lacks the Tertiary or Quaternary volcanism characteristic of the basins of the Rift Valley lakes of East Africa and Ethiopia. Thus relatively recent tectonic events have had much less effect on Chilwa than elsewhere. In Chilwa vertical tectonic movements have been dominantly the result of warping, rather than faulting, in contrast to Lake Naivasha, for example, described by Richardson & Richardson (1972).

The geological composition of the lake basin is reflected in the sediments of the lake and the chemistry of the lake waters. The Chilwa basin lacks the easily eroded recent volcanic materials, ashes and tuffs (consolidated ashes), of other East African lake basins, and thus sedimentation rates are probably lower. Again, the absence of extensive alkaline extrusive rocks in the Chilwa basin is reflected in the much lower carbonate content of the lake waters and the generally poor preservation of molluscs in lacustrine sediments.

In conclusion, Lake Chilwa is unusual in many respects and comparisons with lakes elsewhere in Africa are difficult to make. The distinctiveness of Chilwa is a direct reflection of the nature of the physical environment of its basin, with its generally low relief, developed on ancient crystalline rocks, with locally important igneous intrusions. Climatically the Chilwa basin is not particularly diverse and contrasts in temperatures and precipitation are relatively small. Chilwa is thus very much more sensitive to changes in climatic elements over the basin that many other lakes, a factor of some importance in considering the causes and effects of fluctuations in lake level.

References

Beadle, L. C. 1974. The Inland Waters of Tropical Africa, Longman, London. 365 pp.

Bloomfield, K. 1965. The Geology of the Zomba area. Geol. Surv. Malawi, Bull. 16. Government Printer, Zomba. 193 pp.

Bloomfield, K & Young A. 1961. The geology and geomorphology of Zomba Mountain. Nyasaland Journal, 14:54–80.

Butzer, K. W., Isaac, G. L., Richardson, J. L. & C. Washbourne Kamau, C. 1972. Radiocarbon dating of East African lake levels. Science, 175:1069–1076.

Dixey, F. 1926. The Nyasa section of the Great Rift Valley, Geogr. J. 68:117–140.

Dixey, F. 1933. Water supply of the Zomba District. Geol. Surv. Nyasaland, Water Supply Investigation. Prog. Rep. 2:11–15 Government Printer, Zomba.

Dixey, F., Campbell-Smith, W. & Bisset, C. B. 1955. The Chilwa Series of Nyasaland. Geol. Surv. Nyasaland, Bull. 5. Government Printer, Zomba.

Garson, M. S. 1960. The geology of the Lake Chilwa area. Geol. Surv. Nyasaland. Bull. 12. Government Printer, Zomba. 69 pp.

Garson, M. S. 1965. Carbonatites in Southern Malawi. Geol. Surv. Malawi. Bull. 15. Government Printer, Zomba.

Garsons, M. S. & Walshaw, R. C. 1969. The geology of the Mulanje area. Geol. Surv. Malawi Bull. 21. Government Printer, Zomba. 157 pp.

Hamilton, A. 1976. The significance of patterns of distribution shown by forest animals and plants in tropical Africa for the reconstruction of upper Pleistocene paleo-environments; a review, Paleo-ecology of Africa 9:63-97.

Lancaster, J. N. 1979. Late Quaternary events in the Chilwa Basin, Malawi. Paleo-ecology of Africa II.

Langbein, W. B. 1949. Annual runoff in the United States. U.S. Geog. Soc. Circular 52:1–14.

Morgan, A. & Kalk, M. 1970. Seasonal changes in the waters of Lake Chilwa (Malawi) in a drying phase, 1966–68. Hydrobiologia 36:81–103.

Richardson, J. L. & Richardson, A. E. 1972. History of an African Rift Lake and its Climatic Implications. Ecol. Monogr. 42:499–534.

Stobbs, A. R. 1973. The Physical Environment of Southern Malawi. Government Printer, Zomba, Malawi.

Street, F. A. & Grove, A. T. 1976. Environmental and climatic implications of late Quaternary lake level fluctuations in Africa. Nature, 261:385–390.

Talbot-Edwards, J. 1948. The Zomba Cyclone. Nyasaland Journal 1:53–63.

Van Zinderen Bakker, E. M. & Butzer K. W. 1973. Quaternary environmental changes in Southern Africa. Soil Science, 116:236–248.

Water Resources Division. 1976. Water Resources Assessment of Lake Malawi. UNDP/WMO Project 71518. Lilongwe, Malawi. 167 pp.

3 The changes in the lake level

N. Lancaster

3 The changes in the lake level

Fluctuations in lake level are an important feature of the hydrology of many African lakes, particularly those with no outlet, and Lake Chilwa is no exception to this. In addition to seasonal changes in lake level of about 1 m, the level of Lake Chilwa fluctuates at regular intervals by 2–3 m, which may lead to the partial or complete drying out of the lake with consequent deleterious effects upon fish populations and aquatic life and upon the people who depend upon the lake and the floodplain for their livelihood.

The larger fluctuations are the result of changes in the water balance of the lake and its catchment area. When inflow from rivers draining to the lake exceeds evaporation from the lake and swamp, then the level of the lake rises. But if inflow is less than the total of evaporation from the lake, then its level will fall. The cumulative effects of series of years of high or low rainfall in the catchment result in substantial changes in lake level. These larger changes have been analyzed and, as will be seen, are of a regular nature.

1. Historical changes in the lake level

Records of the levels of Lake Chilwa have been maintained only since 1950. Before this, descriptions of high or low lake levels by local people or visitors to the area provide the main sources of information on the state of the lake. Some of these have been summarized by Chipeta (1972). Considerable caution needs to be used when interpreting these subjective records of lake levels, and comparisons of the descriptions with known events is not always possible. In addition, for an observer on the west shore, the view of the southeast sector which is up to 2 m deeper that the position of the guages, is obscured by Nchisi Island (see Fig. 2.5).

The earliest record of lake levels is in Livingstone's description of the lake when he first visited it in April 1859. From his comments, and the painting by Kirk, reproduced in Wallis (1956) it appears that the level of the lake was at that time very high indeed, and probably somewhat higher that it is at present (May 1978). Significantly, Lake Malawi was also at a very high level at this time, as it is today. The comments of local people recorded in O'Neill (1884) and Buchanan (1893) suggest that lake levels in the 1860s or early 1870s were also high and that the lake level had fallen considerably since then.

Evidence for low lake levels in 1879–1880 is, however, provided by O'Neill (1884) and Buchanan (1893). O'Neill reports that he was told (in 1883) that the western portion of Lake Chilwa was then so shallow that two or three years before, during an exceptionally dry season, a man could have walked from Nchisi Island to the mainland, but people were deterred from doing so by the soft muddy nature of the lake bed. Buchanan states that 'in 1879 the lake very nearly dried out'. It is possible that these descriptions may refer to the same event, but they may represent a period such as 1967–68 when there were two successive years of low lake levels. In 1901, Duff (1906) records that 'last November (1900) there was nothing but black mud cracked and dried by the sun, between Nchisi Island and the mainland'. A rather different picture is given in Chipeta (1968) when local people stated that in 1900 most of the lake was overgrown with tall grass with only occasional pools of water remaining. This situation may have been similar to the rainy season of 1967–68 when the grass *Diplachne fusca* covered some of the lake bed (Chapter 7).

Low levels also appear to have occurred in the period 1913–1916 when there are reports of men hunting antelope in the middle of the lake (Chipeta 1968). just as fishermen hunted an otter in 1967 (see chapter 4). Since then, low levels were recorded in 1920–1922 when, according to Garson & Campbell-Smith (1958), the lake was dry. In 1934 low levels occurred again, with heavy mortalities in hippo populations. The experiences of duck shooters in the late 1930s (B. Burgess pers. comm.) seem to suggest that lake levels were rather high again in those years.

The general picture that emerges from these somewhat scanty records is summarized in Table 3.1. It appears that the lake was at a high level in the late 1850s and 1860s; in 1888 and in the late 1930s. Low levels, comparable with those in 1973, seem to have occurred in 1880, 1900, 1922 and 1934. However, it is possible that the 1900 level was extremely low and the result of a period similar to that of 1966–68, although from the descriptions in Chipeta (1968), it seems that the 1913–16 and 1966–68 periods of low lake level were also comparable. Generally, more data on low lake levels are available than on high levels, due to the greater impact of low levels on the fishing and economy of the lakeshore populations.

Table 3.1 Historical lake level changes.

	High	Low	Very low
1859	Livingstone		
1860	O'Neill 1884		
1870	Buchanan 1893		
1879			Buchanan 1893
1880		O'Neil 1884	
1888	Drummond 1902		
1900			Chipeta 1972
			Duff 1906
1913–15		– – – Chipeta 1972 – – – – – – – –	
1920–22		– – – Garson Campbell-Smith 1958 – –	
1934			
late 1930s	Burgess (pers. comm.)		
1943		Chipeta 1972	
1949		Chipeta 1972	

2. Recorded fluctuations in the lake level since 1950

Fig. 1.4 (Chapter 1) shows the variation in lake level since 1950 at Ngangala Harbour (for location see Fig. 3.5). From this, it will be seen that low lake levels were recorded in 1954, 1960–1, 1967–8 and 1973. Between them are periods when lake level was high, reaching peaks in 1952, 1958, 1963–4, 1971, 1974 and 1976. Significantly, there is also a general decline in lake levels during the 1960s, and it seems that the 1950s and mid 1970s were periods of high lake levels, and the 1960s one of low levels. Statistically the mean lake levels for the 1950s are significantly different from those during the 1960s at the 5 per cent level. Today (1978) lake levels are historically very high, and probably higher than at any time since records began. For shorter periods the data for levels recorded elsewhere on the lake; e.g. at Kachulu (Fig. 3.1) and Chisoni on the east shore (Fig. 3.2) confirm this pattern.

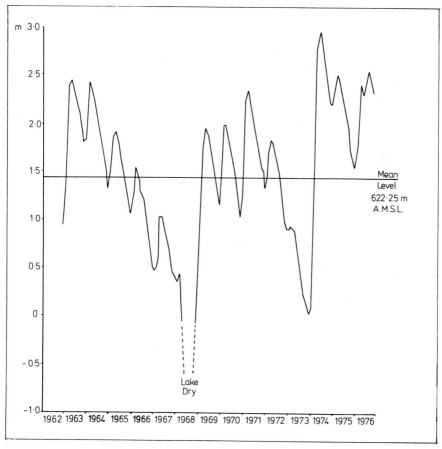

Fig. 3.1 Lake levels at Kachulu, a gently sloping shore on the west, since 1962. For location of station see Fig. 3.5.

3. Seasonal changes of lake level

Each year there is a regular seasonal change in lake levels (Fig. 3.3). The lake is at its lowest in November or December, at the beginning of the rains. It rises thereafter, as runoff from the surrounding catchment reaches the lake, and reaches its peak in March or April. During the dry season, evaporation from the lake increases, whilst inflow gradually declines. As a result the level of the lake falls steadily. If, over the year, inflow is less than evaporation there will be a net fall in lake level as Fig. 3.3a shows. Conversely, when inflow exceeds evaporation, the lake level will rise (Fig. 3.3b). It is the cumulative effects of such surpluses or deficits which give rise to longer period changes in lake level.

45

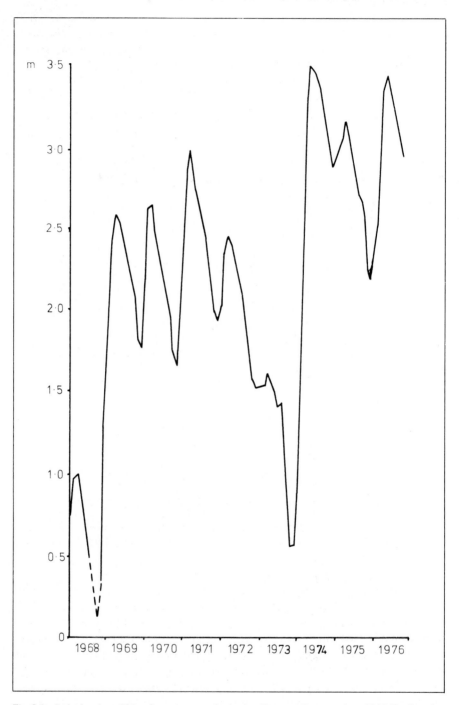

Fig. 3.2 Lake levels at Chisoni, a more steeply sloping shore on the east, since 1968. For location of station see Fig. 3.5.

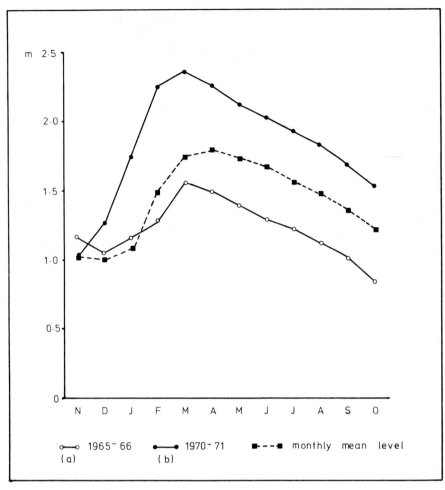

Fig. 3.3 Annual hydrograph of Lake Chilwa at Kachulu: (a) 1965–66 represents the shape of hydrograph when the lake was falling in level; (b) 1970–71 when the lake was rising.

4. Pseudo–periodicities in lake level changes *by* T. G. J. Dyer

The records of the annual maximum depths of Lake Chilwa measured since 1950 have been analyzed using spectral analysis in order to determine whether there are periodicities inherent in the fluctuations of lake level. Such information may enable the minor and major recessions of the lake to be predicted in the future, in time for fishermen and farmers to make provision for one to three or more unfavourable years for fishing and crops.

The mathematical model used (Dyer 1977) detects not only the time periods of the possible oscillations, but also their relative strength or importance. Therefore the amplitude of the oscillations or waves is a measure of the intensity or magnitude of periods of low lake level. The resulting spectrum from this analysis is shown in Fig. 3.4.

For a random time series, i.e. one possessing no organized pattern, the spectrum would theoretically be a straight line. In the case of the lake levels at

47

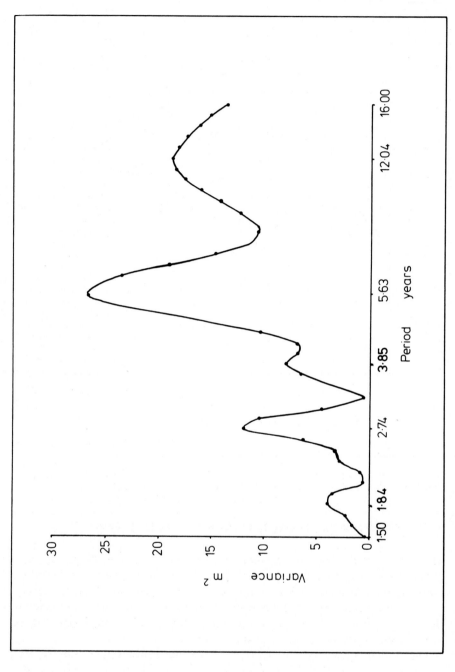

Fig. 3.4　The spectrum for lake levels at Kachulu showing wavelengths of oscillations associated with variations in lake levels.

Kachulu (station 2.C.10) analyzed from 1962–1976, this is not the case. It will be seen that the spectrum detects wave-forms having the periods of 1.84, 2.74, 5.63 and 12.04 years. Of these waves, the one with periodicity at approximately six years is by far the strongest. The 1.84 and 2.74 year periods are multiples of the one at 5.73 years and therefore probably harmonics of it. Harmonics are due to impurities in the basic wave, that is, the period 5.63 is not constant and the distance between the peaks varies slightly over time. Also the highest lake level does not occur in the same month every year. This oscillation with a period of about 6 years has also made an appearance in the variations in the levels of Lake Malawi (Dyer 1976).

In Fig. 3.4, it is seen that the peaks are not of equal amplitude. This situation may be caused by a number of factors. Firstly 'noise' values are probably superimposed on the data, e.g. errors in taking readings of the levels and those caused by onshore winds. Secondly, the various waves coincide from time to time and can have the effect of increasing, or decreasing, the height of the peak, with a similar effect in troughs. These beat frequencies can give rise to peaks in the data that are not observed in this short spectrum. For example, the 5.63 and 12.04 phases will be in harmony and overlap every 5.63 × 12.04, or 68, years and so on.

Analysis of the records of lake level at Ngangala harbour over the period 1950–1976 did not give a well defined spectrum. The noise level here is greater, possibly due to the poor siting of the guage on a very gently sloping shoreline, and possibly due to inaccuracies in the height of the guage.

The accuracy of the 6 year oscillations may be checked against the dates of known low lake levels described above. Thus the six year oscillations correspond quite well to the sequence of guaged low levels in 1973, 1967–8, 1960, 1954 and reports of low levels in 1949, 1943, 1933, 1921–2, 1913–15 and 1900. The harmonization of the 5.63 and 12.04 year pseudo-oscillations every 68 years lends support to Chipeta's account of oral evidence for a partial or complete desiccation of the lake in 1900.

5. Lake level fluctuations elsewhere in Africa

Comparisons of the pattern of fluctuations of Lake Chilwa with those of other lakes in East and Central Africa reveals some interesting features. Direct comparisons are impossible, in view of the varied nature of the catchments of the lakes, but broad similarities in the patterns would tend to suggest regional scale climatic fluctuations. Like Lake Chilwa, Lake Malawi had high levels in the 1860s and 1870s (Pike 1965). It experienced a progressive fall in level in the early part of the 20th century, but in 1915 its level started to rise, reaching a peak in the 1970s. Like Chilwa, it has a 6 year periodicity in levels superimposed on this gradual rise in level (Dyer 1976).

The pattern of lake levels of Lake Rukwa, 150 km north of Lake Malawi, have been summarized by Gunn (1973). The pattern there appears to be out of phase with that of Chilwa. Rukwa was rising from 1963, at a time when the level of Chilwa was falling. Historically, Lake Rukwa was very high in 1880–82, but fairly dry in 1873, just the opposite of Chilwa. The only points of comparison are in 1920, when both lakes were dry, and 1954 when Rukwa was dry and Chilwa at a moderately low level.

Lamb (1966) has summarized the variation in level of the major East

African lakes and notes an abrupt rise in the levels of most lakes in that region in the early 1960s due to a substantial increase in rainfall. Thus, Lake Naivasha underwent a steady fall in level from 1916 to the 1960s, with a rise thereafter (Richardson & Richardson 1972). A similar pattern is reported by Kingham (in discussion in Grove, Street & Goudie 1975) from the Rift Valley lakes of Ethiopia. These began to rise in the early 1960s, apparently as a result not of an increase in rainfall, but as a consequence of reduced evaporation over the catchment.

It appears therefore that the pattern of changes in the level of Lake Chilwa is out of phase with those of the lakes of East Africa. Significantly, rainfall over the north and northeastern parts of South Africa was below normal during the 1960s, and Botswana experienced severe droughts during that time. In view of the fact that the climate of southern Malawi is dominated for much of the year by the position and movements of the Southern African anticyclone, it is more likely that the oscillations in the level of Lake Chilwa would correspond to the pattern of climatic fluctuations in southern rather than eastern Africa. Low levels occurred at Lake Sibaya, a land-locked freshwater lake in the northeast of South Africa, in the 1960s. Very high levels were predicted for the late 1970s from a mathematical model of the rainfall, which exhibited a similar periodicity, and they actually occurred, destroying the research station (Pitman & Hutchison 1975).

Interestingly, the level of Lake Chad declined drastically from 1968–1973 (Chouret 1977), as a result of low rainfall over its catchment at this time, which was one of drought throughout the Sahel region.

6. The water balance of the lake

In the long term, the level of a closed lake, such as Chilwa, represents a balance between inflow to the lake (I) together with rainfall on the lake itself (LR) and evaporation from the surface of the lake (LE). Thus;

$$\text{Inflow} + \text{Rainfall on lake} = \text{Evaporation from the lake.}$$

Consequently changes in lake level are primarily the result of changes in the terms of this equation. Thus, an understanding of the causes of fluctuations in lake level depends upon a knowledge of the components of its water balance.

6.1 *Components of the water balance*

6.1.1 Inflow to the lake

Seven major perennial rivers, together with numerous small seasonal streams, flow into Lake Chilwa and its marginal swamps and marshes. In the Malawi part of the catchment they are the Domasi, Likangala, Thondwe, Namadzi, Phalombe and Sombani rivers. The Mnembo is the principal influent stream from the Mozambique side.

Details of river discharge are available only for the rivers in Malawi and are shown in Table 3.2. Due to deficiencies in the data, the period of record illustrated is restricted to 1961–1971. In addition, many of the gauging stations are at some distance from the lake (see Fig. 3.5) and it is impossible to estimate

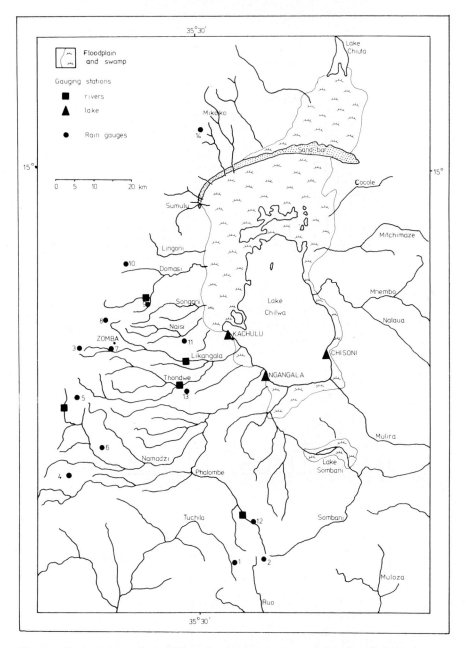

Fig. 3.5 The hydrology of Lake Chilwa; showing drainage network; location of lake level gauges; river gauging stations and raingauges. For names of rain gauge stations see Table 3.4.

how much of the gauged inflow actually reaches the lake and how much is lost, especially at peak flows, by inundation of floodplains and subsequent evaporation. However, it is estimated that the gauged inflow represents some 65 per cent of the total inflow to the lake.

All the rivers have a strongly seasonal regime, as Fig. 3.6 shows. With the exception of the Thondwe, maximum flows occur in February, shortly after the

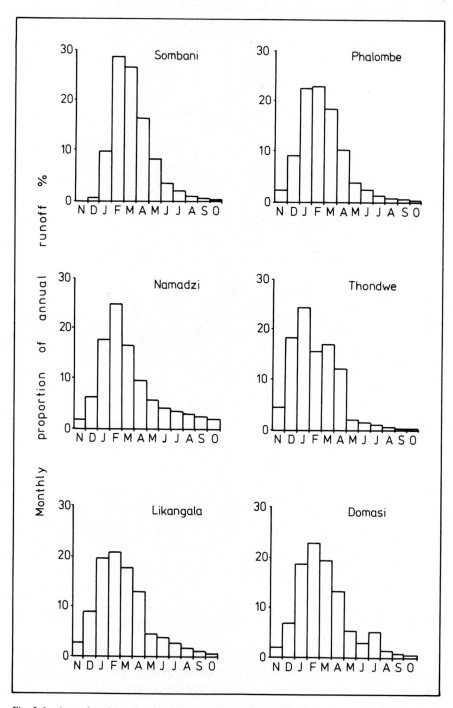

Fig. 3.6 Annual regime of major gauged influent rivers (1961–71).

Table 3.2 Mean monthly and mean annual discharge for main influent rivers (1961–71), m³sec⁻¹.

River	Nov.	Dec.	Jan.	Feb.	Mar.	Apr.	May	June	July	Aug.	Sep.	Oct.	Mean
Sombani	0.12	0.29	3.46	7.31	6.80	3.59	1.95	1.10	0.64	0.37	0.18	0.09	2.23
Phalombe	0.33	1.17	3.41	2.90	2.19	1.04	0.43	0.34	0.25	0.15	0.12	0.08	1.03
Namadzi	0.06	0.17	0.57	0.54	0.50	0.25	0.16	0.13	0.13	0.13	0.09	0.07	0.24
Thondwe	1.36	5.35	7.07	4.60	5.00	3.57	0.69	0.50	0.35	0.25	0.15	0.14	2.56
Likangala	0.79	2.73	6.86	6.16	4.39	2.81	1.50	1.01	0.72	0.46	0.28	0.16	2.42
Domasi	0.76	1.94	4.84	5.71	4.68	3.20	1.33	0.81	0.98	0.39	0.29	0.20	2.10

peak of the rainy season, and fall rapidly thereafter. Dry season flow represents only 5–10 per cent of the total flow for most rivers, except those, like the Likangala and Domasi, which drain well-forested catchments. An important feature of the discharge of all streams is its annual variability. This is greatest in the case of the Sombani and Thondwe rivers, and least in the case of the Phalombe and Domasi rivers.

During the period of records described, discharge was high between 1960 and 1963 and from late 1968 to 1971 and low in the intervening period, as Table 3.3 and Fig. 3.7 show.

Table 3.3 Mean of annual discharge into Lake Chilwa gauged rivers (1961–71).

	m³ sec⁻¹
1961–2	11.67
1962–3	18.26
1963–4	9.43
1964–5	7.68
1965–6	5.90
1966–7	5.44
1967–8	3.30
1968–9	17.94
1969–70	12.30
1970–71	16.01
Mean	10.79

6.1.2   Precipitation

The general features of the rainfall of the Chilwa basin have been described in Chapter 2. Rainfall in the source regions of the major influent rivers averages 1200–1500 mm, rising to 2000 mm or more on Zomba, Malosa and Mulanje mountains. The mean basin rainfall was determined using the Theissen polygons method to offset the over-representation of the wetter highland areas in which more rain gauges are situated. The results of this analysis are given in Table 3.4 and show that mean basin rainfall over the period 1961–1971 was 998 mm per year. The total rainfall varies considerably from year to year, with a coefficient of variability of 19 per cent. Rainfall is much more variable over the Phalombe plain and the lake, with a coefficient of variability of 30 per cent.

Rainfall over the lake was determined by averaging the rainfall for 3 stations, Ntaja, Mpyupyu Prison Farm and Jali, using the Theissen polygons method. The result is a mean lake rainfall over the period 1961–1971 of 893 mm per year.

Table 3.4 Mean catchment rainfall (1961–71).

Station	Weight (1)	Mean annual rainfall (mm) (2)	Weighted mean annual rainfall (mm) (1 × 2)
1 Tuchila Plateau	0.004	2040	8.16
2 Fort Lister	0.054	1389	75.01
3 Changalumi	0.010	1155	11.55
4 Chiradzulu	0.038	1030	39.14
5 Makoka	0.039	1065	41.54
6 Nasawa	0.070	786	55.02
7 Zomba Town	0.033	1365	45.05
8 Zomba Plateau	0.009	2034	18.31
9 Domasi	0.079	1186	93.69
10 Malosa	0.031	1453	45.04
11 Mpyupyu	0.160	827	132.32
12 Phalombe	0.197	958	188.73
13 Jali	0.127	792	100.58
14 Ntaja	0.154	947	145.84
Weighted Mean catchment rainfall			997.97

Numbers locate rainfall stations on Fig. 3.5.
Note: Areal weight for each station is a fraction of the basin area in Malawi covered by Theissen polygons drawn around each station.

6.1.3   Evaporation

Pan evaporation, measured at Khanda near Kachulu, averages 1763 mm per year over the period 1961–1971, with little significant variation from year to year. Due to the 'oasis effect' of an evaporation pan, evaporation from the open water of the lake is rather lower. As a result of studies based upon the dry season fall of Lake Chilwa, the Water Resources Division (1976) calculated a coefficient of 0.88 to be applied to the pan evaporation records. Thus the mean annual evaporation from the open water of Lake Chilwa was 1551 mm per year during the period 1961–1971.

The open water area of Lake Chilwa is surrounded by an almost equal area of swamp and marsh which is hydrologically continuous with the open water area. The amount of evaporation from areas of swamp is a matter of some uncertainty. Using data from experiments carried out by Rijks (1969) and Linacre et al. (1970), the latter on an area of *Typha australis*, a ratio of swamp evaporation to lake evaporation of 60 per cent was estimated and applied to the lake evaporation figure, giving a swamp evaporation total of 931 mm per year.

The areas of swamp and open water vary from year to year as lake level changes, with a mean ratio of areas of 60 per cent open water to 40 per cent swamp. Applying these proportions to the mean evaporation rates, a lake plus swamp evaporation total of 1303 mm per year was calculated.

6.1.4   Computation of the water balance

Considerable, and in the foreseeable future, insoluble difficulties in the provision and reliability of the data prevent computation of an annual water balance for the lake. Chief amongst these problems are the total lack of data on rainfall and inflow from the Mozambique part of the catchment, and uncertainty over the area of the lake and swamp as it fluctuates in level.

As a result only a general water balance of the form indicated above can be calculated. Thus over the period 1961–1971 the annual evaporation from the lake and swamps was calculated as 1303 mm, and annual rainfall on the lake and swamps was calculated as 893 mm. Substituting in the equation below, it is found that water balance is of the form:

Inflow + Rainfall on lake = Evaporation from lake and swamp
410 mm + 893 mm = 1303 mm

The inflow figure derived from this is 130 per cent of that calculated from the gauged rivers, which drain only 18 per cent of the total catchment area (see Fig. 3.5). The contribution from the relatively wet Shire Highlands area to the inflow to the lake is thus very considerable indeed.

7.   The causes of medium term fluctuations in lake level

In the event of being unable to calculate the annual water balances for the lake, it is necessary to try to examine the behaviour of the lake in terms of variation in the amounts of inflow to the lake and precipitation on the lake. The analysis will once again focus on the period 1961–1971.

Fig. 3.7 shows the relationship between annual maximum lake level, mean annual catchment rainfall and mean annual inflow to the lake from the gauged rivers. The striking parallelism of the curves for each of these aspects of the hydrology of the lake is notable, and from this graph it appears that the low lake levels experienced in 1967–68 were principally the result of a cumulative deficit of rainfall during the years 1963 to 1968. In each of these years the mean catchment rainfall was below normal. Fig. 3.8 illustrates this clearly by showing the relationship between lake level and cumulative rainfall deficit.

However, it is interesting to note that the parallelism of the rainfall and inflow curves is not maintained during the years 1965–66 to 1967–68. It is possible that this may be due to other factors, such as higher evapotranspiration during these years, or changes in the intensity and duration of precipitation events, which may have affected the amount of runoff.

An assessment of these factors which affect lake level was made by using a multiple regression analysis. The results of this showed that, during the period 1961–1971, there was a high correlation ($r = + 0.76$) between discharge into the lake and annual maximum lake levels. In view of the difficulties of measuring discharge into Lake Chilwa, this is a surprisingly good correlation. A better correlation coefficient ($r = + 0.83$) occurs between mean catchment rainfall and annual lake level. Similarly good correlations have been obtained for other African lakes, for example, Mörth (1965), quoted in Lamb (1966), obtained a correlation coefficient of $+ 0.96$ between month to month changes in the level of Lake Victoria and rainfall over the catchment area.

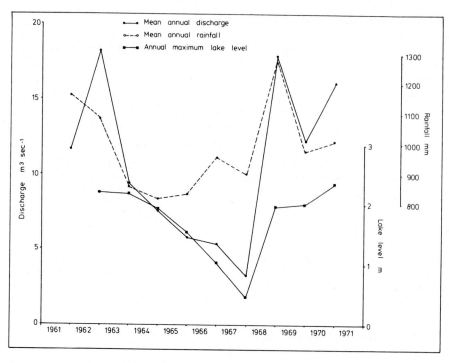

Fig. 3.7   Relationship between annual maximum lake level at Kachulu, discharge into lake and catchment rainfall (1961–71).

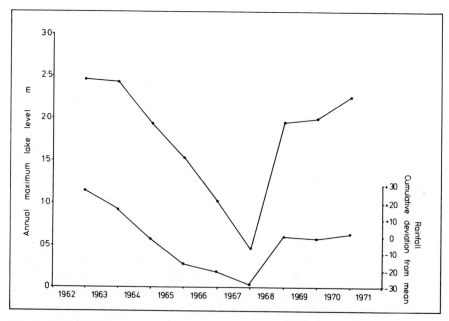

Fig. 3.8   Relationship between annual maximum lake level at Kachulu and cumulative deviation from mean catchment rainfall (1961–71).

56

Although a spectral analysis of rainfall for each of the four stations in the catchment with reliable long term records revealed no significant oscillations, it is interesting to note the close parallelism between the variations in the combined mean rainfall for these stations and lake levels during the period 1954–1975, as shown in Fig. 3.9. Averaging of the combined annual totals of these stations tends to cancel out local factors and shows a clearer pattern of variation. Confirmation of the strong influence of variations in catchment rainfall on lake level may be obtained by examining the rainfall record for Zomba which extends back to 1895. With the exception of 1913–15 all the years of recorded low lake levels are preceded by one or more years of exceptionally low rainfall in Zomba.

Thus it may be concluded that the principal cause of changes in the level of Lake Chilwa is variation in rainfall over the catchment. Although the pattern of variation at individual stations may not be detectable, mean catchment rainfall appears to vary in a regular way. Thus a succession of years of below average rainfall leads to a lowering of lake level. Statistical analysis of the record of lake levels indicate that this will happen approximately every six years. Conversely one or more years of high rainfall lead to a rise in lake level. Thus, it is clear that Lake Chilwa is a very sensitive index of changes in rainfall over its catchment, and lake levels respond very rapidly to periods when rainfall is significantly below or above average.

Fig. 3.9 Relationship between annual maximum lake level at Ngangala Harbour and deviation from mean rainfall at Fort Lister, Chiradzulu, Zomba Plateau and Mpyupyu (1951–75).

References

Buchanan, J. 1893. The Industrial Development of Nyasaland. Geogr. J. 1:245–253.
Chipeta, C. 1968. Cyclical and secular changes in Lake Chilwa water level and its fisheries. Lake Chilwa Co-ordinated Research Project. Economic Report No. 1. (Cyclostyled).
Chipeta, C. 1972. A note on Lake Chilwa cyclic changes. J. Interdisp. Cycle Research. 3:87–90.
Chouret, A. 1977. La persistance des effets de la secheresse sur le Lac Tchad. Food & Agricultural Organization, U.N. CIFA/77/Symp. 28:1–15.

Drummond, H. 1903. Tropical Africa. 11th edn. Hodder and Stoughton, London. 228 pp.

Duff, H. L. 1906. Nyasaland under the Foreign Office, 2nd edn. Bell, London. 422 pp.

Dyer, T. G. 1976. Analysis of the temporal behaviour of the level of Lake Malawi. S. Afr. J. Sci. 72:381–382.

Dyer, T. G. 1977. On the application of some stochastic models to precipitation forecasting. Quart. J. Roy. Met. Soc. 103:177–189.

Garson, M. S. & Campbell-Smith, W. 1958. Chilwa Island. Mem. Geol. Surv. Dept. Nyasaland, 1. Government Printer, Zomba, Malawi. 128 pp.

Grove, A. T., Street, F. A. & Goudie, A. S. 1975. Former lake levels and climatic change in the rift valley of southern Ethiopia. Geogr. J. 141:177–202.

Gunn, D. L. 1973. Consequences of cycles in East African climate. Nature 242:457.

Lamb, H. H. 1966. Climate in the 1960's: world wind circulation revealed in prevailing temperatures, rainfall patterns and the levels of the African lakes. Geogr. J. 132:183–212.

Linacre, E. T., Hicks, B. B., Sainty, G. R. & Gauze, G. 1970. The evaporation from a swamp. Agric. Met. 7:375–386.

O'Neill, H. E. 1844. Journey from Mozambique to Lake Shirwa and Amaramba. Parts 1, 2 and 3. Proc. Roy. Geogr. Soc. 6:632–655; 713–740.

Pike, J. G. 1965. The sunspot/lake level relationship and the control of Lake Nyasa. J. Inst. Water Engineers. 19:221–226.

Pitman, W. V. & Hutchinson, I. P. G. 1975. A preliminary hydrological study of Lake Sibaya. Rep. 4/75. Hydrological Research Unit. University of the Witwatersrand, Johannesburg. 35 pp.

Richardson, J. L. & Richardson, A. E. 1972. History of an African rift lake and its implications. Ecol. Monographs. 42:499–534.

Rijks, D. A. 1969. Evaporation from a papyrus swamp. Q. J. Roy. Met. Soc. 95:643–649.

Wallis, J. C. P. 1956. The Zambesi Expedition of David Livingstone. II. Journals, Letters and Despatches. Central African Archives. Chatto and Windus, London. 215–462.

Water Resources Division. 1976. A Water Resources Investigation of Lake Malawi. Report U.N.D.P./W.M.O. Project 71/518. 166 pp.

4 The aquatic environment: I. Chemical and physical characteristics of Lake Chilwa

A. J. McLachlan

4 The aquatic environment: I. Chemical and physical characteristics of Lake Chilwa

1. Fluctuations in the depth of the lake

Like many other tropical endorheic lakes in Africa, South America and Australia (Talling & Talling 1965, Marlier 1967, Bayly & Williams 1966), Lake Chilwa is characterized by a history of marked seasonal fluctuations in water level, which are accentuated by periodic recessions of the lake. Sometimes the contraction of the open water is 'minor', when up to one fifth of the lake bed may be exposed at intervals of about six years. More rarely the whole lake dries up over a period of three years and refills the following year, i.e. a major recession. The periodicity of the lake's behaviour and the factors contributing to it are fully discussed in Chapter 3. Such an unstable situation is not easy for living things to tolerate. In this chapter an attempt is made to describe the instability in chemical and physical terms. Points for consideration have been selected, bearing in mind relevance to the biota, and this is stressed when appropriate. Details of the living responses of many of the organisms within the ecosystem are considered in the following chapters.

It was exceptionally fortunate that the study was able to include the relatively rare drying phase of the lake (1966–68) and the refilling and recovery phases (1969–72) as well as the later years of high level (1975–76). Conditions over these periods demonstrate the two environmental extremes experienced by the ecosystem and make it possible to put the smaller seasonal fluctuations and recurrent minor recessions into perspective. Physical and chemical phenomena associated with the drying and filling phases are discussed separately below after a consideration of a 'normal year'. At the end of the chapter, the environmental factors are briefly evaluated as forces which limit the occurrence or distribution of the present fauna and flora.

The hydrograph in Fig. 4.1 illustrates the variations in the depth of the lake at Kachulu on the western edge of the lake. The maximum depth at this station at the end of the wet season each year was usually between two and three metres. The annual rise and fall is very uneven and is usually less than one metre, but it is only visible on the few exposed beaches, since the surrounding *Typha* grows in water about 1 metre deep. The effect on the floodplain extends for 10 km or more to the west and south.

The exposure of the lake bed in the last minor recession is shown in the satellite photograph in the frontispiece, taken in October 1973; and an aerial survey in November 1967 reported a similar exposure of 8 km wide around the north of the lake and 2–3 km on the west (F. Nicholson pers. comm.). The water in the southeastern sector of the lake is up to two metres deeper than at Kachulu and this area, from 150–250 km², may remain in some recessions, although it is completely isolated from the incoming rivers. This was the case during the wet season of 1968.

From September to mid-November in 1968, the whole lake bed dried up and then gradually filled to a depth of two metres in the 1968–69 wet season. Various degrees of evaporation in different years, followed by dilution due to rain and river inflow has marked effects on the ionic content and physical properties of the water.

2. Seasonal changes in a 'normal' year

The extremes of variation in the concentration of the major inorganic ions range from that of really fresh water, like Lake Malawi, along the shores after the rains, to that of concentrated brine just before drying out. In a 'normal' year concentrations of salts varies only by a factor of 2–3 (Howard-Williams 1975).

The changes in dissolved salts may be monitored by measuring water conductivity; this method is widely used in tropical limnology because of the ease of obtaining estimates in situations that are often remote from the usual laboratory facilities. It is not ideal, because the results are not always easy to interpret. 'Conductivity' is a measure of the current carrying capacity of water and depends upon the number of dissociated ions and the temperature; it is expressed as the reciprocal of the resistance at 25°C, measured over a distance of 1 cm. In general, the higher the conductivity value, the higher the concentration of ions. However, some ions, notably Cl^-, carry a higher charge than others, so that conductivities are dependent upon the ionic composition as well as the concentration of ions. Since, as will be seen, the proportions of the major ions vary only slightly in Lake Chilwa water, the method is considered to be suited to a short-hand designation of the state of the lake at any time. As a rule of thumb on many inland African lakes, conductivity numerically (as $\mu S\ cm^{-1}$) equals $100 \times$ alkalinity in meq l^{-1} (Talling & Talling 1965) and so may indicate salt concentration in a way which may be familiar to many readers. The relationship in Lake Chilwa is not quite the same, since the chlorinity and alkalinity are almost equal during a range of conductivities; alkalinity surpasses chlorinity only at high conductivities. Table 4.1 illustrates the ionic composition of Lake Chilwa at various conductivities in different phases of its fluctuating regime (McLachlan et al. 1972). We are concerned in this section with the range of about 1000 to 2500 $\mu S\ cm^{-1}$ in a year of normal depth.

Table 4.1 Concentration of the major cations and anions in terms of meq l^{-1} and as percentages of total cations or anions at conductivities of 50 to 12000 $\mu S\ cm^{-1}$. 50 $\mu S\ cm^{-1}$: Likangala River water; 400, 2000 and 12000 $\mu S\ cm^{-1}$: Lake Chilwa water. Values for 50 and 12000 $\mu S\ cm^{-1}$ adapted from Moss & Moss (1969).

	meq l^{-1}				% Total cations or anions			
	50	400	2000	12000	50	400	2000	12000
Na^+	0.3	3.5	24.8	117	29.4	77.3	93.9	98.5
K^+	0.1	0.2	0.4	1.0	7.8	4.0	1.6	0.8
Ca^{2+}	0.4	0.5	0.7	0.5	34.3	10.4	2.7	0.4
Mg^{2+}	0.3	0.4	0.5	0.3	28.4	8.2	1.9	0.3
Cl^-	0.3	2.2	12.7	54.1	30.0	46.0	48.1	46.2
HCO_3^-								
CO_3^{2-}	0.6	1.5	9.2	61.6	60.0	33.3	34.9	52.6
SO_4^{2-}	0.1	0.9	4.5	1.4	10.0	20.7	17.1	1.2

In spite of the strikingly large fluctuations in depths and salt contents of the lake waters, inspection of the hydrograph in Fig. 4.1 shows that the lake depth may be considered 'normal' for about 70 per cent of the 26 years recorded. The major ions in Lake Chilwa water, Na^+, Cl^-, and HCO_3^-, give a slightly alkaline water with pH values varying between 8 and 9. The large proportion of chloride is unusual in inland saline waters and points to some contribution from the three underground hot springs, located in the area. The main sources of the salts are the feldspars and nepheline syenites and the volcanic intrusions of the mountains in the catchment, and the alluvia over which rivers flow on the floodplain (see Chapter 2). The range of ions in a 'normal' year is given in Table 4.2 which are taken from the results of the monthly sampling programme of 1970, in surface waters at an offshore station in Kachulu Bay.

The amount of total salts varied from 1.3 to 2.3 per mille, which is well below the upper limit of about 5 per mille for tolerance by most freshwater organisms (Beadle 1974). An attempt has been made to place Lake Chilwa in the heirarchy of African lakes, listed by Beadle (1974). Lake Chilwa, in a year of normal depth, varies from five to ten times the salinity of Lakes Malawi, George and Chad, and is similar to Lakes Kivu, Rudolf and Rukwa, when their levels are high. But the ionic compositions differ markedly. In contrast to Lakes Kivu, Rudolph and Rukwa, the alkalinity values are relatively low compared with those of chloride (Morgan & Kalk 1970).

The variation in ionic content shown in Table 4.2 shows that the lowest values occurred in February, when the lake level was highest, and highest in November and December when the lake level had fallen to its lowest ebb. In a year of high lake level (1975–76), the conductivity in the dry season did not exceed 2000 μS cm^{-1} (Fig. 4.1) and in 1977 at the same site, the value did not exceed 1600 μS cm^{-1} (Cantrell pers. comm.), showing that in years of high level the upper range of salinity is reduced. The changes in values of the major ions throughout the year can be explained by straightforward physical factors: dilution and evaporation.

Surface water temperatures varied between 20° and 39°C. There is a seasonal variation, when in the cooler months (June to August) the temperature during the day is around 20°C; in hot dry months the average is 25°C–30°C and in hot wet months it is frequently above 30°C. The temperature does not exceed the thermal limits of life in the lake. Its effect is greater in limiting the solubility of oxygen in very shallow water (see below). The oxygen content of the surface water during 1970 is listed in Table 4.2.

The shallowness of the water prohibits any stable thermal stratification. Examination of surface and bottom samples from all stations and on all sampling occasions confirmed that stable stratification was lacking. Variation in surface and bottom values for conductivity, turbidity, temperature, oxygen and pH did occur, but these differences were inconsistent in their distribution and temporary in nature. In calm conditions (which were fairly rare) a predawn decrease in the oxygen content of bottom waters was detected on 24 hour sampling runs. This is shown for one occasion in Table 4.3, which also shows that the bottom waters tended to be of a slightly higher temperature than surface waters throughout the night. Daily temperature inversion was probably associated with bulk vertical water movements due to convection currents, so that even during these calm periods, vertical stratification would tend to be disrupted. The decrease in oxygen concentration of bottom waters was also clearly transient (McLachlan et al. 1972). This temporary oxygen reduction is unlikely to be deleterious to either animals or plants. Indeed on the occasion illustrated, although the drop is conspicuous, there is quite enough oxygen present near the bottom for respiration of the organisms that live there.

Table 4.2 Characteristics of Lake Chilwa surface water in a year of normal water level (1970).

Month	Conductivity µS cm⁻¹	pH	Total alkalinity meq l⁻¹	O₂ mgl⁻¹	Secchi depth cm	Cations and anions as mgl⁻¹							Depth m
						Na⁻	K⁺	Ca²⁺	Mg²⁺	Cl⁻	*PO₄³⁻	NO₃⁻	
January	1000	8.5	6.7	4.2	6	260	13.7	12.0	7.3	280	5.1	0.1	2.0
February	800	8.2	7.15	5.6	7.5	189	14.0	10.8	6.4	182	3.7	–	2.05
March	1250	8.5	7.0	6.4	3.9	324	16.8	10.0	7.1	256	3.5	–	2.0
April	1500	8.6	8.7	7.4	4.2	474	14.8	7.0	5.2	298	3.9	–	1.9
May	1350	8.7	8.0	7.1	6.5	363	11.3	11.4	6.6	288	2.9	–	1.8
June	1350	8.6	8.1	7.8	6.5	345	15.2	18.0	6.2	280	3.25	–	1.7
July	1500	8.6	8.7	10.0	8.5	350	10.5	13.4	6.3	277	3.1	–	1.55
August	1700	8.7	11.1	8.5	7.4	449	16.4	13.6	6.1	333	4.0	0.15	1.4
September	–	8.7	13.4	8.3	7.0	566	20.7	–	–	395	–	0.38	1.3
October	2100	8.8	16.1	7.7	5.7	633	19.5	15.2	6.8	473	6.4	0.10	1.2
November	2400	8.8	17.3	7.6	8.5	713	23.8	15.2	6.2	497	6.8	0.44	1.1
December	2500	8.8	19.0	6.7	11.0	780	23.1	13.2	8.6	515	5.2	0.26	1.0

– = not determined.   * sulphate was not determined, but was always present.

The continuous mixing of the waters results in high turbidity, as seen in Table 4.2. Turbidity was estimated using a Secchi-disc, modified for the turbulent conditions of Lake Chilwa, as described in McLachlan et al. (1972).

Table 4.3 Variation in oxygen concentration and temperature (°C) over 24 hours in May 1970. *Temperature inversion. †Decrease in oxygen concentration (after McLachlan et al. 1972).

Time	Surface water °C	mg/l O$_2$	Bottom water °C	mg/l O$_2$
Noon (12.00)	22.6	7.4	20.9	7.0
Dusk (8.00)	21.4	7.3	22.2*	7.1
Midnight (24.00)	20.2	7.6	21.2*	7.5
Dawn (6.00)	19.4	7.4	20.4*	†4.7

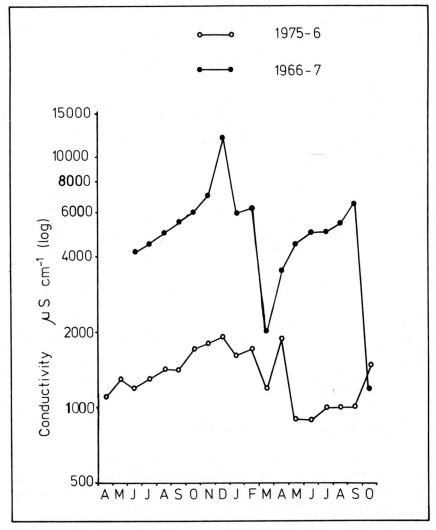

Fig. 4.1 Monthly variations in surface water conductivity (μS cm^{-1}) at an offshore site, beyond Kachulu Bay: open circles 1975–76, a 'normal' year of high lake level; shaded circles 1966–67 the predrying phase (or an example of a 'minor' recession).

As the lake reached higher levels some months after dryness, the turbidity decreased (Fig. 4.6). In the high level year 1975–76 the turbidity varied from 14 to 24 cm. Light transmission at 430 nm measured through a 4 cm light path is normally about 50 per cent.

Table 4.4 gives a list of the chemical and physical properties of the bottom samples from Lake Chilwa in 1970 (McLachlan et al. 1972). A much greater area of the lake bed was covered by mud than sand. The proportion of silt and clay in the mud was very high. This has a high cation exchange capacity, i.e. a strong ability to adsorb cations like Na^+ from the water, which is characteristic of clays. When saturated with cations, clay will precipitate; therefore as the ionic concentrations of the water rises following evaporation, water turbidity decreases. This effect can easily be demonstrated in the laboratory for the particular conditions prevailing in Lake Chilwa, as was done by McLachlan & McLachlan (1976). Organic carbon and nitrogen constitute 4 per cent of the mud, a fairly high level for a shallow lake. The mud in suspension, which accounts for the turbidity, is rich in detritus derived in major part from the decay of the *Typha* plants in the swamp and from the blue-green algae of the lake water.

Table 4.4 Comparison of chemical and physical characteristics of mud samples from Lake Chilwa. A, Mud bottom. B, Sandy bottom. C, Mud aggregates. D, Mud blanket. Values for A, C and D expressed as mean ± standard error (from McLachlan et al. 1972).
*Difference between C and D significant at 5% level (Student's t test).

Analysis	A	B	C	D
Silt + Clay %	87.1±1.1	29.5	80.0±4.0	85.0±4.1
Sand %	12.9±1.1	70.5	10.6±0.2*	1.5±0.1
pH	8.9±0.1	8.7	8.4±0.1	8.5±0.1
Carbon %	1.8±0.1	0.5	2.3±0.1*	2.9±0.1
Nitrogen %	0.3±0.1	0.1	0.3±0.1*	0.4±0.1
Nitrate mg/1000 g	17.7±3.1	13.0	±	±
Phosphate −P mg/1000 g	8 ±	7	10 ±1	10 ±1
Sulphate mg/1000 g	66.0±4.4	46.0	±	±
Na^+	30.8±2.9	11.8	16.0±0.1*	37.3±3.7
K^+	5.3±0.4	1.8	4.0±0.2	3.6±1.2
[1]Extractable Ca^{2+} meq/100 g	32.7±1.2	27.8	42.0±3.1	34.2±2.6
cations Mg^{2+}	11.8±0.9	5.2	14.2±1.4	13.7±3.1
Cation exchange capacity meq/100 g	26.3±0.6	13.8	44.1±4.0	51.5±3.3

[1] Extracted with 1 N ammonium acetate at pH 7.

3. Conditions during the pre-drying and drying phases

As the lake diminishes in size because of evaporative losses and poor rain over three or four years, there is a progressive concentration of dissolved salts in the water and a gradual drying of the lake bottom exposed behind the receding water. This happened throughout 1966, 1967 and 1968.

3.1 *Changes in the lake water*

Data presented in Fig. 4.2 illustrates the inverse relationship that exists be-

66

tween water level and the concentration of dissolved salts, as indicated by conductivity measurements. This is accentuated during the 1966 and 1967 dry seasons; the more 'normal' seasonal relationship is seen in 1970–72. The maximum conductivity of 12 000 μS cm^{-1} recorded in the west in October 1967 (Morgan & Kalk 1970) was greatly exceeded before the last of the lake water disappeared eleven months later. The highest alkalinity in the main lake,

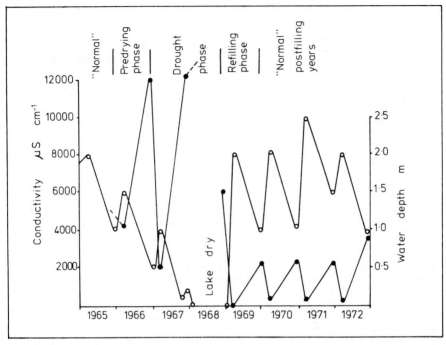

Fig. 4.2 Changes in the concentration of dissoved salts as reflected by conductivity – shaded circles (μS cm^{-1}), and water depth –￩open circles (m), over a seven-year period including a major recession, showing predrying, dry and refilling phases and normal years of recovery.

measured in November 1967, was 143 meq l^{-1} (Moss & Moss 1969) which suggested a conductivity for Lake Chilwa conditions of well over 20,000 μS cm^{-1}. The proportions of the major ions in the drying phase of 1967 are shown in Fig. 4.3. Carbonate ions had increased to a level equal to bicarbonate indicated by the pH of 10.8 (Beadle 1974), and carbonate and bicarbonate together exceeded the concentration of chloride.

Minor ions changed as follows: calcium and magnesium were precipitated at about 6000 μS cm^{-1} in October 1966, an event which was probably associated with physiological changes in organisms described in the following chapters; silicon levels increased after the cyclone in March 1967, probably indicating the large scale death of diatoms; nitrate-nitrogen and phosphate varied by factors of 10 and 5 respectively, levels which were connected with chemical changes in the swamps, discussed in Chapter 5. Unaccountably, ferric iron doubled when the rains started, while after the cyclone only a trace could be detected (Morgan & Kalk 1970) (cf. p. 407).

The monthly changes in conductivity during the period June 1966 to October 1967 are shown in Fig. 4.1. (In the same graph the conductivity (indicating the total ions) of a year of high level, 1975–76, is shown.) The contrast is particularly striking and shows that this fluctuating lake can also experience relatively stable periods, when conductivity is moderate and only small seasonal changes occur.

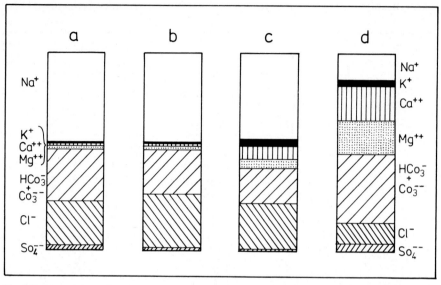

Fig. 4.3 Proportions of major ions in Lake Chilwa at various times in the drying and refilling cycle of the lake and of the Likangala river responsible for some of the input during refilling. (a) conductivity 12,000 μS cm^{-1}, drying phase 1967; (b) conductivity 2000, saline core, refilling phase, 1969; (c) conductivity 400, freshwater ring in the refilling phase, 1969; (d) conductivity 50, incoming river water, Likangala river mouth, filling phase, 1969. (After Moss & Moss 1969 and McLachlan & McLachlan 1976).

The shallow alkaline/saline water was very warm (21°–37°C before noon) and supersaturated with oxygen in the hot wet months by day, (377 per cent oxygen saturation in February), but low in oxygen at night (Morgan & Kalk 1970). Very few organisms except blue-green algae, bacteria and fungi were alive at the end of the predrying year. Kirk (1972) reported that a few catfish, *Clarias gariepinus* and still fewer *Sarotherodon shiranus chilwae*, an endemic cichlid, remained in the lake until June 1967.

The mud was liquid and fine-grained. Deposits of river-borne silt which have accumulated since the late Quaternary have resulted in a considerable depth of mud. During the periods of low water level, the belt of soft deep mud between shore and receding water was virtually impenetrable. This barrier prevented access to the main body of the lake from the west, during the drying phase of 1968. As illustrated in Fig. 4.4a the barrier also has consequences for animals like this otter (exhausted in an attempt to reach land) and for crocodiles, hippopotamuses and fishes cut off from rivers by a drop in lake level.

By energetic and hazardous wading however, Dr. Moss was able to reach the receding water on several occasions in 1968 and was able to observe that, at this stage, the concentration of ions was sufficient to precipitate the clays which had given the water its characteristic turbid appearance. The resulting water clarity and associated light penetration was accompanied in Chilwa by massive blooms of algae (Chapter 6). Indeed the growth was sufficiently intense to be self-shading, thus re-introducing light as a limiting factor. This kind of situation was responsible for the oxygen supersaturation mentioned above. Although we do not have the data, it must be expected that when photosynthetic oxygen production stops at night, the water might rapidly become totally anoxic; this would be a much more severe condition than the mild predawn oxygen reduction in a 'normal' year. This oxygen depletion will be discussed below in relation to limiting factors (section 4.5).

Nitrate and inorganic phosphate were not detected in samples from the main body of water in November 1967. This was probably owing to uptake by abundant algal crops (Moss & Moss 1969). In 1968 the seasonal rise in water level was much smaller than in 1967 and mainly, scattered pools were examined. When the rains came, temporary rain pools had very high concentrations of these and other ions, and their source was undoubtedly the mud which had adsorbed major and minor ions when drying. Another source was also detected: large numbers of corixids decaying in heaps left by the receding water had been noted earlier on the mud flats. Bacterial mineralization of these and many other small organisms was probably responsible for the marked release of the soluble ions from the sediment.

Nitrate levels then fell markedly as water continued to flood over the mud flat and inorganic phosphate and silicate levels also fell. All these decreases were probably due, in part at least, to large growths of epipelic blue-green algae and pennate diatoms on the inundated sediment surface.

The rain pools were connected with the main water mass in February 1968 and lower levels of all ions were recorded, but they were twice as high as the minimum in 1966. During March, evaporation once more started concentrating the ions. The only visible life in the water was abundant blue-green algae. In April 1968 the mud flats dried out and the main body of water became inaccessible except from the eastern shore.

In July 1968 the last conductivity measurement was made in the main body of water, approached from the eastern shore, after travelling overland around the south of the lake. About 150 km² water, 30 cm deep, remained and its conductivity was then 17 000 μS cm^{-1}. All that remained alive was a dense bubbly mat of blue-green algae (Fig. 4.4c). A Land Rover could travel over about 1 km on the sandy bed towards the receding water.

3.2 Changes in the mud

Drying of the exposed lake bottom results first in the appearance of salt on the mud surface. As the clays lose water and shrink, the mud cracks and hardens. Occasional showers of rain redissolve the salt and it is presumably washed into the lake centre. This progressive leaching of the exposed mud must ultimately have resulted in the concentration of much of the total salt there before refilling proper started in late 1968. Both the concentrating of salt at the lake centre and the cracking of the mud into hard blocks of material have a direct bearing on the filling of the lake (see below). The mud produces a noxious odour when exposed to air, which suggests the active decomposition of organic matter, as might be expected under well-aerated conditions of high temperatures that prevailed at the mud surface. For much of the time, the drying process affected only the surface few centimetres of mud which formed a crust over the more liquid deeper layers. The difficulties of moving through this barrier have already been mentioned, nevertheless, as illustrated in Fig. 4.4b, fishermen were undertaking prodigious journeys in the early part of 1967, to the deeper parts of the lake that still held water by pushing their heavy dugout canoes through the mud and climbing aboard periodically for a much needed rest.

Fig. 4.4 (a) Exhausted clawless otter, *Aonyx capensis* about to be killed while attempting to cross the dry lake bed (photo: A. J. McLachlan).

(b) Tracks left by canoes pushed through the mud of the lake bed during the drying phase. Salt crystals on the mud surface and the beginning of the formation of cracks are evident (photo: A. J. McLachlan).

(c) Mats of blue-green algae, mainly *Arthrospira platensis* rising to the surface, full of oxygen bubbles, a few weeks before the lake completely dried (photo: M Kalk).

4. The recovery phase

Conditions during the rapid refilling of the lake basin, when normal rains occurred, are largely the reverse of those during the drying phase except for the time taken. The pre-drying and drying period had taken three years, while the recovery to 'normal' lake level took less than 5 months. Salt, concentrated towards the lake centre, was progressively diluted as the dry, cracked mud of the lake bed was flooded.

4.1 *The water*

Progressive dilution of salts is illustrated in Fig. 4.2 and can be seen to have the same relationship to depth as in the drying phase: alternating dilution and concentration are inversely related to lake level in each succeeding year. In the filling phase, the first record in December 1969 showed a high conductivity. This resulted from a very rapid leaching of salts from the mud similar to that which occurred in the rain pools of the wet season of 1968 described above. The rise was short-lived because water soon poured in from rivers.

The data for Figure 4.2 however, were gathered at a point near the lake edge and reveal nothing about possible discontinuities in salt concentration over the whole lake. Indeed marked discontinuities did occur, especially during filling, when a saline core of water at the lake centre was surrounded by a ring of fresh water from rivers and from run-off from the floodplain (Fig. 4.5.1). The failure of these waters of different salt content (and therefore density) to mix persisted for the entire filling phase. The effect was most pronounced in the first year, but the seasonal change in post-filling years to a lesser extent repeat this pattern of a freshwater ring and more saline body of the lake (Fig. 4.5.4). The freshwater

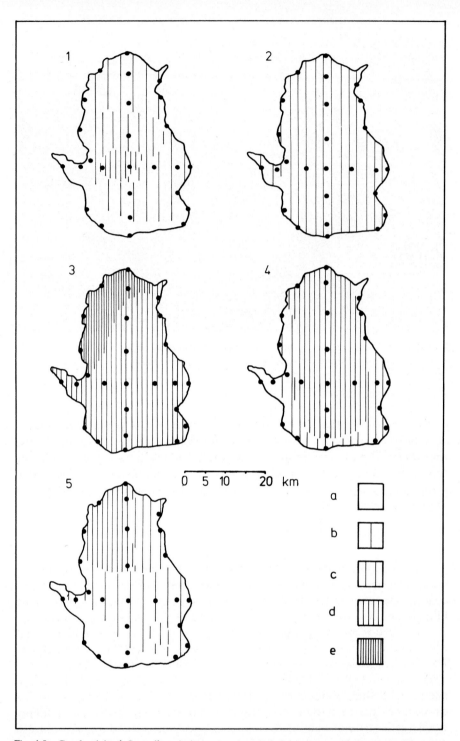

Fig. 4.5 Conductivity (μS cm^{-1}) variations over the whole lake at quarterly intervals, (1) end of rains, March 1969; (2) mid-dry season, June 1969; (3) end of dry season, October 1969; (4) end of rains, February 1970; (5) end of rains, March 1976; (a) less than 500 μS cm^{-1}; (b) 501 to 1000; (c) 1501 to 2000; (d) from 2001 to 2500; (e) over 2500.

Positions of sampling sites are indicated by solid circles. 1–4 after McLachlan et al. (1972); 5 after Kalk (1979).

ring was enhanced in the first flooding because the concentration of salts was at the centre of the lake basin, due to leaching of the peripheral mud during the drying period. Eventually, some months after flooding stopped, the ring disappeared and the salt concentrations became more homogeneous (Fig. 4.5.2).

The pattern for a year of higher lake level, 1975–76 is shown as Fig. 4.5.5 for comparison. The river inflow was very marked in the southern sector of the lake at the time when the northern sector still received water from the swamps, which were still saline.

Examination of the ionic composition of the water confirms that the 'freshwater ring' was of a different type from that in the lake centre. While the percentage composition of salts in the water of the centre was essentially the same as that when the lake was full or during a drying phase, the 'ring' was chemically related to floodplain and river water and was characterized by a greater proportion of K^+, Mg^+ and Ca^{2+} ions (Fig. 4.3). This is the only major exception to the general case that major ions in Lake Chilwa do not vary with concentration.

Nitrate-nitrogen varied from 50 to 250 μg per litre in the lake. 800 μg per litre was recorded on the floodplain during the rains. The interaction of swamp and lake is described in Chapter 13.

Turbid water was especially obvious during filling. Almost opaque at the outset, light penetration, measured by Secchi disc, improved as the lake aged (Fig. 4.6). An annual fluctuation in turbidity accompanied the trend of increasing clarity during the first two years after filling and in later years. 'The swamp-lake edges undergo seasonal change in turbidity due to dilution of the water by rainfall, followed by mixing with the more turbid waters of the open lake' (Howard-Williams 1972).

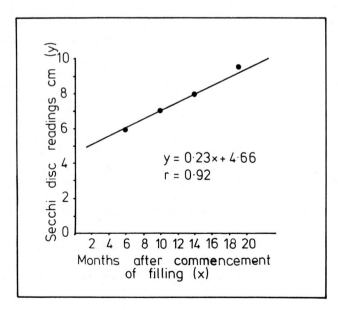

Fig. 4.6 Water turbidity. The correlation between light penetration, as indicated by Secchi-disc readings, and the age of the lake after the commencement of filling (after McLachlan et al. 1972).

It must be expected that species of algae will be influenced by the variation in the amount of light due to changes in turbidity. Restriction of algal photosynthesis to the water surface and to the splash zone on emergent plants was characteristic of the early weeks of the filling phase, after severe drying up. At this time, as was the case in the final stages of drying in the 1968 rain pools, when the algae grew on mud, the ecological emphasis appeared to be on algae growing on surfaces. Production by phytoplankton became more important under 'normal' conditions when turbidity was lower. Turbulence of the water ensured the mixing of the water and distribution of algae throughout the water column. Changes of species from greens to blue-greens and to different species among the latter, occurred in successive years (see Chapter 6).

4.2   *The mud*

The shrinkage of the exposed mud in the dry phase, already referred to, resulted in the development of a pattern of cracks. When the lake refilled, this cracked mud was rapidly eroded and it produced a surface substratum of firm, rounded aggregates, about 3 cm in diameter. The material thus removed during the rounding process was held in suspension by the water and was responsible for the initially high turbidity. When filling was complete, erosion of the lake bottom was reduced because of increased water depth, and particles started to fall out of suspension. This resulted in decrease in turbidity and the deposition of a blanket of very fine material over the mud aggregates. The process was more rapid at the edge due to the shelter from winds provided by emergent vegetation. The depth of this blanket at the lake centre, measured at quarterly intervals, increased progressively from nil to 14 cm; its development was of primary importance to mud-dwelling animals as emphasized in Chapter 9.

There was a reduction in the organic matter in the mud during the dry phase, as noted earlier. Much of what remained became concentrated in the surface sediment layer during filling as a result of the mechanical process of particle size sorting due to wave action (McLachlan & McLachlan 1969). Surface accumulation of organic matter is augmented by the rain of organisms and detritus from the swamp as the recovery phase progresses. The turbulence of the water induced by winds ensures that a proportion of the detritus and mud remain in suspension.

McLachlan (1977) reported that the organic content of the sediment decreased from the filling to the post-filling phases (and consequently the density of the benthos). Subsequently organic matter is derived from the annual growth and decay of the *Typha* swamp, much of which is probably washed into the lake (Howard-Williams & Lenton 1975).

5. Evaluation of changes in the ecosystem in relation to the biota

Two aspects of the effect on the biota of the environmental changes, described in the preceding sections call for comment:

(i) Since all the organisms in the ecosystem, even birds, mammals and some reptiles owe their presence, in greater or less degree to their adaptation to an aquatic way of life, during periods of dryness all the species must be able to make adjustments for survival.

(ii) The fauna and flora of the lake is of low diversity, undoubtedly impoverished by certain factors in the large, gradual or sometimes sudden, changes in the properties of the lake. The species which have been selected in the time since the lake became endorheic, are those which can survive the changes in the long run.

These two aspects are considered at all trophic levels in Chapters 6–14. A general framework for the assessment of the influence of various factors is provided by Beadle (1974). 'Within the freshwater range, mineral composition has little do with the distribution of plant and animal species . . . (since) the biota of freshwater all have ionic regulating mechanisms which normally maintain body fluids hypertonic to the environment.' The range of salinity suitable for most freshwater animals is about 0.5–5 per mille (350–5000 μS cm^{-1}). There is evidence that salinity must remain below about 10 per mille for this mechanism to work. In an environment more saline than this, the regulatory mechanism would have to be a hypotonic type, if the organism is not to succumb.

It has been shown that Lake Chilwa has a seasonal variation in total salinity of 1–2.5 per mille in a normal year, and so is within the range defined for freshwater for seventy per cent of the time. In periods of minor recessions, every six years or so, the salinity reaches 10–12 per mille. Yet all the animal species in Lake Chilwa, as far as is known, have a wide distribution in non-saline water, with the exception of one endemic sub-species of cichlid fish. Presumably all are hypertonic regulators. When salinity approaches 6 per mille (conductivity approx. 6000 μS cm^{-1}) and calcium is precipitated as in 1966, evasive mechanisms come into play, such as succession in algae, diapause in crustaceans and rotifers, migration to the swamp and streams in fishes, aestivation in molluscs even in the bivalve, *Aspatharia* sp. (Cockson 1972) and in some reptiles, migration elsewhere in birds and chironomids, confinement to streams in other reptiles and mammals. These mechanisms are effective during the periods of dryness as well.

There are also glaring cases of mortality, e.g. among diatoms, corixids, molluscs and fishes. The only documented case reported (Morgan, 1972), where ionic regulation appears to have broken down and mass mortalities were reported, was in the endemic cichlid fish *Sarotherodon shiranus chilwae* (Chapter 11).

Osmotic regulation is an energy consuming process which demands an adequate supply of oxygen. Solubility of oxygen decreases with rising temperature and, as stated above, when blue-green algae are so dense as to cause supersaturation of oxygen during the day, the water becomes anoxic at night. In addition, since the regulation is the hypertonic type depending partly on exchange of chloride and bicarbonate in the gill epithelium, the increase in bicarbonate in the water may lower the diffusion gradient of bicarbonate to zero. At some critical level, not investigated physiologically, but considered on experimental and ecological grounds to be at conductivity 12 000 μS cm^{-1} and alkalinity 61 meq l^{-1} regulation by this means would become impossible for this fish (Chapter 11).

No case has been reported in Lake Chilwa of a species with the specialized ionic regulation of the brine shrimp, *Artemia salina* or that of estuarine animals. There are, however, degrees of tolerance of salinity among species. Only three of the thirty nine species of fishes tolerate even the seasonal changes in the lake. Fishes which enter the 'freshwater ring' during the rains, retreat to the swamp and streams when the lake water mixes. No animal species is present in the highly saline and alkaline water of the predrying phase, but some are able to aestivate in the drying mud.

The blue-green algae in this phase are those characteristic of soda lakes, which can tolerate (and may use) carbonate.

The dry lake bed with high salt concentrations and cracking soils proved a very inhospitable habitat for vascular plants and only three species were able to survive these conditions (Chapter 7). The influence of changing salinity during normal years on the rooted vascular plants of the lake appears to be minimal, but it does adversely affect the floating aquatic plants (Chapter 7). Seedling germination of *Typha* was shown to be severely inhibited at salinities approaching those of years of low lake level (Howard-Williams 1975).

Nitrate and phosphate, when scarce are often limiting to plant growth in tropical waters, and Moss (1969) showed in enrichment experiments that *Oscillatoria* could be limited by nitrate. In the post-filling phase *Anabaena* and *Anabaenopsis* occurred (Howard-Williams & Lenton 1975). These have the potential for fixing nitrogen from solution in the water, and they can take the place of nitrate-limited algae during the periods of low nitrate. In time nitrate is built up again, partly by influx of water from the swamp, where it may be released, and different algae *Oscillatoria* sp. and some green algae take over, when the lake is back to normal (Chapter 6).

Moss & Moss (1969) suggested that euplanktonic algae present in the predrying phase could not stand desiccation as well as the epipelic algae, which occurred so plentifully in the drying phase (See Chapter 5).

Calcium ions, which are limiting in some waters, are high in Lake Chilwa, even after precipitations in the drying phase. Silicate ions although variable, depending on uptake by diatom growth, seem to be high. Phosphate is fairly high and also replenished from the wetting of mud in the rains as stated in Chapter 6.

Perhaps the major influence on the biota, which excludes a number of possible candidates for colonization, is the grey, murky, turbid water of Lake Chilwa, which limits the penetration of light, except when salt concentration is very high. Moss & Moss (1969) showed that this was partly due to humic acids released from the swamp, as well as to the high content of silt and detritus. These are the result of large annual and small intermittent exchanges of water with the *Typha* swamp, the decay products of which are flushed out, of the stirring of the bottom muds by water turbulence and also of the high silt load of the rivers. Moss (1969) suggested that 'light may be more important in limiting total primary productivity of the water body than nutrient availability'. The green algal bloom in the filling and post-filling phases were confined to the upper layers of the water and formed an epineustic skin on calm days (McLachlan et al. 1972).

The clogging effect of thick silt probably determines that the robust filter feeders among the zooplankton are more viable (Chapter 8).

Temperatures do not reach lethal levels, except perhaps when the lake has very nearly dried up. They are high enough to facilitate rapid reproduction and extended breeding periods of certain organisms (Chapter 8 and 10).

Many species in swamp lakes are adapted to low content of oxygen. *Clarias gariepinus*, the catfish, has a well-developed respiratory tree in the branchial cavity, which can function in air. Cockson (1972) demonstrated for the Chilwa species, that it has a general distribution of mucous cells which keep the tree moist when exposed to air. The aquatic gills have fewer mucous cells, confined to the tips of the lamellae. Midge larvae of the genera *Nilodorum* and *Chironomus*, which are dominant in Lake Chilwa, and the schistosome vector *Biomphalaria* in the swamp carry a physiological store of oxygen combined with haemoglobin, which enables them to withstand conditions of oxygen shortage for several hours each day. Crustacean zooplankton and other molluscs have haemocyanin which serves a similar function, although the molluscs are mainly air breathers. Nocturnal oxygen reduction alone may not therefore be a serious factor for determining the disappearance of the well-adapted fauna in the drying lake. In combination with ionic stress, it is probably important.

The factors precipitating the evasive responses are discussed in the relevant chapters on organisms.

References

Bayly, I. A. E. & Williams, W. D. 1966. Chemical and biological studies of some saline lakes of South-East Australia. Aust. J. mar. Freshwat. Res. 17:177–228.

Beadle, L. C. 1974. The inland waters of tropical Africa: An introduction to tropical limnology. Longmans, New York. 365 pp.

Cockson, A. 1972. Notes on the anatomy, histology and histochemistry of the respiratory tree of *Clarias mossambicus* (*gariepinus*). Zool. Beitrage 18:101–107.

Howard-Williams, C. 1972. Limnological studies in an African swamp: Seasonal and spatial changes in the swamps of Lake Chilwa, Malawi. Arch. Hydrobiol. 70:379–391.

Howard-Williams, C. 1975. Vegetation changes in a shallow African lake. Response of the vegetation to a recent dry period. Hydrobiologia 47:381–398.

Howard-Williams C. & Lenton, G. M. 1975. The role of the littoral zone in the functioning of a shallow tropical lake ecosystem. Freshwat. Biol. 5:445–459.

Kalk, M. (1979). Zooplankton in a quasi-stable phase in Lake Chilwa, Malawi. Hydrobiologia.

Kirk, R. G. 1972. Economic Fishes of Lake Chilwa. Fisheries Bulletin 4:1–13. Ministry of Agriculture and Natural Resources, Zomba, Malawi.

Marlier, G. 1967. Ecological studies on some lakes of the Amazon valley. Amazoniana. 1:91–115.

McLachlan, A. J. 1977. The changing role of terrestrial and autochthonous organic matter in newly flooded lakes. Hydrobiologia 54:215–217.

McLachlan, A. J. & McLachlan, S. M. 1969. The bottom fauna and the sediments in a drying phase of a saline African lake (L. Chilwa, Malawi). Hydrobiologia 34:401–413.

McLachlan, A. J., Morgan, P. R., Howard-Williams, C., McLachlan, S. M. & Bourn, D. 1972. Aspects of the recovery of a saline African lake following a dry period. Arch Hydrobiol. 70:325–340.

McLachlan, A. J. & McLachlan, S. M. 1976. Development of the mud habitat during the filling of two new lakes. Freshwat. Biol. 6:59–67.

Morgan A. & Kalk M. 1970. Seasonal changes in the waters of Lake Chilwa (Malawi) in a drying phase, 1966–68. Hydrobiologia 36:81–103.

Morgan, P. R. 1972. Causes of mortality in the endemic *Tilapia* (*Sarotherodon*) of Lake Chilwa, Malawi. Hydrobiologia 40:101–119.

Moss, B. 1969. Limitation of algal growth in some Central African Waters. Limnol. & Oceanogr. 14(4):591–601.

Moss, B. & Moss, J. 1969. Aspects of the limnology of an endorheic African lake (L. Chilwa, Malawi). Ecology 50:109–118.

Talling, J. F. & Talling, I. B. 1965. The chemical composition of African lake waters. Int. Rev. ges. Hydrobiol. Hydrogr. 50:421–463.

5 The aquatic environment: II. Chemical and physical characteristics of the Lake Chilwa swamps

C. Howard-Williams

5 The aquatic environment: II. Chemical and physical characteristics of the Lake Chilwa swamps

The form of many large tropical African lakes favours extensive development of marginal vegetation, often as swamps. Lakes such as Chad, Chioga, Bang-weulu, Mweru, Upemba, George, Naivasha, Malombe and Chilwa as well, are wholly or partially surrounded by extensive swamp areas. Beadle (1974) mentions that the area occupied by swamps in tropical Africa may be greater than that of the open waters of all the lakes of the region. Lake Chilwa is therefore not unique in this respect, although as pointed out in Chapter 7 the swamp here is made up of *Typha domingensis* (Fig. 5.1), rather than *Cyperus papyrus* as in most other tropical African lakes. The general structure of swamp communities is similar irrespective of the dominant plant species but, as will be discussed in this chapter, the physical environment of a *Typha* swamp differs in certain important respects from that of a papyrus swamp. These differences make the Chilwa swamps unlike those of other lakes.

Fig. 5.1   The entrance to a canoe channel which extends through the dense *Typha* swamp to shore. These channels may be up to 11 km long (photo: A. J. McLachlan).

1. General features of the swamp habitat

Swamps occupy an ecological position between purely aquatic and purely terrestrial environments and the structure of the swamp ecosystem is rather different from either. In a swamp the autotrophic and heterotrophic layers of the ecosystem (see Odum 1971) are not only functionally different, they occur

81

in different physical media. The bulk of the primary production occurs in the air, and secondary production and consumption generally occur in the water. This results in a distinct spatial separation of most of the producer and consumer organisms.

In a dense stand, swamp vegetation generally forms a closed canopy (Beadle 1974), thus distinct vertical differences in wind, air temperature and light occur through a swamp, and the swamp waters are subjected to an environment totally different from that of an open lake. Distinct variations in the swamp environment occur both spatially and temporally. Spatially we can show both vertical and horizontal differences, whilst temporally, distinct short term (hourly and diurnal) and seasonal changes occur which are very different to those of an open lake.

2. Variations in the Lake Chilwa swamps

2.1 *Spatial variations*

2.1.1 Vertical stratification

Fig. 5.2 shows the vertical differences in midday air temperature, relative wind speed, and light levels from above the swamp vegetation down to the water level in the *Typha* swamp. There is, on average, a 3°C drop in air temperature from just above the top of the vegetation to the water. Light penetration to the water surface through the swamp vegetation varies depending on the water level and the growth of the plants (Howard-Williams 1972), but Fig. 5.2 shows a mean general light profile from a variety of stands on the land and lake edges and in the central swamp. Light levels at the water surface seldom rise above 50 per cent and are often below 10 per cent of those above the vegetation.

Air flow is often reduced to almost zero on the water surface of a *Typha* swamp, and Fig. 5.2 shows that the mean surface wind speed for a variety of stands in the swamp was 18 per cent of that above the vegetation.

2.1.2 Horizontal stratification

Horizontal variations from land through to the lake side of the swamp vary with the seasons, often breaking down completely during the rains when the 'freshwater ring' effect occurs (Chapter 4), and then re-establishing themselves when the rain stops.

Fig. 5.3 shows some typical horizontal gradients in the swamp on the southwest side of Lake Chilwa in May. Gradients are often most pronounced at the end of the dry season (September), but then the water level has dropped such that no water would be present at stations 4 and 5 in Fig. 5.3.

The sand fraction of the soil decreases from land into the lake, and the fine sediments (silt + clay) increase considerably as the water depth increases. The pH of both soils and waters decreases from the lake through to the land edges of the swamp. Although this is the general rule there are some exceptional areas in the south occupied by alkaline marsh vegetation where this does not occur (Chapter 7). The typical decrease in pH shown in Fig. 5.3 is found in most tropical swamps (Carter 1955, Beadle 1974) and many temperate swamp

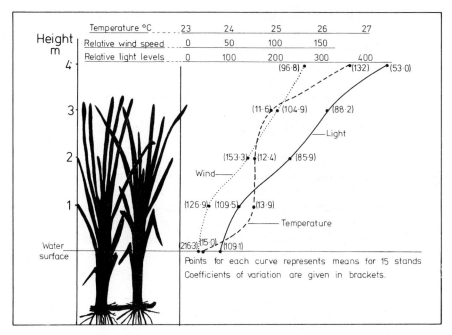

Fig 5.2 Vertical profiles of temperature, wind speed and light through a *Typha* swamp. Points for each curve represent means for 15 stands.

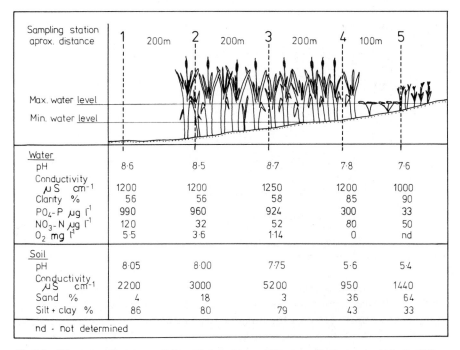

Fig 5.3 Diagrammatic representation of a long transect through a *Typha* swamp in Lake Chilwa. Spatial changes in physical and chemical factors in May 1971 are given across the swamp from lake to land.

communities (Gorham 1953, Dokulil 1973, Přibil & Dvorak 1973). The drop in pH from lake to swamp is due largely to the CO_2 formed from organic decomposition in the swamp and the release of H^+ ions (Gorham 1953, Beadle 1974). Visser (1962) working on Uganda papyrus swamps also suggested that the acid papyrus swamp peat acted as an ion exchange system, exchanging H^+ ions for other ions in solution. This would also tend to lower the pH of the water, but as there is almost no peat formation in the *Typha* swamps of Lake Chilwa (Chapter 13) this H^+ ion exchange system could not operate to any extent here.

Water and soil conductivity increase into the swamp centre from the land edges and then decrease slightly again in the open lake. The central regions of the Chilwa swamps show a typically higher water and soil conductivity than the lake itself. In the extensive swamps on the northwest side of Lake Chilwa (Fig. 7.1), this high dry season conductivity extends up to 3 km inside the swamp from the lake edge, and then decreases rapidly (Howard-Williams 1973). An example of pH and conductivity changes in an 11 km long transect from land to lake through the northwest swamp of Lake Chilwa is given in Fig. 5.4.

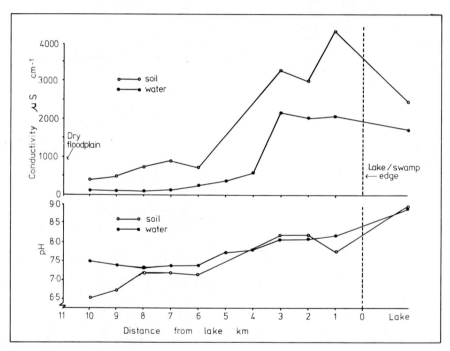

Fig. 5.4   Changes in soil and water conductivity and pH in an 11 km long transect through the northern swamp of Lake Chilwa.

If we consider again the description of the 'freshwater ring' in Chapter 4, when fresh river water moves through the swamps into the lake, it seems somewhat surprising that by the time the dry season comes, the concentrations of ions should be higher in the central swamp than in the lake itself. This can be explained by a number of factors. Firstly, following the rains, the freshwater ring in the lake is rapidly broken down by wind and wave action, and the concentrated lake water then begins to penetrate the swamp as a result of winds. Weisser (in press) gives a full description of this process in shallow lakes. Fig. 5.3 shows how the turbid lake water (53–57 per cent light transmission through a 4 cm light path) has penetrated into the swamp centre. In the extreme northern swamps, this concentrated lake water eventually penetrates several kilometres into the swamp. This process will be described later in more detail (Fig. 5.5), but for the present discussion it can be seen that these wind-induced water movements can rapidly increase the concentration of salts in parts of the swamp. However, they could only result in ionic concentrations equal to, not

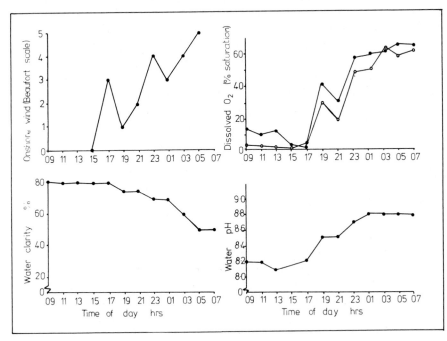

Fig 5.5 Short term temporal changes in the northwest swamps of Lake Chilwa. The sampling site was in the swamp 200 m from the open lake. Changes in dissolved oxygen, water clarity and pH are shown in relation to an onshore wind (after Howard-Williams & Lenton 1975).

higher than, those of the lake. In addition to this process, concentration of salts to levels above those in lake water can only occur as a result of the release of ions from swamp sediments and from the decomposition of swamp plant material. The central swamp muds are richer in total ions (Figs. 5.3 and 5.4) than the lake muds, and the water in the swamp is shallower than in the open lake, so the ratio of water volume to mud exchange surface is lower in the swamp than in the lake. Both these facts would allow for a further increase in the ionic concentration of the swamp water. In addition, as will be described shortly, the central swamp waters become deoxygenated and the now classic work of Mortimer (1941) shows conclusively how nutrient release from muds into water is enhanced in anaerobic conditions. Thus in the dry season the central swamp waters reach higher conductivities than those of the lake, but only in areas where lake water can penetrate and where the muds are as rich or richer in total ions than those of the lake.

Perhaps one of the most important factors influencing aquatic organisms in swamps is the availability of oxygen in the water. The waters of very dense tropical swamps are usually devoid of oxygen (Carter 1955, Beadle 1958, 1974). This deoxygenation is due to three factors. The vegetation cover hinders air movement across the water surface and reduces turbulence and aeration. Secondly, the low light intensity reaching the water surface retards algal growth and oxygen production. Thirdly and most important, there is a continual rapid uptake of oxygen by organisms decomposing the abundant organic material produced by the macrophytes. The waters of the landward edges and much of the centre of the Lake Chilwa swamps are thus devoid of oxygen except during the rainy season when oxygenated water from the catchment enters the swamps. The significance of this to the fauna is discussed in Chapter 13. However, on the lake edges of the swamp, and in some central areas, mixing with oxygenated lake water occurs as described previously, and in sheltered lagoons within the swamps (natural and man made) where high light intensities reach the water surface (Chapter 7), dense growths of epiphytic algae, benthic algal mats and submerged macrophytes such as *Utricularia* spp., *Ceratophyllum demersum* and *Ottelia* spp. occur. In these lagoons the daytime oxygen values can reach 188 per cent saturation (Howard-Williams & Lenton 1975).

2.2 Temporal variations

2.2.1 Short term changes

The area in Lake Chilwa which is subjected to the greatest short term change is the lake edge of the swamp which mixes with the lake water. This is because, due to wind drift, it only takes a few hours for lake water, with its very different properties, to penetrate the swamp margins. This effect can be clearly seen in Fig. 5.5 which shows the influence of an onshore wind on dissolved oxygen, pH and water clarity in a swamp sampling point 200 m from the open lake edge. The swamp waters were deoxygenated from morning until 17.00 hours. At this time a strong onshore wind started to blow on the lake, and within two hours lake water had penetrated the 200 m of swamp and had reached the sampling site. The dissolved oxygen and pH suddenly rose and the water clarity dropped. The wind continued to blow for the remainder of this sampling period, the lake water penetrated further into the swamp and more and more oxygenated turbid lake water entered the sampling area. Because of the very dense floating mat formed by *Cyperus papyrus*, such extensive water movements do not occur in papyrus swamps. This is a major difference between papyrus swamps and those dominated by *Typha* or *Phragmites*.

Since strong short lived winds can blow from any direction throughout the year, mixing of peripheral swamp and lake water often occurs around the lake. This mixing is a very important phenomenon in the ecology of Lake Chilwa and more attention will be given to it in Chapter 13.

2.2.2 Seasonal changes

Generally, seasonal changes in the physical and chemical environment in the swamps are related to annual changes in water level which in turn reflect the climatic changes in the catchment (Howard-Williams 1972). The general pattern of change in swamp waters can be seen from conductivity changes (Fig. 5.6) as the changes of all the major ions in the swamp waters closely followed conductivity (Howard-Williams 1972).

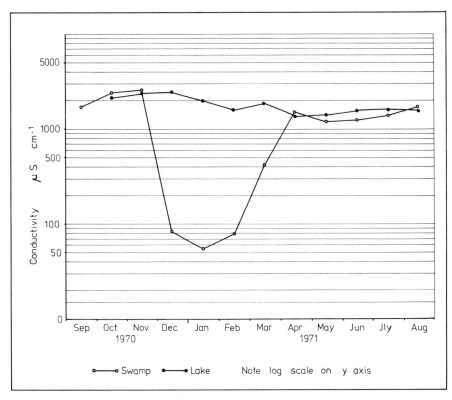

Fig. 5.6 Seasonal changes in conductivity in the swamps when compared with those in the open lake (after Howard-Williams & Lenton 1975).

Towards the end of the dry season in September and October, the conductivity values were the highest for the year. After the rains began in November the conductivity dropped by 2000 μS cm^{-1} in one month. This was due to flooding of the swamp by freshwater runoff (conductivity 80 μS cm^{-1}) from the catchment area. However, the very sharp drop in conductivity from November to December cannot be accounted for merely by dilution of the existing swamp water, as the water level had only just begun to rise. The swamp water volume had increased two to three times whilst the conductivity had dropped twenty five times. What appears to happen in the swamps is that the swamp water is completely replaced by fresh water from the catchment. The inflowing fresh waters seem to push the swamp water out into the lake ahead of them (Chapter 13, Howard-Williams & Lenton 1975). The conductivity then slowly rises again due to wind induced mixing with lake water and, later in the year, concentration due to a drop in water level caused by evaporation.

While conductivity reflects changes in the major ions and in Lake Chilwa, phosphate also (Chapter 4) it does not always reflect changes in nitrate, the other very important minor ion necessary for plant growth. Fig. 5.7 shows changes in water level, conductivity and nitrate-nitrogen during 1971 on the

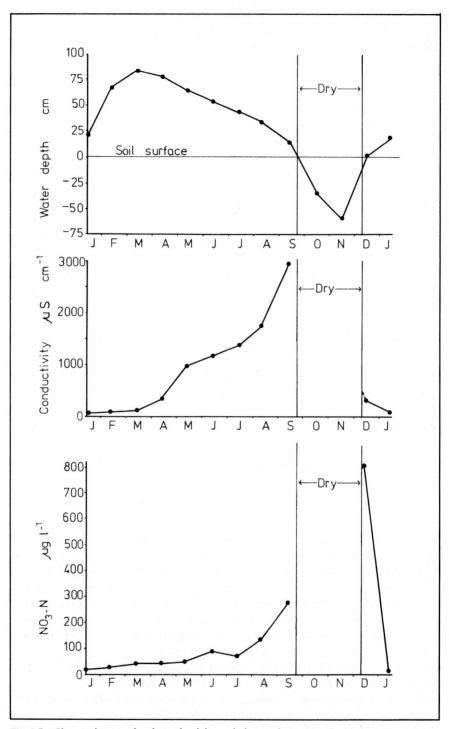

Fig. 5.7 Changes in water level, conductivity and nitrate-nitrogen on the landward edges of the *Typha* swamp at station 5 in Fig. 5.3. Note the high nitrate pulse following the dry period.

seasonally flooded marsh behind the *Typha* swamp. The conductivity shows an almost exactly opposite trend to water level, the highest conductivity occurring just before the marsh dried out completely. The marsh remained dry for only about two months before flooding with water of low conductivity occurred again in December. During the dry period the marsh area was extensively grazed and trampled by cattle. The nitrate-nitrogen level did not show any definite trend until just before final drying when it rose sharply. Immediately after flooding, however, a massive release of nitrate-nitrogen occurred (Fig. 5.7). Such releases of nitrate following flooding and subsequent rewetting of tropical soils are well documented (Birch 1960, Hickling 1961, Viner 1975 etc.). Alternate wetting and drying of soils appears to stimulate, temporarily, microbial activity and enhance the mineralization of carbon and nitrogen (Agarwal et al. 1971, Patrick & Tusneem 1972). McLachlan S. M. (1970, 1971) attributed a release of nutrients, including nitrogen, from the gently sloping shorelines of Lake Kariba following rains, to a release from dry grass and animal dung. In this respect the large herds of cattle on the Chilwa floodplain almost certainly play a part.

3. Summary

In conclusion, the Lake Chilwa swamps provide a habitat in which water turbulence is reduced, and light levels, dissolved oxygen and pH are reduced generally although there are variations within the swamp. Enrichment of major ions in the central swamp waters occurs through wind mixing of lake and swamp water and also from release of ions from the rich muds of the central swamp. On the landward edges, enrichment of nitrate-nitrogen occurs after a dry spell. Very marked changes occur both on a short term and a seasonal basis, and when compared with the open lake there is far greater spatial variation in environmental factors.

References

Agarwal, A. S., Singh, B. R. & Kanchiro, Y. 1971. Soil nitrogen and carbon mineralization as affected by drying-rewetting cycles. Proc. Soil Sci. Am. 35:96–100.
Beadle, L. C. 1958. Hydrobiological investigations on tropical swamps. Verh. internat. Ver. Limnol. 13:855–857.
Beadle, L. C. 1974. The inland waters of tropical Africa. Longmans, London. 365 pp.
Birch, H. F. 1960. Nitrification in soils after different periods of dryness. Plant and Soil 12:81–96.
Carter, G. S. 1955. The papyrus swamps of Uganda. W. Heffer and Sons, Cambridge. 25 pp.
Dokulil, M. 1973. Planktonic primary production within the *Phragmites* community of Lake Neusiedlersee (Austria). Pol. Arch. Hydrobiol. 20:175–180.
Gorham, E. 1953. Chemical studies on the soils and vegetation of waterlogged habitats in the English Lake District. J. Ecol. 41:345–360.
Hickling, C. F. 1961. Tropical inland fisheries. Longmans, London. 287 pp.
Howard-Williams, C. 1972. Limnological studies in an African swamp: seasonal and spatial changes in the swamps of Lake Chilwa, Malawi, Arch. Hydrobiol. 70:379–391.
Howard-Williams, C. 1973. Vegetation and environment in the marginal areas of a tropical African lake (L. Chilwa, Malawi). Ph.D. Thesis, University of London. 312 pp.
Howard-Williams, C. & Lenton, G. M. 1975. The role of the littoral zone in the functioning of a shallow tropical lake ecosystem. Freshwat. Biology 5:445–459.
McLachlan, S. M. 1970. The influence of lake level fluctuations and the thermocline on water

chemistry in two gradually shelving areas in Lake Kariba, Central Africa. Arch. Hydrobiol. 66:499–510.

McLachlan, S. M. 1971. The rate of nutrient release from grass and dung following immersion in lake water. Hydrobiologia 37:521–530.

Mortimer, C. H. 1941. The exchange of dissolved substances between mud and water in lakes. I and II. J. Ecol. 29:280–329.

Odum, E. P. 1971. Fundamentals of ecology. W. B. Saunders Company, Philadelphia. 574 pp.

Patrick, W. H. & Tusneem, M. E. 1972. Nitrogen loss from a flooded soil. Ecology 53:735–737.

Přibil, P. & Dvorak, J. 1973. Variation in some physical and chemical properties of the water in the stand of *Phragmites communis*. In: S. Hejný, (ed.) Ecosystem study on wetland biome in Czechoslovakia. Czechoslovak IBP/PT–PP Report No. 3, Trebon. 71–78.

Viner, A. 1975. The supply of minerals to tropical rivers and lakes (Uganda). In: G. Olson (ed.) An introduction to land-water relationships. Academic Press, London. pp. 227–261.

Visser, S. A. 1962. Chemical investigations into a system of lakes, rivers and swamps in S.W. Kigezi, Uganda. East Afr. Agric. and Forestry Journal 28:81–86.

Weisser, P. 1978. A conceptual model of siltation system in shallow lakes with littoral vegetation. J. Limnol. Soc. South. Afr. 4:145–149.

Part 2. The response of plants and animals to changes

6 Algae in Lake Chilwa and the waters of its catchment area

Brian Moss

Algae in Lake Chilwa and the waters of its catchment area

All the genera and many of the species of algae have a world-wide distribution, unlike most other groups of organisms. Though only rarely has dispersal been observed in progress, the microscopic algae seem to have been moved rather readily by air currents and animals associated with water. There may be endemic tropical freshwater micro-algae, and careful workers have indeed recognized some on the basis of descriptive studies. However many algae are now known to be extremely plastic, and in the absence of very thorough studies in culture, it is unwise to draw conclusions about the biogeography of species. However, understanding of some of the physiological ecology of the algal phyla and sub-taxa has increased recently and the algal flora of Lake Chilwa and its catchment is best considered in the light of these rather than in that of biogeography.

The habitats of the catchment area will be considered as follows: (1) dilute oligotrophic waters of upland streams and man-made dams, which ultimately feed the lowland rivers; the Mlungusi dam and river on Zomba mountain are representative of these; (2) lowland rivers immediately feeding Lake Chilwa the Likangala River draining the Shire Highlands is typical; (3) Lake Chilwa itself in both its low and high water level phases, and (4) temporary pools forming on the Chilwa plain after heavy rain.

Some indication of the features of water chemistry at the sites examined in 1967/1968, particularly as they relate to the algal flora are shown in Table 6.1, and further details will be found in Moss & Moss (1969), Moss (1970) and elsewhere in this volume (Chapters 4 and 5).

Table 6.1 Characteristics of sites sampled for algae in the Lake Chilwa Catchment area, November 1967 – April 1968.

	Temperature °C	Alkalinity meq l^{-1}	Max. PO_4–P $\mu g\ l^{-1}$	Max. NO_3–N $\mu g\ l^{-1}$	Max. SiO_3–Si mg l^{-1}
Mlungusi R.	10.4–17.6	0.076–0.139	Trace	68	2.11
Mlungusi Dam	14.0–22.3	0.091–0.129	Undetectable	39.6	4.8
Likangala R.	19.6–24.5	0.16– 1.10	24	171	7.5
L. Chilwa (drying phase)	25.0–35.5	15.2–200	7520	39 000	540

The dissolved ion content of the water increases markedly from the upland headwaters on the hard rock mountain massifs, through the lowland plain where a softer rock catchment of increasing area provides additional ions. Finally the evaporation-dominated water regime and local soluble carbonatites combine to give high ionic levels in the lake itself, particularly during its drying phase, but comparatively so also during its flooded phase (McLachlan et al. 1972). The temperature range of the water also increases both in amplitude and level from the headwaters to the lake basin. In the former it overlaps that of many temperate streams, but is clearly higher in the lake basin. The tempera-ture range over which many common green (*Chlorophyta*) and blue-green

95

(*Cyanophyta*) algae will grow, however, easily encompasses the range found in L. Chilwa (Fogg et al. 1973, Moss 1973), and temperature is unlikely to exert as decisive an effect on algal distribution within the catchment as water chemistry (Moss 1972).

1. The upland waters

The limnology of the Mlungusi dam and river has been described by Moss (1970). The dam is poor in phytoplankton, partly because of the infertility of the water, and partly because of the high flushing rate, but supports an interesting epipelic (sediment-living) flora (Fig. 6.1). This comprises pennate diatoms, a few blue-green algae and euglenoids, flocs of *Spirogyra* and Mougeotia and a range of those desmid genera (e.g. *Micrasterias, Penium, Euastrum, Pleurotaenium*), associated with waters of alkalinity below 1.5 meq/l (Moss 1972). Among the diatoms are *Eunotia* and *Frustulia*, which also are characteristic of waters of low alkalinity elsewhere in the world (Round 1964).

2. The lowland river

The Likangala River was regularly sampled in 1967–68 at a point about halfway between Zomba and Lake Chilwa. Though the water of the Likangala River is more fertile than that of the Mlungusi dam, vigorous water flow prevents the epipelic flora from developing to the potential maximum, and sediment was often washed completely away from the sampling station. The flora differed from that of the Mlungusi dam and river in its greatly increased representation of diatoms and in its few desmid species (Fig. 6.2). This is of interest since the water chemistry was not such as to prevent development of a rich desmid flora. However, the desmids are often associated with quiet backwater and lake sediments rather than rushing rivers, and diatoms, with their heavier, silica-walled cells and their greater powers of attachment to sediment grains may have a selective advantage in the latter.

3. Algae in the lake

3.1 *Drying phase*

As river water enters Lake Chilwa, the many chemical changes it experiences are reflected also in changes in the algal flora. In the pre-drying phase of Lake Chilwa, 1966 to early 1967, there were few green algae; *Scenedesmus quadricauda* (Turp.) Breb. was present, although scarce. The water was dominated by a dense growth of blue-green algae, consisting of *Oscillatoria planctonica* Wol. and *Anabaena torulosa* (Carm.) Lagh.

In its drying phase when very alkaline and saline, Lake Chilwa supported very few species of algae, though each was very abundant. The lake water then contained only the planktonic filamentous blue-green algae (*Arthrospira, Spirulina* and *Anabaenopsis*) (Fig. 6.3). Predominant was *Arthrospira platensis* (Nordst.) Gom., which is characteristic of carbonate-rich warm waters elsewhere in Africa (Rich 1932, Talling & Talling 1965) and in South and Central America. The maximum density recorded was 92,000 filaments per ml.

96

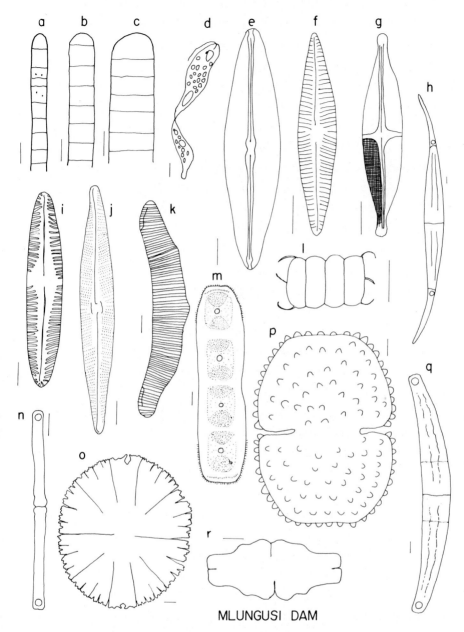

MLUNGUSI DAM

Fig. 6.1 Representative epipelic algae from the Mlungusi dam, drawn from live specimens or acid cleaned frustules (diatoms). Scale lines represent 5 μm. a–c, Cyanophyta (blue-green algae): (a) *Oscillatoria terebriformis* Ag; (b) *Oscillatoria subbrevis* Schmidle; (c) *Oscillatoria nigra* Vaucher; (d) *Euglena elastica* Prescott (Euglenophyta); e–g, i–k, Bacillariophyta (diatoms); (e) *Frustulia rhomboides* Breb; (f) *Navicula cryptocephala* Kuetzing; (g) *Stauroneis lauenbergiana* Hust; (i) *Pinnularia brebissonii* (Kuetz.) Rab; (j) *Navicula viridula* Kutz; (k) *Eunotia triodon* Ehr.; h, l–r, Chlorophyta (Green algae, all are desmids except 1), (h) *Closterium ralfsii* (Bréb.); (l) *Scenedesmus spinosus* Chod; (m) *Penium margaritaceum* (Ehr.) Bréb; (n) *Pleurotaenium trabecula* (Ehr.) Naeg; (o) *Micrasterias denticulata* Bréb; (p) *Cosmarium supraspeciosum* Wolle; (q) *Closterium striolatum* Erh; (r) *Euastrum ansatum* (Ehr.).

97

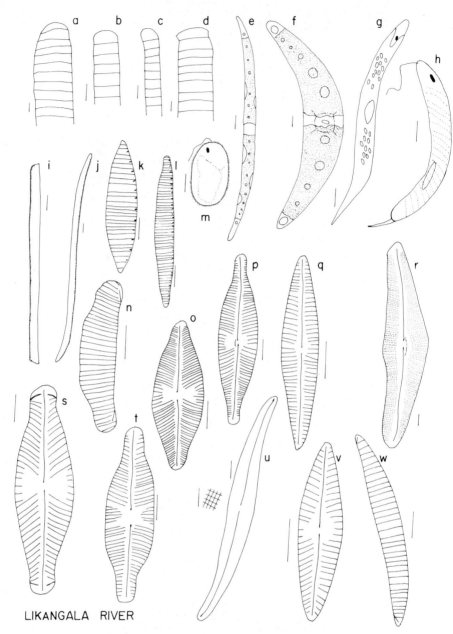

LIKANGALA RIVER

Fig. 6.2 Representative epipelic algae from the Likangala River, drawn from live specimens or acid cleaned frustules (diatoms). Scale lines represent 5 μm, except that for the detail of striae in (u) which represents 2.5 μm. a–d, Cyanophyta: (a) *Oscillatoria curviceps* Ag; (b, c, d) *Oscillatoria* spp; e, f, Chlorophyta, both desmids; (e, f) *Closterium* spp; g, h, m, Euglenophyta; (g) *Euglena* sp; (h) *Euglena spirogyra* Ehr; (m) *Trachelomonas* sp; i–l, n–w, Bacillariophyta; (i) *Nitzachia* sp; (j) *Nitzschia sigma* (Kütz) W.Sm; (k) *Nitzschia fonticola* Grun; (l) *Nitzschia palea* Kutz; (n) *Eunotia pectinalis* var. *minor* (Kütz.) Rabh; (o) *Navicula placentula* Ehr; (p) *Navicula virdula* var. minor, Kg. K. B; (q) *Navicula gracilis* Ehr; (r) *Cymbella cistula* (Hemprich) Grun; (s) *Navicula pupula* Kütz; (t) *Capartogramma* sp; (u) *Gyrosigma acuminatum* (Kütz.) Rabh; (v) *Navicula cryptocephala* var. *veneta* (Kütz.) Grun; (w) *Eunotia veneris* (Kütz) O. Hull.

Anabaena and *Anabaenopsis* have heterocysts and almost certainly actively fix nitrogen in the lake. This may be important in the nitrogen economy of the lake since nitrogen may be limiting (Moss & Moss 1969) to phytoplankton growth, particularly at the end of the dry season when supplies from the catchment area are at their lowest. These genera were also abundant at this season in 1971/72 (Howard-Williams & Lenton 1975). On sediment underlying water with dense blue-green algal populations epipelic algae were scarce, but as the water retreated in 1968 leaving wet sediment, sometimes temporarily flooded with rainwater, extensive crops of *Oscillatoria* spp., (maximum number of filaments $7 \times 10^6/cm^2$) and two pennate diatoms, *Nitzschia palea* (Kutz.) W.Sm. and *Anomoeoneis sphaerophora* (Kutz.) were patchy but often very large (maxima $1 \times 10^6/cm^2$ and $5 \times 10^4/cm^2$ respectively). Further details of crop sizes are given in Moss & Moss (1969).

Table 6.2 provides a comparison of features of the epipelic algal populations from the Mlungusi dam, the Likangala and Lake Chilwa on the same dates in

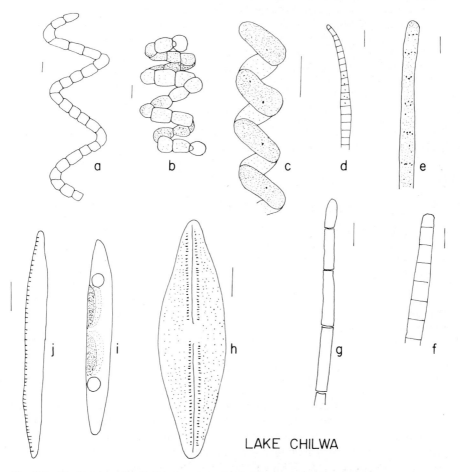

LAKE CHILWA

Fig. 6.3 Planktonic and epipelic algae from Lake Chilwa, drawn from live specimens or acid cleaned frustules (diatoms). Scale lines represent 5 μm. a–c, planktonic, d–j, epipelic, a–g Cyanophyta; (a) *Arthrospira platensis* (Nordst.) Gom; (b) *Anabaenopsis circularis* G. S. W. Wol. et Miller; (c) *Spirulina major* Kg; (d, e, f, g) *Oscillatoria* spp; h–j, Bacillariophyta; (h) *Anomoeoneis sphaerophora* (Kütz.); (i, j) *Nitzschia palea* Kütz.

1967–68. The extreme fertility and relatively low diversity of the last compared with the former sites are well shown. The Chilwa wet mud flat was a hot, salty, probably highly reducing environment, judging from the abundant organic matter dissolved in the water, the decomposing bodies of stranded invertebrates and the odour which emerged from it. In an extreme environment such as this the generally thermophilic (Fogg et al. 1973) and microaerophilic (Stewart & Pearson, 1970) blue-green algae tolerating higher temperatures and exposure to air were able to increase in the absence of competition from groups other than the two species of diatoms found there.

Table 6.2 Comparative measurements of the epipelic algae of the Mlungusi Dam, Likangala R. and Lake Chilwa (drying phase). (Based on means of at least two replicates in each case). Diversities are calculated from the Shannon-Weaver formula.

	Mlungusi	Likangala	Chilwa
December 18, 1967			
Mean standing crop (mg chlorophyll *a* m^{-2})	2.8	4.3	15.5
Total number of algae cm^{-2}	13.4×10^3	9.5×10^3	No data
% diatoms	36	83.5	
% blue-green algae	7.4	13.9	
% green algae	52.6	0.8	
% others	4.0	1.8	
Diversity (bits per individual)	2.2	3.7	
January 3, 1968			
Mean standing crop (mg chlorophyll *a* m^{-2}	3.7	1.4	17.4
Total number of algae cm^{-2}	11.99×10^3	5×10^3	0.9×10^6
% diatoms	18.8	92.9	0
% blue-green algae	15.4	0	100
% green algae	39.7	6.2	0
% others	26.1	0.9	0
Diversity (bits per individual)	3.1	3.4	0.58
January 9, 1968			
Mean standing crop (mg chlorophyll *a* m^{-2})	2.1	1.9	56.5
Total number of algae cm^{-2}	5.2×10^3	3.9×10^3	6.5×10^6
% diatoms	33.6	93	0.01
% blue-green algae	21.7	0	99.99
% green algae	33.7	1.3	0
% others	11.0	5.7	0
Diversities (bits per individual)	3.2	2.8	0.72

3.2 Early filling phase

As the lake itself filled in late 1968/1969 it became very turbid with suspended inorganic matter (McLachlan et al. 1972), and phytoplankton was scarce, though some organisms were noted in a surface film. It is of interest that during the 1967–68 rainy season, pools of water, up to 50 cm deep, lying temporarily on the soil surface of the Chilwa Plain, had a comparable algal community, which may perhaps be regarded as a microcosm of what happened in the lake itself. This algal community is called the neuston.

The pools were extremely turbid and light absorption measurements showed that light would not penetrate at intensities capable of supporting net photosynthesis to depths below 0.7 mm. Algae were thus confined to the surface film, where a striking red scum of *Euglena sanguinea* Ehr. was often characteristic (Fig. 6.4). (I have seen similar neustonic scums of this species on pools during summer in Michigan, U.S.A.). The red colour is given by red (probably carotenoid) pigments which are enclosed in movable granules in the cell. The scums appear red during the higher light intensities of the late morning and mid afternoon, but change to green late in the afternoon. The red pigments may thus protect the photosynthetic apparatus from extremely high light intensities. Cysts of *E. sanguinea* were found and these may be the means by which it survives the intervening dry periods. Other species (Fig. 6.4) were also present in the neuston scums. These were largely flagellate species, including other *Euglenas*, *Trachelomonas*, *Chlamydomonas*, *Eudorina*, *Platydorina*, and *Pandorina*. An occasional cell of *Nitzschia palea* or *Closterium* was sometimes present, and nematodes, rotifers and protozoa (*Arcella*) were found grazing in the film.

When the lake reached its high level of the filling phase during and after rains of 1969, a surface film of green alga was present. This consisted of Cholorophyceae, especially Volvocales and Chrysophyceae, including *Synura* sp. (Mwanza 1970). In the swamps Chlorophyta were also present, and *Spirogyra*, *Pediastrum* and *Staurastrum* were recorded by Mwanza (1970), and undoubtedly represent a flora of considerable richness at the time.

3.3 *High water phase*

When the filling phase came to an end with the decrease in lake level of the dry season, the flora changed. *Anabaena* and *Anabaenopsis* played a major part in the algal populations of 1971 and 1972 (Howard-Williams & Lenton 1975). The post-filling phase thus displayed yet another change in the nature of the algae. At the end of the rains when the lake was almost full in a year of especially high level, in March 1977, samples provided by Dr. M. A. Cantrell, from Kachulu Bay, which is usually typical of the lake, displayed a much richer flora. *Oscillatoria* sp. was most abundant in the phytoplankton, though *Trachelomonas* spp., *Euglena spirogyra*, and *Phacus* sp. (Euglenophyta), *Cyclotella* sp. and *Nitzschia* sp. (Bacillariophyta), *Anabaena* sp. (Cyanophyta), *Scenedesmus quadricauda* (Chlorophyta) and *Peridinium* sp. (Pyrrophyta) were also present.

Scrapings from the dominant plant of the swamp, *Typha domingensis*, showed the periphyton community to be a mass of largely blue-green algae (*Phormidium*, *Oscillatoria* and *Anabaena*) with Chlorophyta (*Spirogyra* and *Stigeoclonium*) and diatoms (*Cymbella*, *Nitzschia palea*, *Synedra* and *Eunotia*).

4. Summary

The algae of Lake Chilwa and its inflowing waters demonstrated sharply the general correlation between algal communities and water chemistry observed widely in fresh and brackish waters. Soft water highland streams and dams have desmid-dominated floras, whilst a lowland river flowing into the lake showed a predominance of diatoms on its sediments. The lake itself in its drying and most saline phases had the dense population of Cyanophyta, particularly *Spirulina*

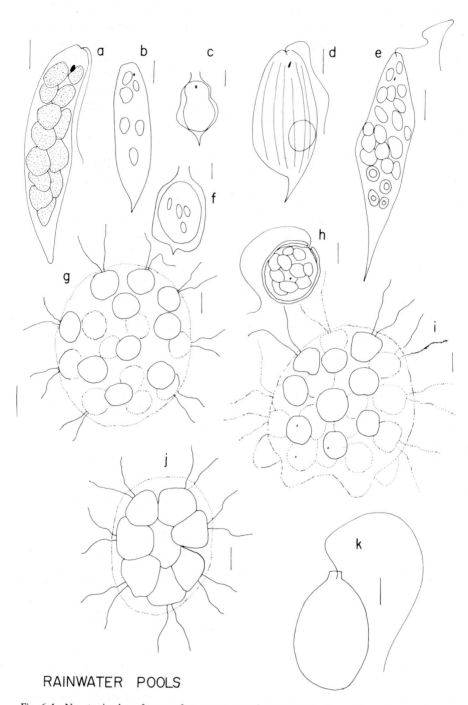

RAINWATER POOLS

Fig. 6.4 Neustonic algae from surface scums on rainwater pools. Drawn from live specimens. Scale lines represent 5 μm. a–f, h, k, Euglenophyta; (a, b), *Euglena sanguinea* Ehr; (c, f, h, k) *Trachelomonas* spp; (d) *Phacus caudatus* Hueb; (e) *Euglena* sp; g, i, j Chlorophyta (Volvocales); (g) *Eudorina elegans* Ehr; (i) *Platydorina caudata* Kofoid; (j) *Pandorina morum* Bory.

platensis, observed in other endorheic lakes, but the phytoplankton greatly diversified as water levels rose and salinities fell. Various species of Cyanophyta predominated at different times.

An unusual flora of the surface water film was associated with the lake as its level rose after a period of complete desiccation. This neustonic flora was paralleled by communities of algal flagellates which developed on the temporary puddles in the catchment area during the wet season.

References

Fogg, G. E., Stewart, W. D. P., Fay, P. & Walsby, A. E. 1973. The blue-green algae. Academic Press, London. 459 pp.

Howard-Williams C. & Lenton, G. 1975. The role of the littoral zone in the functioning of a shallow lake ecosystem. Freshw. Biol. 5:445–459.

McLachlan, A. J., Morgan, P. R., Howard-Williams, C., McLachlan, S. M. & Bourn, D. 1972. Aspects of the recovery of a saline African lake following a dry period. Arch. Hydrobiol. 70:325–340.

Moss, B. 1970. The algal biology of a tropical montane reservoir (Mlungusi Dam, Malawi). Brit. phyc. J. 5:19–28.

Moss, B. 1972. The influence of environmental factors on the distribution of freshwater algae – an experimental study. Part I. Introduction and the influence of calcium concentration. J. Ecol. 60:917–932.

Moss, B. 1973. The influence of environmental factors on the distribution of freshwater algae – an experimental study. Part 3. Effects of temperature, vitamin requirements and inorganic nitrogen compounds on growth. J. Ecol. 61:179–192.

Moss, B. & Moss, J. 1969. Aspects of the limnology of an endorheic African lake (Lake Chilwa, Malawi). Ecology 50:109–118.

Mwanza, N. P. 1970. Algal ecology: Recovery phase. In: M. Kalk, (ed.) Decline and Recovery of a Lake. Government Printer, Zomba, Malawi. 28–29.

Rich, F. 1932. Reports on the Percy Sladen Expedition to some Rift Valley lakes in Kenya in 1929 – IV. Phytoplankton from the Rift Valley lakes in Kenya. Ann. Mag. Nat. Hist. Ser. 10:233–262.

Round, F. E. 1964. The ecology of benthic algae. In: D. F. Jackson, (ed.) Algae and Man. Plenum Press N.Y. 138–184.

Stewart, W. D. P. & Pearson, H. W. 1970. Effects of aerobic and anaerobic conditions on growth and metabolism of blue-green algae. Proc. R. Soc. B. 175:293–311.

Talling, J. F. & Talling, I. B. 1965. The chemical composition of African lake waters. Int. Rev. ges. Hydrobiol. 50:421–463.

See checklist for additional diatoms on *Scirpus littoralis* in open water [Ed.].

7 The distribution of aquatic macrophytes in Lake Chilwa: Annual and long-term environmental fluctuations

C. Howard-Williams

7 The distribution of aquatic macrophytes in Lake Chilwa: Annual and long-term environmental fluctuations

One of the most striking features of all temperate water bodies is the conspicuous zonation of vegetation, usually associated either directly or indirectly with changes in water depth. However, in Lake Chilwa, as in other tropical African lakes particularly the shallow ones, the relationship between zonation of vegetation and water depth is not as simple. This is due to the unusually large water level fluctuations relative to lake depth which occur here. Indeed, Beadle (1974) pointed out that the type of vegetation in tropical Africa seems to depend on water level fluctuations rather than merely a water depth gradient.

Because it is a closed drainage basin, changes in water level in Lake Chilwa result in particularly noticeable changes in other environmental factors (McLachlan et al. 1972, Howard-Williams 1972, and Chapters 3 & 4), particularly those relating to water chemistry. The latter can also have an effect on the vegetation although, as will be described later in this chapter, it is only during exceptionally dry conditions that such changes (mostly reflected as increases in salt concentration) appear to have a direct measurable effect on the presence or absence of various plant species.

In this chapter the vegetation of Lake Chilwa is described in outline and the major environmental influences on the vegetation are discussed. Particular stress is laid on the fluctuations in water level and concentration of sodium chloride and bicarbonate, and the effects of these on the growth of various plant species. Of particular interest in the African context is the presence of *Typha domingensis* rather than *Cyperus papyrus* as the dominant swamp plant in Lake Chilwa. Species such as *Salvinia hastata*, present in other water bodies in the region, are conspicuous by their absence in Lake Chilwa.

The object is therefore not only to describe the vegetation but to stress the dynamic nature of the plant communities in a lake with a fluctuating environment.

1. Classification of the vegetation of Lake Chilwa

Seven aquatic and semi-aquatic vegetation types have been recognized in the Chilwa basin, each type comprising one or several plant associations (Howard-Williams & Walker 1974). Of these seven types, one is characteristic of areas disturbed by man's activities, and one is of minor importance, so I will be considering here only five major natural vegetation types. These are called: floodplain grassland, neutral to acid marsh, alkaline marsh, swamp transition and swamp. The distribution of species with naming authorities in these vegetation types is shown in Table 1.

1.1 *Floodplain grassland*

On the periphery of the Chilwa basin, in areas which are dry for more than nine

Table 7.1 Average frequency and presence values of the species in the vegetation types in Lake Chilwa. F, % frequency of the species in stands in which they occurred; P, % of the stands in each vegetation type in which the species occurred (Howard-Williams & Walker 1974; by courtesy of J. Ecol.)

	Floodplain grassland		Neutral to acidic marsh		Alkaline marsh		Swamp transition		Swamp	
	F	P	F	P	F	P	F	P	F	P
Scirpus littoralis Schrad.	–	–	–	–	–	–	–	–	52	9
Paspalidium geminatum (Forsk.) Stapf	–	–	–	–	–	–	–	–	23	48
Pseudowolffia hyalina (Del.) den Hartog & van der Plas	–	–	–	–	–	–	–	–	58	9
Ceratophyllum demersum L.	–	–	–	–	–	–	78	11	38	55
Pistia stratiotes L.	–	–	–	–	–	–	68	11	39	10
Lemna perpusilla Torr.	–	–	–	–	–	–	44	11	38	6
Panicum repens L.	–	–	–	–	–	–	62	89	32	15
Utricularia inflexa Forsk.	–	–	–	–	–	–	18	22	14	42
Diplachne fusca (L.) Beauv.	–	–	–	–	–	–	35	56	22	27
Nymphaea caerulea Savigny	–	–	–	–	–	–	17	22	35	36
Cyperus laevigatus L.	–	–	–	–	–	–	20	11	12	3
Spirodela polyrhiza (L.) Schleid.	–	–	–	–	–	–	28	11	2	6
Utricularia reflexa Oliv.	–	–	–	–	–	–	35	33	3	3
Cyperus alopecuroides Rottb.	–	–	–	–	–	–	31	22	49	6
Alternanthera sessilis (L.) DC.	–	–	–	–	–	–	5	22	–	–
Typha domingensis Pers.	–	–	35	13	47	40	95	33	75	61
Ipomoea aquatica Forsk.	–	–	16	38	5	40	35	44	20	33
Vossia cuspidata Griff.	–	–	52	38	50	80	31	44	31	42

Cyperus articulatus L.	–	–	56	25	–	–	23	56	2	6
Phragmites mauritianus Kunth	–	–	27	44	–	–	–	–	39	15
Echinochloa pyramidalis (Lam.) Hutch.	–	42	71	44	100	20	1	11	20	3
Aeschynomene pfundii Taub.	10	100	64	25	12	20	–	–	11	25
Cynodon dactylon (L.) Pers.	75	14	50	50	68	60	–	–	33	6
Cyperus procerus Rottb.	8	14	53	13	–	–	51	33	–	–
Scirpus maritimus L.	56	–	13	17	–	–	49	44	–	–
Utricularia gibba L.	–	–	–	–	–	–	34	11	–	–
Wolffiopsis welwitschii (Hegelm.) den Hartog & van der Plas	–	–	–	–	–	100	–	–	–	–
Cyperus longus L.	6	14	6	13	45	–	–	–	–	–
Aeschynomene nilotica Taub.	–	–	3	19	–	–	–	–	–	–
Oryza sativa L.	–	–	14	13	–	–	–	–	–	–
Cyperus digitatus Roxb.	–	–	20	13	–	–	–	–	–	–
C. papyrus L.	–	–	69	19	–	–	–	–	–	–
Pycreus mundtii Nees	–	–	30	6	4	60	–	–	–	–
Polygonum limbatum Meisn.	42	14	8	13	–	20	–	–	–	–
Leersia hexandra Swartz	18	14	82	94	82	–	–	–	–	–
Commelina africana L.	35	57	56	6	–	–	–	–	–	–
Eragrostis gangetica (Roxb.) Steud.	24	43	64	13	22	40	–	–	–	–
Sporobolus pyramidalis Beauv.	24	14	18	13	–	–	–	–	–	–
Eriochloa borumensis Stapf	78	100	2	20	–	–	–	–	–	–
Hyparrhenia rufa (Nees) Stapf	39	43	–	–	–	–	–	–	–	–
Hygrophila auriculata (Schumach.) Heine	22	–	–	–	–	–	–	–	–	–
Chloris virgata Swartz	–	28	–	–	–	–	–	–	–	–

months of the year, a grass-dominated vegetation type occurs. *Hyparrhenia rufa*, *Cynodon dactylon*, *Sporobolus pyramidalis* and *Eragrostis gangetica* are the principal species. This vegetation type can be equated partially with the valley grassland described by Vesey-Fitzgerald (1970).

1.2   Neutral to acid marsh

This vegetation type consists of a number of plant associations in soils where pH levels varied from less than 5.0 to 7.0. Opposite the perennial river mouths in the most acid soils, *Cyperus papyrus* occurs. This association is surrounded by a zone of tall grasses – *Phragmites mauritianus*, *Echinochloa pyramidalis* and *Vossia cuspidata*. Extending down the west side of the lake between the rivers is the other neutral to acid marsh association, that dominated by *Cyperus procerus*. The marsh grass *Leersia hexandra* commonly occurs with *C. procerus*.

1.3   Alkaline marsh

This occurs widely at the southern end of Lake Chilwa on heavy clay soils with a pH exceeding 7.5. *Vossia cuspidata* and *Cyperus longus* occur in a mixed sward interspersed with large clumps of *Aeschynomene pfundii*.

1.4   Swamp transition or alkaline grassland

This vegetation type occurs extensively around the northern half of the lake on alkaline soils which, although slightly above lake level during the dry season, are kept moist by wind generated surges of water (wind tides) which flood the northern plains during strong southeast winds. The grasses *Diplachne fusca* and *Panicum repens* usually form the bulk of the plant biomass. *Cyperus laevigatus* and *Scirpus maritimus* are also common. Scattered *Typha* clones may also occur.

1.5   Swamp

By far the largest area of vegetation is swamp, dominated by *Typha domingensis* occurring largely as pure stands. Swamp transition and marsh species may occur in the *Typha* on the landward edges, whilst several free floating species such as *Pistia stratiotes*, *Ceratophyllum demersum*, members of the Lemnaceae, and *Utricularia* spp. are found on the lake edges of the swamp. The large sedge *Scirpus littoralis* and the aquatic grass *Paspalidium geminatum* occur in patches out in the lake itself. These latter two appear to be the only species able to live in fairly deep water and to tolerate heavy wind and wave action. The very turbid lake water keeps the lake bed free from submerged vegetation.

Fig. 7.1 is a vegetation map of Lake Chilwa, drawn up after extensive ground and aerial surveys and utilizing the classification derived from a cluster analysis (Howard-Williams & Walker 1974). The map shows clearly the relative importance of the various vegetation types. *Typha* swamp occupies by far the largest area, with floodplain grassland the next largest. Marsh vegetation for convenience has been subdivided according to the dominant species. *Cyperus papyrus* occurs only opposite the Likangala and Domasi River mouths.

110

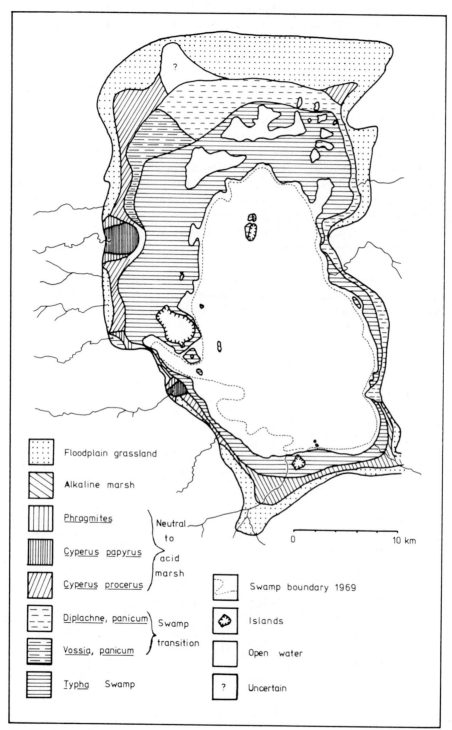

Fig. 7.1 Vegetation map of Lake Chilwa.

Floodplain grassland

Alkaline marsh

Phragmites

Cyperus papyrus

Cyperus procerus

Diplachne, panicum

Vossia, panicum

Typha Swamp

Neutral
to
acid
marsh

Swamp
transition

Swamp boundary 1969

Islands

Open water

? Uncertain

0   10 km

The vegetation of the lake islands, which occupy a total area of some 10 km²
is very interesting. However, as it is essentially a terrestial vegetation it will not
be discussed in detail here. Almost half the total number of plant species found
in the Chilwa basin occur on the lake islands, and I have ascribed this rich flora
to the complex mineralogy of the islands (see Chapter 2), resulting in very
varied soil patterns (Howard-Williams 1977). These in turn give rise to a rich
and varied flora.

Soil conductivity (which can, for the purposes of this discussion, be equated
with soil salinity) and water depth were the two factors shown to cause most of
the variation in the environment around Lake Chilwa (Howard-Williams &
Walker 1974). However, it is unfortunately not possible to separate these two
effects because in the Chilwa system soil conductivity is closely related to water
depth, so the effects of either of these two factor complexes on the vegetation is
masked by the other one in a field analysis.

After water depth and conductivity, Howard-Williams & Walker (1974)
showed that pH seemed to be the next most important factor correlated with
changes in the vegetation.

Fig. 7.2 shows the relationship between the various vegetation types around
Lake Chilwa and a few selected environmental factors. Values for the open
lake (19 stands widely scattered throughout the open water) where no vegetation occurs are also shown.

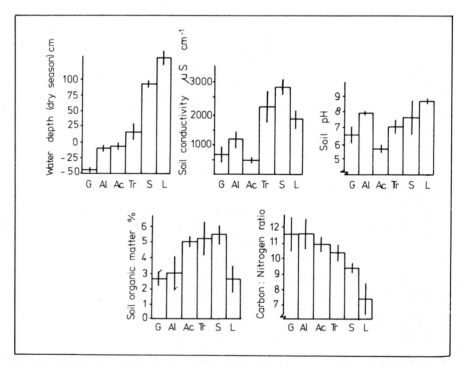

Fig. 7.2 Levels of selected environmental factors associated with the various vegetation types in
Lake Chilwa. Vertical bar = 2 Standard Errors. G = Floodplain grassland, A1 = Alkaline marsh,
Ac = Neutral to acid marsh, Tr = Swamp transition (or alkaline grassland), S = Swamp,
L = Open lake.

There is a steady increase in water depth as the vegetation progresses from grassland through to open lake. Soil conductivity increases markedly in the swamp transition and swamp vegetation types, and is higher in alkaline than in acid marsh. Lake sediments have a lower conductivity than swamp sediments. This is probably due to lower organic matter content (and hence lower exchange capacity) in the lake sediments when compared with those of the swamp. Other possible reasons for this are discussed in Chapter 5. With respect to soil pH, however, the alkaline marsh and open lake have the highest values, the sediments of the open lake being very alkaline. Soil organic matter (decomposed material) increases steadily with water depth to the swamp, but then there is a sharp drop into the lake. However, the C/N ratio decreases from grassland to swamp and then drops significantly into the lake where there are high nitrogen residues, due largely to the blue-green algae which make up the bulk of the lake's phytoplankton (Chapter 6). The high C/N ratio on the floodplain grassland soils can be attributed to the conversion of organic nitrogen to nitrate during the dry season, and subsequent flushing out of the nitrate during reflooding (Howard-Williams 1972 and Chapter 5). However, the essential feature of Fig. 7.2 is the marked relationship between the vegetation types and both water depth and soil conductivity (salinity). The effects of these two factors will now be discussed in more detail.

2. Changes in vegetation with changes in water level

2.1 *Seasonal changes*

Table 7.1 shows the species composition of the vegetation, Fig. 7.1 shows how this vegetation is distributed around Lake Chilwa, and Fig. 7.2 shows how the vegetation is affected by various factors in the environment. However, this presentation gives a rather static appearance to the vegetation itself and to vegetation-environment interactions.

In order to overcome this, a series of fixed sampling localities was laid out in a transect from deep to shallow water across the swamp in the southwest region of the lake. Fifty quadrats at each of these stations were examined at monthly intervals and the percentage frequency of each species noted each month (Howard-Williams 1975a). Fig. 7.3 shows how the species composition of the vegetation at a stand situated on the landward edge of the *Typha* swamp varied during the year in relation to water level changes. With the exception of *Typha*, the plants occur in varying frequencies throughout the year, some being absent during high water level (*Cyperus* and *Hibiscus*), while some increase as the water depth increases and then fall again as the water level drops.

Few species were present in the community in late October at the end of the dry season. Then, following rains in late October, moist ephemeral species such as *Cyperus esculentis* and *Hibiscus cannabinus* germinated rapidly on the wet mud (Fig. 7.3). These grew, flowered and died within a few months as the water level (Fig. 7.3a) rose. However, with the rising water a further set of species such as *Panicum repens*, *Nymphaea caerulea* and *Utricularia* spp. (Fig. 7.3 c, d, e) grew rapidly. These then declined as the water level dropped again.

The marginal vegetation of Lake Chilwa is thus in a constant state of flux depending on changes in water level. However, the occurrence of the dominant perennial species *Typha* (Fig. 7.3) does not fluctuate to any marked extent over the year. The vegetation classification (Table 7.1) and map (Fig. 7.1) are based almost entirely on these dominant perennials.

The mechanism by which water level changes affect vegetation is complex; it is not merely a case of increasing depth as one might infer from Fig. 7.2a. Walker & Coupland (1968) showed that the effect of water depth on species composition was not linear, as a change in water level at a shallow depth would have a greater effect on the vegetation than the same change in deep water. Segal (1971) suggests that it is the presence or absence of a temporary drying-up period rather than the degree of fluctuation which determines the vegetation type, whilst Harris & Marshall (1963) showed that the speed at which a soil dried out and the length of time it remained dry were important to vegetation in a fluctuating water regime.

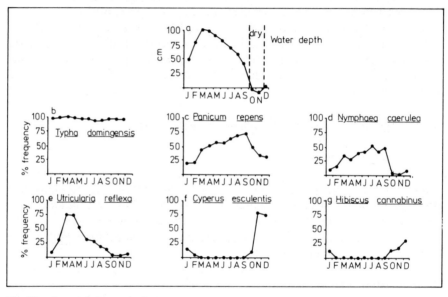

Fig. 7.3 Seasonal changes in the per cent frequency of various species from the landward edges of the *Typha* swamp.

Certainly in Lake Chilwa, the presence or absence of a dry period has a major effect on the vegetation. The primary division of the Chilwa vegetation into grassland and marsh on the one hand, and swamp transition and swamp on the other, coincided with the presence and absence of an annual dry period (Howard-Williams & Walker 1974). In areas which do dry out, the length of the dry period is the next most important factor influencing the Chilwa vegetation. Many grassland species can tolerate limited periods of flooding, whilst many aquatic plants tolerate limited drying out. *Panicum repens*, for instance, is generally found in permanent water, but occurs extensively on the north Chilwa plains which are dry for much of the year, but where wind tides keep the soil moist. *Diplachne fusca* also occurs on these north plains but can grow in deep water; however, it dies after one year in permanent standing water (Howard-Williams 1975b).

In areas where the water level fluctuations result in a dry and a wet period each year, two distinct habitats are created. It was pointed out by van der Maarel & Leertouwer (1967) that fluctuations in habitat factors will affect species composition and diversity. In an area of very gently sloping ground such as in the marginal areas of Lake Chilwa, a very gradual gradient exists between permanently wet and permanently dry regions. Conditions here are such that it is possible for hygrophilous species to tolerate the dry periods, whilst hydrophobic species can tolerate (or are adapted to) the wet periods as seeds. This results in two or more sets of species living in the same area at different times of the year. Fig. 7.3f and g, for instance, show two such species which are abundant in the early rainy season but which die during the high water level, while Fig. 7.3c and d shows how other species may occur later in the year. Others (Fig. 7.3b) occur throughout the year. This not only leads to greater species diversity in the areas of alternate dry and wet conditions than in areas of permanent standing water, but also leads to changes in this diversity during the year.

2.2 *Long-term changes*

As would be expected, there were significant changes in the vegetation of the Lake Chilwa basin when the lake dried out in 1968. Fig. 7.4 illustrates broadly what happened during and after the dry phase. Before the lake dried, the margins were defined entirely by *Typha domingensis*. By 1968 an enormous area of exposed mud existed which was potentially available for colonization by plants from the surrounding areas. However, as described by Moss & Moss (1969), McLachlan et al. (1972) and Howard-Williams (1975b) this mud proved to be a very unfavourable habitat because of high concentrations of sodium chloride and bicarbonate on the mud surface, high temperatures (over 40°C) and the extensive system of large cracks which developed in the mud which would have the effect of tearing the roots of small plants (Chapter 4). In these rather extreme conditions, only three species survived. These were the alkaline grass species *Diplachne fusca*, the sedge *Cyperus laevigatus* and the large legume *Aeschynomene pfundii* (Fig. 7.5). Not all the lake soils were unfavourable during this phase, however. Restricted areas where sand occurred, or where perennial rivers enter the lake and a certain amount of fresh water was available, had lower conductivities, temperatures and little or no soil cracking. In these relatively small areas, a fairly rich flora developed as the water level rose in 1969. Here a sward of *Diplachne fusca* interspersed with other aquatic grasses such as *Paspalidium geminatum*, *Vossia cuspidata* and *Echinochloa pyramidalis* developed. *Aeschynomene pfundii* and *A. nilotica* grew up in clumps through the grasses. Water lilies (*Nymphaea* spp.) and the floating-stemmed *Ipomoea aquatica* and *Ludwigia stolonifera* spread out across the water (Fig 7.5). Submerged plants such as *Ottelia* spp. and *Nitella* sp. were also common in the clear rising waters. Fig. 7.4 shows a general outline of the lake margins in 1969 when the water level rose again (Chapter 3 & 4). Eighty-four km² were colonized by new vegetation. *Diplachne fusca* was found almost all around the lake, while *Aeschynomene pfundii* was restricted to the southwest shore. As the lake refilled, a number of environmental changes occurred within a few months which adversely affected the colonizing vegetation. Firstly the water depth increased to 2 m on the lake side of the *Typha* zone, wind action mixed the water column disturbing the bottom muds and causing a sharp decrease in water clarity to a Secchi disc value of only 6 cm (Chapter 4), the water conductivity increased from 200 to $1500\,\mu$ S cm^{-1}, and wave action resumed. These factors are described more fully in Howard-Williams (1975b). They caused most of the newly colonizing species to disappear within the first year of refilling. By 1970 the distribution of *Diplachne fusca* had diminished considerably (Fig. 7.4) and by 1972 it had disappeared completely. *Aeschynomene pfundii* had gone by February 1973 (Fig. 9.6b) when the margins of the lake were once again made up of *Typha* as in the pre-drying phase.

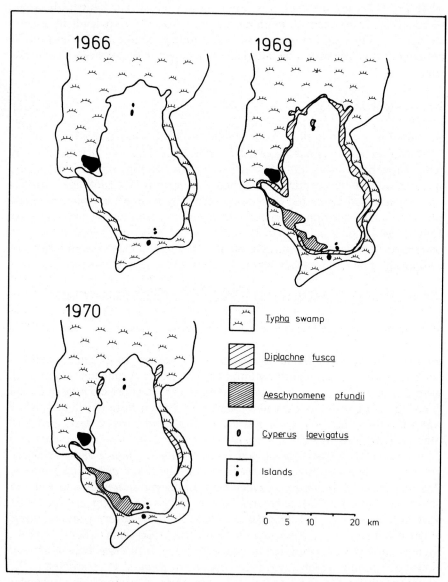

Fig. 7.4 Changes in the swamp-lake margin as a result of the dry phase. 1966 is the pre-drying phase, 1969 shows the filling phase and 1970 shows the initial dieback of the vegetation during the post-filling phase. By 1972 the margin resembled that in 1966.

116

Experiments with seedlings of *Aeschynomene pfundii*, showed clearly that the changing salinities after lake filling would not have influenced the species, but that the seedlings were not able to grow under water and the seeds require a dry period to germinate (Howard-Williams 1975b). Thus no sexual reproduction in the lake population could occur, and once the colonizing individuals, with a life-span of 3–4 years had died, there was no further chance of the population growing again. However, large numbers of seeds were set by this population, and more are washed in each year with the rivers. Many of these will almost certainly germinate when Lake Chilwa dries again, and the growth of *A. pfundii* in the lake will be repeated.

As the water level of Lake Chilwa fluctuates, so does the ionic composition of the lake waters, and of particular importance to plant growth and species composition are changes in salinity.

Fig. 7.5   The temporary swamp community which developed in the lake following refilling after the dry phase. The tall legume *Aeschynomene* can be seen with the *Typha* swamp in the background. *Ipomoea aquatica* and *Nymphaea* spp. occur on the water surface. (photo: C. M. Schulten-Senden).

3. Changes in the vegetation with changes in water chemistry

Here again the fact that carbonate and bicarbonate ions make up the major proportion of the Chilwa anions (Chapter 5) is very important, as many organisms which may be tolerant of high salinity where chloride is the only major anion cannot survive high bicarbonate concentrations. Beadle (1974) gives an account of the available evidence which illustrates this feature.

117

Plants which would be most affected by seasonal changes in the salt concentration in the waters of Lake Chilwa are obviously the free floating species rather than the rooted ones. In addition, the seedling stages of rooted aquatic plants would be sensitive to changes in water chemistry, particularly those such as *Typha* which often germinate on the water surface.

Fig. 7.6 shows seasonal changes in conductivity, sodium, alkalinity (carbonate plus bicarbonate) and chloride in the centre of a swamp area in Lake Chilwa. Sodium concentrations rise to almost 800 mg l⁻¹, and chloride and bicarbonate ions are present in about equal proportions.

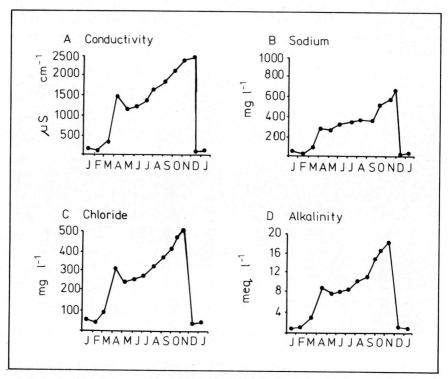

Fig. 7.6 Seasonal changes in conductivity, sodium, chloride and alkalinity (carbonate + bicarbonate) during a 'normal' year in the southwestern swamps.

In order to assess the effects of seasonal salinity changes on the growth of floating plants in Lake Chilwa their tolerance to increasing levels of sodium chloride and sodium bicarbonate was measured under laboratory conditions. A culture solution was used in which the levels and proportions of the major ions approximate those prevailing in Lake Chilwa (Howard-Williams 1975b). It was clearly shown by McLachlan et al. (1972) that as the lake dried out and the salinity increased, the relative importance of Na^+, Cl^- and $CO_3^{2-} + HCO_3^-$ increased. Sodium chloride and sodium bicarbonate were therefore the only salts used to vary the salinity of the culture solution. The treatments contained equal amounts of NaCl and Na_2CO_3 in the following concentrations: 5, 20, 100, 250, 500, 750 and 1000 mg l⁻¹ of Na^+. The final treatment was much higher than is normally found in Chilwa during a seasonal cycle, and would only be expected when the lake entered a dry phase, as shown in Chapters 4 & 5.

For these specific experiments, the following species from Lake Chilwa were studied: *Ceratophyllum demersum*, *Spirodela polyrhiza*, *Pseudowolffia hyalina* and *Lemna perpusilla* and seedlings of *Typha domingensis*. Details of the experimental layout are given in Howard-Williams (1973). The result of these experiments are shown in Fig. 7.7.

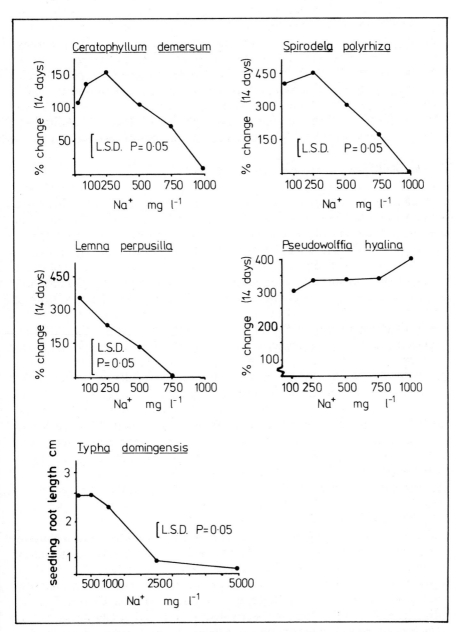

Fig. 7.7 The influence of increasing salinity on the growth of floating aquatic macrophytes from Lake Chilwa. Compare salinity values with those in the field (Fig. 7.5b). Note that *Typha* was grown in much higher salinity treatments to simulate dry lake periods. L.S.D. = Least significant difference between treatments at P = 0.05. See text for further details.

It can be seen that salinities of 1000 mg l^{-1} Na$^+$ (above that in a 'normal' year, see Fig. 7.6) have little effect on *Typha* and *Pseudowolffia*, but the other three common floating aquatic species showed significant growth inhibition at salinities which would be encountered during the 'normal' seasonal changes in Lake Chilwa.

As the lake enters a dry phase these species would die out, and presumably following re-filling, recruitment occurs from the rivers, or from seeds shed before the populations died out. Although *Typha* seedlings can survive a 'normal' year's salinity maximum, they do not survive salinities during periods when the lake dries out (Fig. 7.7). This is of some considerable significance, as it means colonization of the dry lake bed by *Typha* seedlings is unlikely to occur when the lake dries out. This is perhaps very fortunate, as otherwise the many millions of *Typha* seeds which blow in clouds around the lake basin each year would quickly establish themselves across the entire lake bed during dry periods, thereby considerably speeding up the process of lake infilling.

4. The significance of environmental influences

As is pointed out throughout this book, changes in water level mean changes in a number of other environmental factors, particularly salinity (Chapter 4). In the field it is not possible to separate these two effects when examining spatial or temporal changes in the biota. However, with respect to the higher plants, the salinity, or more likely the alkalinity, is probably of over-riding significance in determining the species composition of the swamp community. For instance, why is *Cyperus papyrus* which is so common in the lake basins of most of tropical Africa relatively scarce in Lake Chilwa? It is confined only to the fairly acid low conductivity regions around the Likangala and Domasi river mouths (Fig. 7.1). It does not occur on the higher conductivity, alkaline marshes in the south around the Sombani and Mulira river mouths. As the Sombani river is strongly perennial, the absence of papyrus from this area is hardly explained by the presence of dry periods as Beadle (1974) suggested. On the contrary, Gaudet (1977) showed that a drawdown on Lake Naivasha actually resulted in the extension of a papyrus swamp. Howard-Williams & Walker (1974) reviewed evidence which indicates that papyrus is generally restricted to areas where the pH varies between 5.5 and 6.4, and is at a competitive disadvantage at conductivities much above 600 μ S cm^{-1}. In a recent extensive review of the biology of *Cyperus papyrus*, Thompson & Gaudet (in press) show that papyrus can grow in water with a pH of 9.0 (Lake George, Uganda) but that high conductivities can limit its growth.

The presence of a *Typha* swamp around Lake Chilwa rather than a papyrus swamp can almost certainly be ascribed to the high conductivity and alkalinity of the lake muds, which give *Typha* a competitive advantage over *Cyperus papyrus* in this region. This may also be the reason why *Salvinia hastata*, common in the nearby Shire River, does not occur in Lake Chilwa. Fig. 7.7 shows that even in the floating species which are found in Chilwa, growth is often considerably retarded in salinities which can occur during a 'normal' year. Changes in salinity brought about by water level changes have an important effect on the presence or absence of species, and on the distribution of species in Lake Chilwa. Here again, the studies on *Cyperus papyrus* by Gaudet (1977) make a useful comparison. In Lake Naivasha, when the water level dropped during a drying phase, swamp development was promoted. However, in Lake Chilwa, due to unfavourably high salinity regimes when the lake was dry, colonization of the mud flats by the *Typha* swamp was inhibited. Thus the vegetation responds to the interacting effect of water levels and associated changes in the water and soil chemistry.

Salt concentrations in the drier soils around the swamps themselves are generally much lower than in deeper water (Fig. 7.2) so here water level fluctuations alone are the major controlling factor. Available evidence points to the presence of a dry period as the factor which causes the greatest vegetation change on the land-lake gradient. Here, water level fluctuations are of major importance as discussed earlier, and a highly dynamic state exists in the vegetation on these regions with the continual response to water level changes (Fig. 7.3). It must be stressed, however, that water level changes in Chilwa during a 'normal' year are relatively small (about 1 m each year) on a vertical scale, but because of the terrain, this relatively small change affects large areas ($\pm$ 500 km^2). In places where fluctuations in water level are much greater than 1 m, a very adverse effect on the vegetation can result. Such regions are found in the várzea, the floodplain of the Central Amazon River, where enormous areas are alternately flooded and dried each year, but the water level fluctuations are in the region of 11 m in height. Few species can tolerate these drastic changes (Junk 1970).

The response by one particular species to different degrees of water level change can also vary, as Gaudet (1977) points out. For instance, *Cyperus papyrus* on the Nile river is restricted to areas with a 20 cm range in water level, but on Lake Naivasha, due to a different reproductive strategy it can tolerate 130 cm changes in water level.

It is obvious that an increase in water depth alone does result in changes in vegetation along the depth gradient. However, where a fluctuating water level with corresponding changes in chemical environment is superimposed on a depth gradient in a complex vegetation, there are both temporal and spatial changes. Lake Chilwa provides an excellent site for the study of such vegetation dynamics.

References

Beadle, L. C. 1974. The inland waters of tropical Africa. Longmans, London. 365 pp.

Gaudet, J. J. 1977. Natural drawdown on Lake Naivasha, Kenya, and the formation of papyrus swamps. Aquatic Botany 3:1–47.

Harris, S. W. & Marshall, D. H. 1963. Ecology of water level manipulations on a north western marsh. Ecology 44:331–343.

Howard-Williams, C. 1972. Limnological studies in an African swamp: seasonal and spatial changes in the swamps of Lake Chilwa, Malawi. Arch. Hydrobiol. 70:379–391.

Howard-Williams, C. 1973. Vegetation and environment in the marginal areas of a tropical African lake (Lake Chilwa, Malawi). Ph.D. thesis, University of London. 312 pp.

Howard-Williams, C. 1975a. Seasonal and spatial changes in the composition of the aquatic and semi-aquatic vegetation of Lake Chilwa, Malawi. Vegetatio. 30:33–39.

Howard-Williams, C. 1975b. Vegetation changes in a shallow African lake: response of the vegetation to a recent dry period. Hydrobiologia 47:381–398.

Howard-Williams, C. 1977. A checklist of the vascular plants of Lake Chilwa, Malawi, with special reference to the influence of environmental factors on the distribution of taxa. Kirkia. 10:563–579.

Howard-Williams, C. & Walker, B. H. 1974. The vegetation of a tropical African lake: classification and ordination of the vegetation of Lake Chilwa (Malawi). J. Ecol. 62:831–854.

Junk, W. J. 1970. Investigations on the ecology and production biology of the 'floating meadows' (Paspalo-Echinochloetum) on the Middle Amazon. I. The floating vegetation and its ecology. Amazoniana 2:449–495.

Maarel, E. van der & Leertouwer, J. 1967. Variation in vegetation and species diversity along a local environmental gradient. Acta Bot. Neerl. 16:211–221.

McLachlan, A. J., Morgan, P. R., Howard-Williams, C., McLachlan, S. M. & Bourn, D. 1972. Aspects of the recovery phase of a saline African lake following a dry period. Arch. Hydrobiol. 70:325–340.

Moss, B. & Moss, J. 1969. Aspects of the limnology of an endorheic African lake (L. Chilwa, Malawi). Ecology 50:109–118.

Segal, S. 1971. Principles on structure, zonation and succession of aquatic macrophytes. Hidrobiologia Bucur. 12:89–96.

Thomson, K. & Gaudet, J. J. in press. A review of papyrus and its role in tropical swamps. Arch. Hydrobiol.

Vesey-FitzGerald, D. F. 1970. The origin and distribution of valley grasslands in East Africa. J. Ecol. 58:51–75.

Walker, B. H. & Coupland, R. T. 1968. An analysis of vegetation-environment relationships in Saskatchewan sloughs. Can. J. Bot. 46:509–522.

8 Zooplankton in Lake Chilwa: adaptations to changes

M. Kalk

8 Zooplankton in Lake Chilwa: adaptations to changes

Zooplankton occupy a prominent place in the food web of both temperate and tropical lakes. They are small, transparent organisms with a rapid rate of reproduction (eight or more generations a year in the tropics) and with resistant resting stages in times of stress. The most abundant in fresh water are crustaceans (cladocerans and copepods) and rotifers. Few are predators and most feed on algae and bacteria and on the organic material derived from both dead aquatic organisms and from the detritus which is formed from decaying plants brought in to a lake by inflow from rivers, swamps or seasonally flooded shores (Mann 1972). Zooplankton, in turn, are included in the foods of larger invertebrates and fishes.

In Lake Chilwa the zooplankton plays a significant part in the predominantly detritus food web in which primary production depends mainly on the contribution to the lake from the decomposition products in the very large *Typha* swamp (Chapter 13). Their predators are the three dominant fishes of the lake, which are largely planktivorous when juvenile and partly so when adult (Chapter 11).

Lake Chilwa exhibits a major seasonal rise and fall in lake level and, in addition, there are minor and major recessions. The changes in the chemical and physical factors of the lake and swamp have been described in Chapters 4 and 5 and the responses of the biota which are relevant to the life of zooplankton have been recorded: algae in Chapter 6, swamp vegetation in Chapter 7 and fishes in Chapter 11.

In this chapter, after considering the species persistently present, (i.e. the same species throughout the whole period), the characteristics of the zooplankton in a 'normal' year are described from data collected during a year of fairly high lake level, 1975–76 (Kalk 1979). Some of the behavioural patterns of the zooplankton in the pre-drying, drying, filling, post-filling and recovery phases (Kalk & Schulten-Senden 1977) are then briefly summarized. Finally, the environmental parameters which may influence the zooplankton pattern and its changes are discussed in relation to the species present and the kinds which are absent, in order to characterize the nature of the zooplankton in Lake Chilwa and its place in the economy of the lake.

1. Species composition

In all the phases of the lake, there were only three abundant species of zooplankton in Lake Chilwa, and they were all crustaceans: the cladoceran *Diaphanosoma excisum* Sars and the calanoid copepod, *Tropodiaptomus kraepelini* Poppe & Mrazek, both of which breed throughout the whole year. In the cooler months, the number of the cladoceran, *Daphnia barbata* (Weltner) became about equal to each of the other two species, which at that time were declining. These are the largest zooplankters in Lake Chilwa, being 1.0 to 1.3 mm in length when mature (Fig. 8.1). In comparison with species of cladocerans and copepods in other lakes, they may be considered medium-sized.

Three smaller cladocerans were also present: *Moina micrura* de Guerne &

Richards, *Ceriodaphnia cornuta* Sars and *Alona* sp., which altogether comprised less than 20 per cent of the total number (Fig. 8.1). *Alona* (Chydoridea), which is reputed to be associated with aquatic macrophytes (Burgis 1969, Keen 1976) was ubiquitous, but contributed less than 1 per cent of the total throughout the year. There is no vegetation in the open lake, but *Alona* may be derived from the swamp vegetation by interchange of water when wind tides occur and when floods bring swamp water into the lake (see Chapter 13).

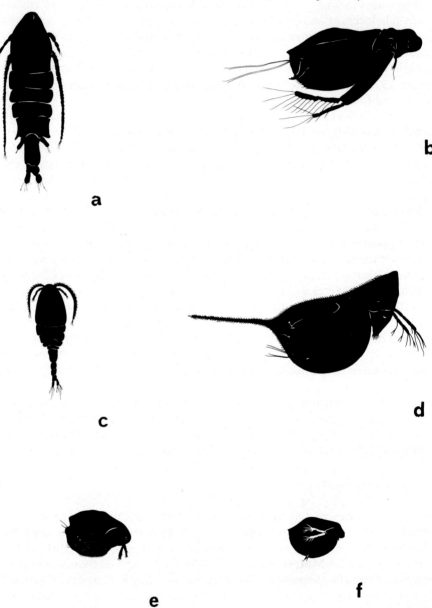

Fig. 8.1 The three larger and three smaller species of zooplankton in Lake Chilwa: (a) *Tropodiaptomus kraepelini*; (b) *Diaphanosoma excisum*; (c) *Mesocyclops leukarti*; (d) *Daphnia barbata*; (e) *Moina micrura*; (f) *Ceriodaphnia cornuta*. Range of size 1.3 mm to 0.4 mm (after C. Schulten-Senden).

The carnivorous cyclopoid copepod, *Mesocyclops leukarti* (Claus) usually contributed 3–10 per cent of the total populations of zooplankton. They feed on rotifers when available, on nauplius larvae of copepods and on young instars of cladocerans (Fryer 1975).

Rotifers were even less well represented than the smaller cladocerans in our catches. This was partly due to the method of sampling with a coarse net – the only possible way in the excessively turbid water – which precluded exact knowledge of the numbers of the smaller organisms, rotifers and nauplii, in the water. Several methods of sampling had been tried (Kalk & Schulten-Senden 1977), but the silt interfered with sedimentation, a bubble pump and fine nets. Hodgkin (1977) compared both net sampling and bottle sedimentation methods and showed that the use of vertical net hauls gave a fair indication of comparative density and seasonal changes. The extreme turbidity of Lake Chilwa, limiting light penetration and therefore green algal growth, is unfavourable for rotifers, since they prefer green flagellates as food (Hutchinson 1967). Only three species, *Keratella tropica* (Apstein), *Brachionus calyciflorus* Pallas and *Filina* (*Tetramastix*) *opoliensis* Zacharias were usually found and then mainly in the wet months, when they formed 1–5 per cent of our samples.

This list of ten species enumerated above, in which only three are abundant, is smaller than that of many other lakes, but the species are certainly not peculiar to waters of the Lake Chilwa type, which are turbid, vary in salinity and alkalinity and are subject to complete drying up at times. On the contrary, their distribution is very wide, and they occur in many different kinds of lakes, as is shown below in order to indicate their tolerance of a wide range of environmental conditions. A significant feature is, however, the absence of many other species which occur in African lakes. Among the Lake Chilwa zooplankton:

(i) the smaller crusteaceans, the cyclopoid copepod and the rotifer species are cosmopolitan (Hutchinson 1967).

(ii) the larger, dominant cladocerans are exclusively tropical and subtropical in distribution. *Diaphanosoma excisum*, the most successful species in Lake Chilwa, has the widest distribution and occurs in Africa, Asia and Australia (Green 1962). It is, however, absent from the small clear, crater lakes of Uganda (Green 1965) and from the large Lake George, which appears turbid only from the dense growth of the blue-green alga, *Microcystis* (Burgis 1969). In general this species seems to prefer turbid water.

(iii) *Daphnia barbata* is exclusively African in distribution, and it extends from the highveld of South Africa to the Nile (Rzóśka 1968). It does well in alkaline or saline, often temporary waters (Hutchinson 1967), but it is also present in the permanent, equatorial Lake George (Burgis 1969).

(iv) It has frequently been observed that one or two cladocerans and one copepod dominate a lake zooplankton fauna, irrespective of species or places, although there are exceptions. Cladocerans are absent from Lake Tanganyika except near river mouths (Harding 1966) and are a very minor component in

Lake George (Burgis 1969). This general pattern is, however, followed in Lake Chilwa, but in contrast to the wide geographical range of cladoceran species, calanoid copepods have a narrower distribution (Green 1962). *Tropodiaptomus kraepelini* has been reported in Lake Malawi (Jackson et al. 1963), a lake of a very different kind from Lake Chilwa in size, depth, clarity and chemical composition. The more distant Lake Chad, which more closely resembles Lake Chilwa in turbidity, but not salinity or pH, has the same two dominant cladocerans and *T. incognito*, closely related to *T. kraepelini*, abundant in the zooplankton (Robinson & Robinson 1971).

Briefly, the outstanding features of the species composition of Lake Chilwa are the following:

There are ten broadly tolerant species, of which three robust, filter-feeding species are numerous.

Species larger than 1.3 mm in length are absent.

Smaller cyclopoid copepods are absent.

The proportion of small cladocerans is very low.

Rotifers are not numerous in species or individuals.

2. Zooplankton in a 'normal' year of high level, 1975–6

The rate of breeding in the zooplankton species in Lake Chilwa appears to be temperature dependent. In the hot months of October, November and December (24°–28°C), when the lake is more shallow, numbers of all species increase (except *Daphnia barbata*) and tends to remain high until the cooler months of April, May and June (24°–21°C) when the lake is at its highest level and numbers decline. Reproduction does not cease in the cooler months, since eggs, nauplii of the copepods and embryos of the parthenogenetic cladocerans are seen throughout the year, but it has been observed that the numbers declined in winter. The total numbers of zooplankton soon recover in the cold months of July and August, since *D. barbata* (the species that extends furthest south to cooler latitudes) reproduces most intensively from May to August. Fig. 8.2 shows the typical total population density variations from April 1975 to July 1976 in shallow sites near the swamp in the northeast, northwest, in water over 1 m deeper in the southeastern sector and near the mouth of a perennial river. The seasonal increases to peaks and the general troughs are discernible, but, in addition, there are sudden drops in numbers for single months, followed by recovery in the next month to a level expected from the reproductive potential during the hot months of December, January, February or March.

In order to explain the non-conformities in the histograms (i.e. sudden drops in the general summer high level and the troughs before the coolest months), fish predation must be considered since the juvenile fishes of the open lake are planktivorous (Chapter 11). During the hot dry season the dominant planktivorous fishes, the minnow, *Barbus paludinosus* and the catfish *Clarias gariepinus* migrate to the swamp to breed and *Sarotherodon shiranus chilwae*, a

128

mouth-brooding cichlid, congregates on sandy patches to nest on the lake periphery. These movements lower the fish population in the lake considerably (see Chapter 11). This allows the zooplankton population to build up to peaks by natural increase. Then numbers begin to fall coincidentally with the return of juvenile and adult fishes to the lake in the hot wetter months. Sudden decreases in these months are seen in the histograms (Fig. 8.2), more especially near the swamp, when the fish are in transit. But the zooplankton is still breeding at a high rate and can recover to high numbers. Predation reaches its greatest intensity when the number of juvenile fishes is largest and the fishes are increasing in size; partly for this reason, numbers of zooplankton are lowest in April, May and June. Since the fishes broaden their diets as they mature in the first year, to include detritus from plant material, insects and smaller fish (see Chapter 11), by July and August populations of zooplankton build up again. During the cooler months when the temperature is lowest (July and August) the increase is due mainly, but not solely, to the reproductive activity of the cooler water species *D. barbata*.

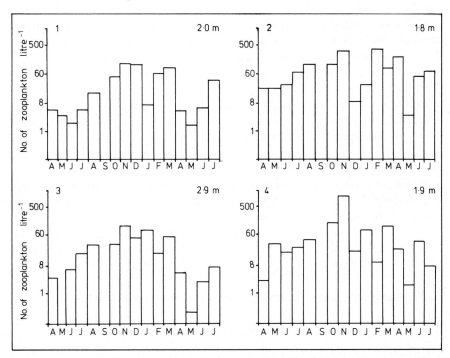

Fig. 8.2 Variation in density of total zooplankton numbers, monthly from April 1975 to July 1976 at sites: (a) near a river mouth; (b) northwest; (c) in deeper water in the southeast sector; (d) northeast swamp fringe. Note vertical scale expressed as $\log_e$. Mean depth m. (After Kalk 1979).

Compared with other African lakes, the zooplankton was fairly dense: maxima of 800 l^{-1} were reached in the northeast, 200 l^{-1} in the northwest and 100 l^{-1} in the deeper water of the southeast and in the south near a river mouth (Kalk 1979). Higher densities were shown in proximity to the large expanses of swamps. The lowest densities occurred in all parts of the lake in May, when numbers fell to 1–2 l^{-1}. The quick recovery to 14 l^{-1} in July 1976, the number of the previous July, which is the coldest month, shows that there is a stable reservoir of zooplankton in the 'normal' lake, from which numbers can multiply quite rapidly. This, as we shall see, is in contrast to the situation described below for the years just after the lake had recovered from dryness. Unlike the habit of some of the zooplankton in temperate waters, in Lake Chilwa no diapause (resting) stages have been seen in a 'normal' year.

The percentage of species composition was variable in different months. In the six hotter months, *Diaphanosoma* and *Tropodiaptomus* combined to form 80 per cent of the total although their proportions varied reciprocally from 10 to 70 per cent. In the cooler months, *D. barbata* increased at the expense of *Diaphanosoma*. After the low population density in May, the gain was due mainly to the reproduction of the larger species with *D. barbata* in the lead in June and July. *Tropodiaptomus* overtook it in August, and *Diaphanosoma* reached higher densities only after *D. barbata* had declined in October. Thus the two cladocerans change in numbers reciprocally in the different seasons, as though they occupied very similar feeding niches – the explanation given in the very similar case described in some Australian waters for two species of *Daphnia* (Herbert 1977).

From the analyses of percentages of species it was observed that whenever numbers suddenly declined at any site, there was a relatively greater decrease in *Tropodiaptomus* than in other species. In several instances this was definitely associated with reports of enormous numbers of *Sarotherodon* juveniles (Fisheries Dept. 1978). Subsequent increases in population were usually initiated by an increase in *Diaphanosoma*. This type of recovery points to the greater efficiency of the parthenogenetic mode of reproduction with protected embryos, compared with the bisexual mode of the copepod, involving a longer generation time and greater susceptibility to mortality in its larval stages from the predation of *Mesocyclops*.

In brief, it has been shown that the population patterns in a 'normal' year can be interpreted by a knowledge of the interaction of specific reproductive potential and predation by fishes.

No vertical migration of the zooplanktons in this very shallow lake was detected. The wind-induced turbulence, high turbidity with very low light penetration would militate against photosensitive responses.

3. Zooplankton during the drying, filling and recovery of the lake

3.1 *Zooplankton in the 'pre-drying phase' (August to October 1966)*

In the first half of 1966 when the water level was not unusually low, i.e. 1 m deep at the inner edge of the swamp at Kachulu (see Fig. 2.5), the zooplankton was relatively dense. The same species occurred as have been described above in a 'normal' year, with *Tropodiaptomus* most abundant. However, in the three critical dry months of the 'pre-drying phase', August to October 1966, when the lake level at the inner swamp edge rapidly fell from 70 cm to 10 cm, numbers of zooplankton began to decline. This was unexpected since a population peak had been thought to be characteristic of the increasing temperature shown. Males of cladocerans had started appearing in the preceding month and then, during this period, every species of cladoceran produced ephippia (thick-shelled resting eggs), each with coat and float characteristics of the species, and some with oil as food store. Next appeared *Tropodiaptomus* carrying smaller thicker-shelled eggs without floats; and later came the ostracods in the shallow water samples, producing similar thick-shelled eggs. *Moina* now became the most abundant zooplankter, and many individuals carried two or three very large late-instar embryos in the brood pouch, which are said to feed on maternal blood plasma (Weismann 1877 in Hutchinson 1967). *Mesocyclops* did not produce resting eggs; and at this time copepodites (late immature stages) were still being developed from nauplii derived from apparently ordinary eggs. Circumstantial evidence suggests that copepodites later went into diapause and aestivated in the mud in cysts since, in the filling stage, *Mesocyclops* was the first crustacean to appear in large numbers as mature adults.

No nauplii, nor young stages of cladocerans, nor rotifers were seen after October 1966. Adult populations declined markedly in the November samples, and was probably the result of anoxia for a few days, during and after high winds, described in Chapter 11. Rotifers were not studied closely, and it is likely that the large resting eggs were overlooked. In normal circumstances rotifers reproduce parthenogenetically, which develop without fertilization into parthenogenetic females. Under certain circumstances the females produce small eggs that become males. These fertilize the females, which then produce large resting eggs that undergo a prolonged diapause. These have thickened ornamented walls. It has been shown for *Brachionus caclyciflorus* (the most abundant rotifer in Lake Chilwa), that fertilized eggs are produced when there is overcrowding, as when, in particular, the volume of the water in which the organisms live is reduced by intensive evaporation (Hutchinson 1967).

In general, in the Lake Chilwa zooplankton, diapause was initiated while the following events were occurring (see Chapter 4):

(i) The water depth dropped from 70 cm to 10 cm;

(ii) the conductivity rose from 5000 to 6000 μS cm^{-1};

(iii) the chlorinity exceeded 30 meq l^{-1};

(iv) alkalinity had begun to exceed chlorinity;

(v) calcium and magnesium were being precipitated;

(vi) the water was highly supersaturated with oxygen in the day time;

(vii) temperatures were rising above 35°C in the afternoon;

(viii) the pH was rising above 9 during the day;

(ix) the species of blue-green algae were changing (Chapter 6).

Hutchinson (1967) discusses laboratory investigations to establish which factors determine the formation of males, the production of resting eggs and their times of hatching in various species of zooplankton. The factors were similar to some of those listed above. They are all inter-related in Lake Chilwa, since they depend on evaporation of water and concentration of salts in the water, which becoming shallower, increases in temperature. Critical factors have not been isolated in the laboratory.

Different species of the zooplankton reacted to the changing situation at slightly different times, but all of them had entered a diapause state before the end of the dry season, *two years* before complete drying of the lake took place.

3.2 Zooplankton in the 'drying phase' (December 1966–November 1968)

At the end of the dry season of 1966, when the conductivity rose in one month from 600 to 1200 μS cm^{-1}, conditions were lethal to adult zooplankton largely resulting from anoxia at night and ionic stress described in connection with fish mortality in Chapter 11. It is not known how far diapause was broken in the 1967 wet season. A very few *Moina*, *Mesocyclops* and *Keratella* were collected in January but none later. Conductivity came down below the 'critical' level in March with the rains. The wide belt of soft mud prevented regular sampling. There was a cyclone over the lake in March 1967, which brought the surface conductivity down to 500 μS cm^{-1} after 200 mm rain in 24 hours, but no zooplankton was found in April. Samples of water from the eastern shore in July 1968, when conductivity was 17 000 μS cm^{-1} and the water was 30 cm deep, contained no zooplankton. Thick, bubbly mats of blue-green algae filled the shallow water at this time (see Chapter 6).

3.3 Zooplankton in the 'filling phase' (December 1968–July 1969)

In the first few months of the 'filling phase', zooplankton was very sparse. At first only single specimens of *Mesocyclops*, *Brachionus* and *Ceriodaphnia* were found in samples, collected in very shallow undisturbed water of high salinity.

During the year of refilling, rotifers multiplied much more quickly than crustaceans. In July 1969, 78 per cent of the zooplankton was *Keratella*, 17 per cent was composed of nauplii and 5 per cent only were adult crustaceans, namely, *Moina* and *D. barbata*. Near the swamp edge, there was 52 per cent *Keratella*, 22 per cent nauplii, 22 per cent *Mesocyclops* and 4 per cent *D. barbata*. Ephippia were seen floating on the surface of the water for many months, but failed to develop in the laboratory.

In the very dilute period of the first few months of 1970 when the lake continued to fill at the end of the first dry season after recovery, rotifers were more diversified in species at an inshore station. *Filina* spp., *Testudinella patina* (Hermann) and the carnivorous species *Asplanchna brightwelli* Gosse were recorded, although the number of rotifers caught was at that time only 4 per cent of the total number of zooplankton in the samples. At this time on calm days, there was a thin surface film of green algae present. No such diversity of rotifers was recorded at any other time, nor was such a high proportion of rotifers ever found again after the 'filling phase'. Rotifers are thus characteristic of this period of Lake Chilwa. They owe their predominance to the shorter hatching time of fertilized eggs and quicker turnover of rotifer generations than crustacean (Hutchinson 1967). It is known that some rotifers survive desiccation as adults (the phenomenon of cryptobiosis) and are reanimated on first wetting. The availability of green flagellates or very small particles of detritus, which must be under 10 μm, allowed them to flourish (Hutchinson 1967). The rotifers provided plentiful food for the large populations of *Mesocyclops* which occurred at that time.

3.4   Zooplankton in the 'post-filling phase' (August–December 1969)

The hot dry season of 1969 when the lake level was undergoing its seasonal fall in level may be considered to be a post-filling phase of the lake. *Moina* was dominant and reached 82 per cent of the total zooplankton, which by this time was composed mostly of crustacean species. *Daphnia* did not become prominent in the post-filling phase and its ephippia were still seen floating on the surface of the water. By November, *Diaphanosoma* had overtaken *Moina* and reached 50 per cent of the total zooplankton numbers. It fell to a steady level around 40 per cent of the total as *Tropodiaptomus* increased more slowly in numbers. During the later rains when water flowed in from the streams and swamp, *Tropodiaptomus* became dominant as it had been in the early months of 1966, before the pre-drying phase occurred (Schulten-Senden 1972).

The early supremecy of *Mesocyclops* among crustaceans in the filling phase and *Moina* in the post-filling phase is probably related to their aestivating habits. *Mesocyclops* is known to aestivate in copepodite form (Fryer & Smyley 1954) and *Moina* has a very short hatching time from resting eggs, containing comparatively large, late instar embryos. Rzoska (1961) reported that *Moina* was present in temporary pools in the Sudan within two days of the first rains. They were also among the first crustaceans to be detected in Lake Chilwa. The availability of suitable food must also have played a part: rotifers and nauplii for *Mesocyclops* and small particles for *Moina*. The diet of *Moina* will be discussed further below.

3.5   Zooplankton in the 'recovery phase' (1970–71)

The repopulation of the lake by zooplankton was not uniform and it was not sustained in the second and third years after filling. Table 8.1 shows the data obtained from sampling in those years. They indicate that: repopulation took place more successfully and earlier in the west, east and south than in the northern part and centre of the lake; and that the numbers declined in the second and third years after filling.

From the conductivity data presented and discussed in Chapter 4, one might suggest an explanation for this. The 'saline core' present during the first months of filling may have delayed hatching of the ephippia of the cladocera; the greater salinity in the north in the next dry season might have produced a similar effect. The influx of freshwater from the rivers in the south, west and east might have stimulated the breaking of diapause.

Although the numbers of zooplankton in the first post-filling year were higher than in the second and third years after filling, the zooplankton was not as dense as found in a 'normal' year. The maximum from vertical hauls (comparable with the 1975–76 data) at Kachulu Bay was 110 l^{-1} in February 1970. The decrease in February of the following year is noteworthy since it was the hot season, when numbers would be expected to be high. From McLachlan's (1974) review of invertebrates in African lakes one would have expected higher numbers in a 'new' lake, i.e. a population explosion. But in one sense, Lake Chilwa does not conform to the pattern. Large numbers of the juvenile planktivorous fish, *Clarias*, took severe toll of the crustacea even in the first year. *Clarias* had been 'conserved' in the natural swamp

133

lagoons and streams for two years or had aestivated in the mud part of the time. It bred in the swamps in the 'filling phase' of the first year of recovery so that 3000 tonnes were caught in 1969, a figure not far short of the peak of fishing before the lake dried up (Chapter 12). In other new lakes, fish suffered from the effect of widespread de-oxygenation in the first year, for example in Kariba (Harding 1966), Volta (Ewer 1966); and for other reasons fish were not so numerous as later in Kainji, and Nasser-Nubia (Beadle 1974). Explosive populations of fish built up later in the man-made lakes, due largely to the great extension of shoreline sources of food.

Table 8.1 Numbers of zooplankton in Lake Chilwa during the post-filling period; subsample of horizontal haul (approximately 40 litres).

	Oct. 1969	Feb. 1970	Aug. 1970	Feb. 1971
Inshore sites				
1. W	2945	1155	540	140
7. E	1235	1865	3235	1015
8. N	320	1765	–	77
14. S	2110	806	1170	141
Offshore sites				
2. W	1860	1505	95	56
6. E	2295	2296	725	340
9. N	351	289	1110	–
10. N	1260	226	235	102
12. S	1270	1290	284	60
13. S	2800	3245	905	141
Open water				
3. Centre	368	–	348	110
4. Centre	487	1150	725	24
5. Centre	975	1710	475	23
11. N	–	68	340	255
Average depth (m)	13	20	24	22
Average conductivity				
(μS cm^{-1})	2000	1180	2000	1300
Average temperature (°C)	27	31	32	27
Average Secchi disc (cm)	3.7	5.7	6.4	7.5
Average pH	8.6	8.8	8.7	8.4
Season	hot dry	hot wet	cool dry	hot wet

In the second year after filling, planktivorous juveniles of *Barbus paludinosus* were added in significant numbers to the predators on zooplankton since the adults had survived in small numbers in the streams (see Chapter 11). In the third year, the mouth-brooder *Sarotherodon*, which has many fewer eggs, had increased in the lake. It seems therefore reasonable to suppose that fish predation was thus responsible for the failure of the zooplankton to maintain high densities in three post-filling years of the lake. It was noted in experimental trawling in 1971–72 that the density of fishes was greater in the 'zone of swamp influence' (Chapter 12). This was most probably dependent on the zooplankton there as well as on the greater volume of detritus being discharged from the swamp.

134

In the post-recovery year, 1973, a minor recession occurred when the periphery of the lake bed was again exposed in a belt from 2–8 km wide in the west and north, as in 1967, but only for a few months. Fishing was not good at that time. No zooplankton studies were carried out again until 1975–76, when the populations had built up to a much higher level than in the years after dryness. Data from the latter period have been used above in assessing the zooplankton in a 'normal' year. The fish landings at this time were twice as high as ever recorded before. Both zooplankton and fish populations were more evenly spread over the whole lake, except in times of fish migration to breeding sites.

Table. 8.2 summarizes the changes that have occurred in the species composition in the Lake Chilwa zooplankton during this study.

Table 8.2 Percentage composition of zooplankton species in Kachulu Bay:
(a) Pre-drying phase   (b) Drying phase
(c) Filling phase end   (d) Post-filling phase
(e) High level phase   (f) Recovery phase

Zooplankton species	(a) cool	(b) hot	(c) cool	(d) hot	(e) cool	(f) hot
Crustacea						
Tropodiaptomus kraepelini	32	0	0	5	10–55	20–70
Mesocyclops leukarti	8	10	14	4	15–5	7–1
Nauplii	+	0	16	+	++	+++
Diaphanosoma excisum	15	0	0	3	40–20	70–20
Daphnia barbata	15	0	4	2	24–10	0–1
Moina micrura	25	80	1	82	10–5	0–1
Ceriodaphnia cornuta	1	0	0	0	1–5	1–0
Alona	1	0	0	0	0–1	1–0
Rotifera						
Keratella tropica						
Brachionus calyciflorus	3	10	65		0	1–7
Filina opoliensis						
Filina sp.				4		
Testudinella patina						
Asplanchna brightwelli						

4. Fitness of zooplankton species in the variability of Lake Chilwa

It has been shown that Lake Chilwa has relatively few species of zooplankton but quite large numbers are present. The few species that have persisted through 'natural selection' during the time since the lake became endorheic, about 10,000 years ago (Chapter 2), all have in common tolerance of changes in many environmental factors which enables them to survive in the fairly frequent times of stress in Lake Chilwa.

4.1 Temperature

The difference in geographical range (wide in cladocerans, cyclopoid copepods and rotifers, narrower in the calanoid copepod, mentioned above under species composition), suggests a tolerance of wider temperature variation in the former groups. In fact, the extremes in Lake Chilwa (18°C and 40°C) are not limiting to any tropical species. Temperature change certainly influences reproduction, as is usual, and it cannot be ruled out as a partial stimulus to diapause initiation and termination.

4.2 *Salinity and alkalinity*

During normal years the salinity and alkalinity may increase or decrease by a factor of 2 or 3 (Chapter 4). The highest density occurred in the months showing the highest conductivity of the year, largely because the lake is then most shallow and hot. The species present are euryhaline and did not die until the conductivity increased above $6000 \mu S cm^{-1}$. The limit suggested by Beadle (1974) as a practical definition of fresh water was five parts per thousand (i.e. with a conductivity approximately $5000 \mu S cm^{-1}$). He says 'The apparently insignificant effects of ionic difference should not surprise us because, in contrast to the sea, inland waters are chemically unstable and variable by nature, and could thus only have been colonized by organisms with ionic regulating mechanisms that can function under a wide range of chemical conditions'. The Cladocera, as a group, are confined to freshwater and regulate osmotic pressure hypertonically. It is known that the subgenus *Ctenodaphnia*, to which *D. barbata* belongs, is more tolerant of 'mineralized' water than others, and species of *Moina* occur in salt lakes (Hutchinson 1967).

The response to increasing alkalinity and salinity in the drying lake may, by forming diapause stages, be one of the crucial properties of the limited number of species present, since the influence of natural selection would have been present every thirty years or so, at least, in the last hundred years. If the chemical events of the 'critical' months August to October 1966 are indicative of minor recessions in the periodic cycle, then diapause may be precipitated more often (see Chapter 3–4). Adults would not necessarily be subjected to lethal conditions as they were in late 1966, if the evaporation effect were to occur later in the year and if there were no high winds to stir up the anoxic bottom mud. In retrospect, one can deduce whether a minor recession would have induced diapause, since the conductivity is a reflection of the water depth. The depth of the lake at the edge near the gauge at Kachulu was 43 cm in 1966. According to this criterion of depth, the 1954 recession was not severe enough to provoke diapause, but the 1960 recession certainly was (cf. hydrograph in Fig. 1.5). The 1973 recession was probably low enough towards the end of the dry season to have conditions that would prompt diapause in some species, but the resting eggs would have hatched about two months later when the level of the lake rose again. There were no fish mortalities in 1973 from anoxia.

4.3 *Fish predation*

Macan (1977) has reviewed the influence of predation on the composition of freshwater animal communities and shown that there is good evidence from several sources that species of Cladocera and Copepoda, which are larger than 1.5 mm, are selected out by intensive fish predation. He concludes that 'Planktivores feed on the largest specimens and frequently cause the absence of

136

larger species . . .'. According to Galbraith (1967) larger species of zooplankton take longer to achieve maturity and are eaten before they can breed, thus the smaller species have a selective advantage. Green (1967) also observed that a reduction in numbers can sometimes be attributed to the effect of the smaller invertebrate carnivores on the smaller zooplankton species. In Lake Chilwa fish predation on zooplankton is very high, as described in Chapter 11 and so might account for the absence of larger species of zooplankton from the lake. *Mesocyclops* may play a part in keeping the numbers of the smaller crustacea and rotifers low. The dominance of the 'medium-sized' zooplankters is discussed below.

4.4 *Food*

The very small raptorial copepods (i.e. those which seize their prey with their mouthparts) which dominate Lake George, feed on *Microcystis* and nauplii (Burgis 1969). This species of blue-green alga has not been recorded in Lake Chilwa, where only filamentous species occur, nor is there much green algae (Chapter 6). This may account for the lack of small copepods of similar habit in Lake Chilwa.

The species of ciliary-feeding rotifers, which predominated only in the filling phase and which repeatedly reached higher densities with the inflow of river water each successive year, require food particles under $10 \mu m$ in size and select green flagellates (Lewkowicz 1974). In the rainpools on the lake bed in 1968, which Moss (Chapter 6) describes as a microcosm of the lake, which simulated the filling of the whole lake, green flagellate species predominated among algae, and rotifers were found grazing. Similarly, the green 'epineustic skin' or thin algal film of the filling period reflected the same algal composition, and rotifers were the dominant zooplankters. Then, in later years during the months of maximum inflow of river water, an algal bloom developed. In 1971 and 1972 this largely comprised *Anabaena* sp. and, in 1977, *Oscillatoria* sp. but as well as the blue-greens, there are, at these times, also flagellates listed by Moss (Chapter 6). *Scenedesmus* was also significant and this is reported to be selected by *Brachionus calyciflorus* (Hutchinson 1967) one of the three rotifer species present.

The crustacean zooplankters will consume flagellates and other small particulate green algae when they are present, although some species reported in Lake Chilwa, such as the flagellates *Trachelomonas* and *Phacus* (Chapter 6) have been shown to be rejected by *Moina* and *Ceriodaphnia*, but *Euglena* species are apparently palatable (Hutchinson 1967). In the first filling phase crustaceans could not successfully compete with rotifers, because the latter had the advantage of shorter hatching time and generation times, as mentioned above. In the normal year 1976, *Tropodiaptomus* exhibited a surge in reproduction in March which might have been related to the earlier consumption of green algae by adults in the 'dilute' months of that year, which would have increased the production of eggs.

Crustacean zooplankters are able to change their diet, (i.e. they are facultative feeders) although they may select green algae preferentially when detritus is also present, they can take whatever type of food that is available within a specific selected range of particle size. Using differential radio-active labelling for algae, bacteria and detritus, Lampert (1974) and Gerber & Marshall (1974) have demonstrated that bacteria and detritus are assimilated as well as algae by cladocerans and calanoids. Ferrante (1976) in a field situation, similarly showed that a *Daphnia* and a *Diaptomus* species fed mainly on allochthonous organic matter (i.e. of external origin) and bacteria in the season when detritus exceeded algae in density and *vice versa*. There is reason to suppose that the zooplankters of Lake Chilwa are also facultative feeders, since Nyirenda (1975) was able to breed *Tropodiaptomus*, *Diaphanosoma* and *Moina* in both bacterial cultures and algal cultures in Lake Chilwa water.

In a fundamental paper on Lake Chilwa, Howard-Williams and Lenton (1975) demonstrated that 'the swamp is the major energy source for the food web'. They calculated that about 0.8×10^6 tonnes of plant material was produced in the littoral swamp every year; and much of it is flushed out and utilized by the organisms in the open lake. On the other hand, measurements of chlorophyll *a* in 1971–72 in the open lake, in the years of recovery after refilling, were low (4.5–9 mg m^{-2}), except at the time of the algal bloom in January, when 239 mg m^{-2} were registered (Howard-Williams et al. 1972). The values may be compared with the range of 35–100 mg m^{-2} in the euphotic zone of Lake Victoria and 200–600 mg m^{-2} in Lake George (Beadle 1974). Obviously algal food is usually in limited supply in Lake Chilwa.

One may now consider why the middle-sized species of zooplankters are usually dominant, though in the post-filling phase a smaller one, *Moina*, predominated. Burns (1968) and Boyd (1976) have reviewed the relationships between the sizes of filter-feeders and the sizes of particles. Larger particles are preferred by organisms and a range of up to 100 μm is used 'according to the pore size of the rather leaky sieve of the mouthparts . . . and their raptorial (seizing) power . . .". Fryer (1954) demonstrated that of two calanoid copepods in Lake Windermere, one larger and robust (length 1.5 mm) and one smaller and slender (length 1.15 mm), the former selected larger particles (diatoms such as *Melosira*) and the latter selected little spherical green algae or small detritus particles. *Tropodiaptomus* is medium-sized (1.3 mm in length) and has robust appendages, so it may be that, in the absence of *Melosira* or algae of comparable size, it selects the larger detritus particles.

Diaphanosoma excisum and *Daphnia barbata* are similarly medium-sized and it may be inferred (from Burns 1968) that they too select larger rather than smaller particles. In this they differ from *Moina* which has a faster feeding rate than the larger cladocerans and does well on bacteria and very small similar-sized detritus particles (Hutchinson 1967).

The dominance of *Moina* in the post-filling phase may be related partly to the breakdown of detritus on the lake bed into very fine particles, as described in Chapter 4. The sediment at the time of filling was richer in organic material than later (Chapter 4), so, by implication, was the material in suspension derived from that sediment, and which made the new lake so very turbid. In the pre-drying phase when the blue-green algal species were dying in very large quantities and being replaced by others as described in Chapter 6, the lake was very rich in bacteria (Mwanza 1969). This too may have favoured *Moina* at that time.

The predominance of *Diaphanosoma* and *D. barbata* in normal years would be explained if it could be shown that the detritus particles in normal years were larger than in the post-filling phase, since then they would have a feeding advantage.

Particulate organic decomposition products of *Typha* would most probably be available in a range of sizes suitable for the larger zooplankters. It would be most interesting to measure the sizes of particles in suspension at various times of the year and in different phases of the lake. Could it be that the success of the three dominant zooplankters is determined by the rate of decomposition of the *Typha* from the swamp?

The niche differentiation between *Diaphanosoma excisum* and *Daphnia barbata* is temporal, in that the former is a summer form and the latter a winter form, so that they may consume without competition the same food resources, as described for two cladocerans in Australia (Herbert 1976). It has been postulated by Angino et al. (1973) that the simultaneous occurrence of a diaptomid and a cladoceran in ponds and lakes could be accounted for by their differential reproductive responses to environmental components, due to their different generation times. The effect of a change would be seen more quickly in the number of parthenogenetic cladocerans and only after some weeks would an increase or decrease in the number of adult copepods occur. Thus, would one species be prevented from maintaining dominance over the other and there would be no mutual exclusion, although their feeding niches might overlap. The inverse variations in the proportions of the two species in Lake Chilwa described above in Section 2 would illustrate this postulate quite well, if one knew what minor factors were involved.

The role of detritus and bacteria in aquatic food chains has recently received considerable emphasis (Melchiorri-Santolini & Hopton 1972; Anderson & Mcfadyen 1976). Saunders (1972) has demonstrated that there is considerable excess of detritus over biomass of phytoplankton in most lakes, and after the initial high rate of decay, the rate of decomposition is slow compared with the turnover of phytoplankton. The ecosystem of Lake Chilwa as a whole, with its fluctuating phases of water depth, thus seems to provide an example *par excellence* of the 'generally important functional role of detritus in aquatic ecosystems . . . providing a greater stability in the total dynamics of the system . . . (where) the detritus serves as an energy store that is dissipated at relatively slow rates . . . and may operate as a buffer to stabilize energy flow'. (Saunders 1972).

Two properties of the zooplankton indicate that Lake Chilwa has a certain degree of stability: firstly, the 'persistence' of the zooplankton species after the major and minor recessions, in the sense of long-term adaptedness; secondly, the increase in their numbers and their broader distribution in the lake in the intervening years between catastrophes. In contrast to Lake George, which is an example of a delicately poised equilibrium in an equatorial lake (Ganf–Viner 1973), Lake Chilwa can return to the *status quo* after radical disturbance and exhibits a dynamic equilibrium over the long term, in the sense of Margalef (1975) in his 'Perspectives of Ecological Theory'.

References

Anderson, J. M. & Macfadyen, A. (eds.) The Role of Terrestial and Aquatic Organisms in Decomposition Processes. 17th Symp. Britt. Ecol. Soc. Blackwell, Oxford.

Angino, E. E., Armitage, K. B. & Saxone, B. 1973. Population dynamics of pond zooplankton II. Hydrobiologia 42:491–507.

Beadle, L. C. 1974. The Inland Waters of Tropical Africa. An introduction to tropical limnology. Longman, London. 365 pp.

Boyd, C. M. 1976. Selection of particle sizes by filter-feeding copepods; a plea for reason. Limnol. Oceanogr. 5(21):175–180.

Burgis, M. J. 1969. A preliminary study of the ecology of zooplankton in Lake George, Uganda. Verh. internat. Verein. Limnol. 17:297–302.

Burns, C. W. 1968. The relationship between body size of filter-feeding Cladocera and maximum size of particle ingested. Limnol. Oceanogr. 13:675–678.

Ewer, D. W. 1966. Biological investigations in the Volta Lake, May 1964 to May 1965. In: R. H. Lowe-McConnell (ed.), Man-made Lakes. Symp. Inst. Biol. Lond. (1965) Academic Press, London. 21–30.

Ferrante, J. G. 1976. The role of zooplankton in the intrabiocoenotic phosphorous cycle and factors affecting P excretion in a lake. Hydrobiologia 49:203–214.

Fisheries Dept. 1978. Data from experimental trawling in Lake Chilwa 1975–1977. Fisheries Dept. Lilongwe, Malawi (unpub.).

Fryer, G. 1954. Contributions to our knowledge of the biology and systematics of the freshwater Copepods. Schweiz Z. Hydrol. 16:64–77.

Fryer, G. 1957. Food of some cyclopoid copepods and its ecological significance. J. Anim. Ecol. 26:263–286.

Galbraith, M. G. 1967. Size selection predation on Daphnia by Rainbow Trout and Yellow Perch. Trans. Amer. Fish. Soc. 96:1–10.

Ganf, G. G. & Viner A. B. 1973. Ecological stability in a shallow equatorial lake (Lake George, Uganda). Proc. Roy. Soc. Lond. B. 184:321–323.

Gerber, R. P. & Marshall, N. 1974. Ingestion of detritus by the lagoon pelagic community at Eniwetok atoll. Limnol. Oceanogr. 19:815–824.

Green, J. 1962. Zooplankton of the River Sokoto. The Crustacea. Proc. Zool. Lond. 135:491–523.

Green, J. 1965. Zooplankton in three crater lakes in Uganda. Proc. Zool. Soc. Lond. 138:383–402.

Green, J. 1967. The distribution and variation of Daphnia lumholtzi (Crustacea: Cladocera) in relation to fish predation in Lake Albert, East Africa. J. Zool. Lond. 151:181–197.

Harding, D. 1966. Lake Kariba. The hydrology and development of Fisheries, In: Lowe-McConnel, R. H. (ed.) Man made Lakes. Symp. Inst. Biol. Lond. (1965) Academic Press, London. 7–18.

Herbert, P. D. N. 1976. Niche overlap among species in the Daphnia carinata complex. J. Anim. Ecol. 46:399–409.

Hodgkin, I. J. 1977. The use of simultaneous sampling bottle and vertical net collections to describe the dynamics of a zooplankton population. Hydrobiologia. 52(2):197–205.

Howard-Williams, C., Furse, M. T., Schulten-Senden, C., Bourn, D. & Lenton G. 1972. Lake Chilwa, Malawi. Studies on a tropical freshwater ecosystem. Rep. IBP/UNESCO Symp. Reading, England. 40 pp.

Howard-Williams, C. & Lenton G. 1975. The role of the littoral zone in the functioning of a shallow lake ecosystem. Freshw. Biol. 5:445–459.

Hutchinson, G. E. 1967. A Treatise on Limnology, Vol. II. Introduction to Lake Biology and Limnoplankton. Wiley, 1115 pp.

Jackson, P. B. N., Iles, T. D., Harding, D. & Fryer G. 1963. Report on a survey of Northern Lake Nyasa by the Joint Fisheries Research Organization, 1953–55. Gov. Printer, Zomba, Malawi.

Kalk, M. (1979). Zooplankton in a quasi-stable phase in Lake Chilwa, Malawi. Hydrobiologia.

Kalk, M. & Schulten-Senden, C. M. 1977. Zooplankton in a tropical endorheic lake (Lake Chilwa, Malawi) during drying and recovery phases. J. Limnol. sth. Afr. 3:1–7.

Keen, R. 1976. Population dynamics of the Chydorid Cladocera of a southern Michigan marl lake. Hydrobiologia 48:269–276.

Lampert, W. 1974. A method for determining food selection by zooplankton. Limnol. Oceanogr. 19:995–997.

Lewkowicz, M. 1974. The communities of zooplankton in fish ponds. Acta Hydrobiol. Krakow 16:139–172.

Macan, T. T. 1977. The influence of predation on the communities of freshwater zooplankton. Biol. Rev. 52:45–70.

Mann, K. H. 1972. Detritus and its role in aquatic ecostystems. Mem. Istituto Ital. Idrobiol. 29: suppl. 13–16.

Margalef, R. 1975. Perspectives in Ecological Theory. Univ. Chicago Press, 4th edn. 111 pp.

McLachlan, A. J. 1974. Development of some lake ecosystems in tropical Africa, with special reference to the invertebrates. Biol. Rev. 49:365–397.

Melchiorri-Santolini, U. & Hopton, J. W. 1972. Detritus and its role in aquatic systems. Mem. Istituto Ital. Idrobiol. 29: suppl. 540 pp.

Mwanza, N. P. 1960. Preliminary studies on the role of bacteria in the food cycle in Lake Chilwa. In: N. P. Mwanza & M. Kalk (eds.). Problems of Natural Resources in Malawi. Proc. IBP Symposium, Blantyre. 24–26.

Nyirenda, G. M. 1975. The feeding behaviour of some cladocerans and copepods from Lake Chilwa, Malawi. Cyclostyled Report, University of Malawi.

Rzóśka, J. 1961. Observations on tropical rainpools and general remarks on temporary waters. Hydrobiologia 17:268–286.

Rzóśka, J. 1968. Observations on zooplankton distribution in a tropical river dam basin (Gebel Aulia, White Nile, Sudan). J. Anim. Ecol. 37:185–198.

Robinson, A. H. & Robinson, P. K. 1971. Seasonal distribution of zooplankton in the northern basin of Lake Chad. J. Zool. Lond. 163:25–61.

Saunders, G. W. 1972. The transformation of artificial detritus in lake water. Mem. Istituto Ital. Idrobiol. 29: suppl. 261–288.

Schulten-Senden, C. M. 1972. Zooplankton. In: Howard-Williams, C., Furse, M. T., Schulten-Senden, C. M., Bourn, D. & Lenton, G. Lake Chilwa, Malawi. Studies on a tropical ecosystem. Rep. IBP/UNESCO Symp. Reading, England. 25–26.

9 Decline and recovery of the benthic invertebrate communities

A. J. McLachlan

9 Decline and recovery of the benthic invertebrate communities

The term 'benthic' may be applied to animals living on a solid substratum of any kind in a lake or swamp, as well as in the ocean. There are generally two substrata of interest in lakes: the mud and the surfaces of the aquatic vascular plants.

Although mud is probably the most available habitat, it is not always the most satisfactory. This is frequently so in the tropics where high temperatures often result in the mud being anaerobic due to the accelerated oxygen demand of mud dwelling micro-organisms. Classical examples are provided by mud in deeper tropical waters like the great African lakes (reviewed by Beadle 1974), the Nile River (reviewed by Rzóśka 1976) and the Amazon River (Marlier 1976). In shallow lakes it may be inimical for other reasons, as in the African lakes Bangweulu (Bowmaker 1964) and George (Darlington 1977), where the mud is too liquid to support animals. In the English bog lakes like Blaxter Lough, the fauna inhabit only a very small part of the lake bottom where a critical balance is found between peat deposition and erosion (McLachlan & McLachlan 1975). In all such situations the submerged parts of aquatic vascular plants take on a special significance, particularly in 'swamp lakes' like Chilwa where there is a very substantial area of aquatic plants.

In common with most other lakes, the benthos in Chilwa is composed principally of midge larvae (chironomids) although in some situations, for example on the vegetation, other taxa like snails, oligochaetes, leeches, beetles and other fly larvae may be important. Relatively few species, however, are able to tolerate the rigorous conditions imposed by this lake (Chapter 4). The resulting simplified ecosystem is similar to 'stress' environments in Australian lakes (Williams 1972). To the ecologist interested in invertebrates in the tropics, the existence of only a few species is an advantage, for the difficulty of identification is a severe handicap in work on the more numerous species in other less harsh environments. Indeed, partly because of the complexity of most other types of lakes, benthic invertebrates have received less than their fair share of attention in Africa. Apart from Lake Chilwa, mud dwelling animal communities in four tropical African lakes have been studied in detail: Chad (Dejoux 1969), Volta (Petr 1971), Kariba (McLachlan 1970) and George (Darlington 1977). This is a very small sample from a continent with an exceptional number and variety of water bodies. Of these, Chilwa is, at present, the only example of a large unstable body of water studied, which may be considered as typical of much of those regions of the tropics that have a high evaporation rate and highly variable rainfall.

The figures and tables to support the story of benthos on Chilwa have been drawn, in varying proportions from my own published work, previously unpublished data and material from a recent visit to Africa. The chronological order of observations has been re-arranged in order to reconstruct the probable sequence of events during typical drying and recovery phases. The common genera of insects and molluscs discussed in this chapter are illustrated in Figs. 9.1 and 9.2.

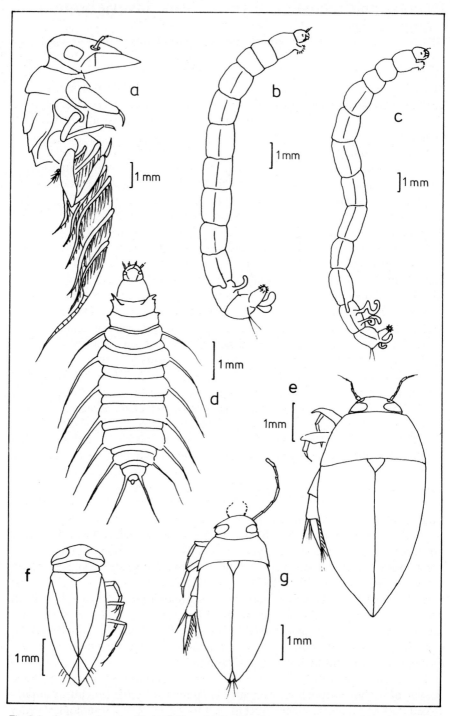

Fig. 9.1 Common insects of Lake Chilwa: (a) *Povilla adusta* Navas., nymph; (b) *Nilodorum* sp., larva; (c) *Chironomus* sp., larva; (d) *Berosus* sp. (from Peterson, A. 1960. Larvae of Insects II. Edwards Bros. Michigan); (e) *Hydrocanthus* sp.; (f) *Micronecta* sp.; (g) *Synchortus* sp.

1. Drying phase

There appear to be three critical periods during the drying phase. The first is the contraction of the shore-line from the swamp. Secondly, as drying proceeds, the decrease in lake volume is accompanied by more pronounced physical and chemical changes in the water. Finally, the lake dries up completely.

Contraction of the shore-line means the loss of a major substratum, the aquatic vascular vegetation, which normally supports many more animal species than the alternative mud habitat (Table 9.1). None of these species is unique to Lake Chilwa – all are widespread in tropical Africa. Particularly striking is the loss of *Nilodorum brevipalpis* Kieffer, (Figs. 9.1b, 9.4), a midge larva which dominates the swamp community when the lake is full. This first result of drought is the most frequently encountered. In a major recession such as that of 1967–68 it occurs for two years before drying and for two years after, for a short period at least, at the end of the dry season. It did not occur in the years of high level such as 1975–77.

Further contraction of the lake volume accompanied by substantial chemical

Table 9.1 Benthic animals recorded in shallow water at the lake edge when full and when the shore-line had just receded from the swamp during a drying phase. Values are the mean percentage dry weight contributed by each species to the community weight. Only species accounting for more than 1% of the weight appear but rarer species are included in the totals* at the bottom of the table and will be found in the checklist in Appendix A.

Benthic Animal Species	Lake Full Swamp Flooded	Lake Drying Swamp Stranded
Chironomidae (Midges)		
Chironomus formosipennis Kieffer	2	
Nilodorum brevipalpis Kieffer	30	
Nilodorum brevibucca Kieffer	31	90
Cryptochironomus neonilicola Freeman		1
Cryptochironomus stylifer Freeman		1
Dicrotendipes fusconotatus Kieffer		1
Clinotanypus claripennis Kieffer		2
Other Diptera (Flies)		
Ceratopogonidae spp.	5	
Tipulidae sp.	4	
Trichoptera (Caddis Flies)		
Dipseudopsis sp.	3	
Ecnomus sp.	4	
Hemiptera (Bugs)		
Micronecta scutellaris Stal	2	2
Coleoptera (Beetles)		
Berosus vitticollis Boh		2
Annelida (Segmented worms)		
Oligochaeta	2	
Hirudinea	1	
Mollusca (Snails)		
Lanistes ovum Troschel	4	
Bulinus (Physopsis) globosus Morelet	3	
Biomphalaria sp.	3	
* Total Number of Species Recorded	45	11

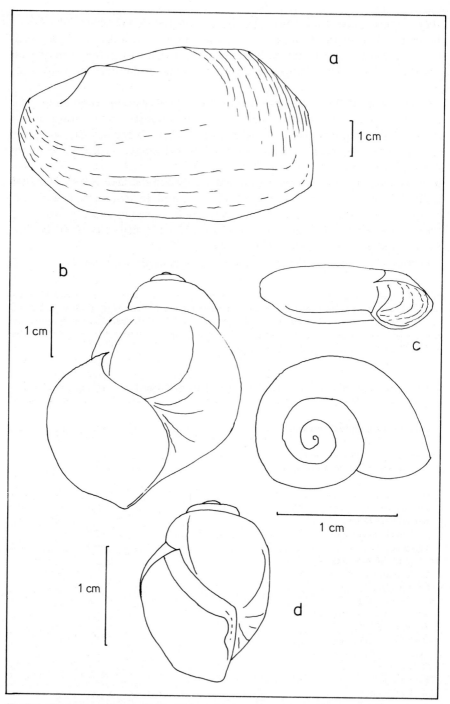

Fig. 9.2 Common molluscs of Lake Chilwa: (a) *Aspatharia* sp.; (b) *Lanistes ovum* Trosch; (c) *Biomphalaria pfeifferi* Krauss; (d) *Bulinus* (*Physopsis*) *globosus* Morelet.

changes in the water (see Chapter 4), occurs less often. We were hampered in our study of these changes by the impenetrable barrier of deep soft mud, progressively exposed between the original shore and the receding water. Conductivities of 5000 to 12 000 μS cm^{-1} were recorded in the water at this time and were associated with the disappearance of at least two further species of animal. The midge *Nilodorum brevibucca* Kieffer, (Figs. 9.1b, 9.4) which persisted for a while at the water's edge, was the first to go. Then the water boatman, *Micronecta scutellaris* Stal. (Figs. 9.1f, 9.4) was affected; carcasses accumulated at the water's edge in banks about 30 centimetres high stretching for many kilometres on windward shores. Soon after this, large numbers of shells of the snail *Lanistes ovum* Troschel, (Fig. 9.2b) conspicuous because of their size and sun bleached appearance were visible in the mud flats (Fig. 9.3). The fact that many shells contained decomposing tissue indicates that a large number of snails died in a short time rather than as the result of normal mortality. However, it is not clear whether they died because of the loss of suitable substratum (possibly the vegetation), because of the effects of chemical changes in the water, or simply because of their inability to keep up with the receding water.

Fig. 9.3 Snail shells (*Lanistes ovum*) stranded behind the receding water during a drying phase (After McLachlan & McLachlan 1969).

In any event, a progressive reduction of species occurred well before the lake dried up, the mud being inhabited for some time near the end almost entirely by a single species, the larvae of the beetle *Berosus vitticollis* Boh. (Fig. 9.1d). These animals were scarce, less than 100 mg m^{-2} compared with up to 22 000 mg m^{-2} dry weight of fauna of other species when the lake was full.

The entire sequence of events between the full lake and the dry lake basin is reconstructed in Fig. 9.4.

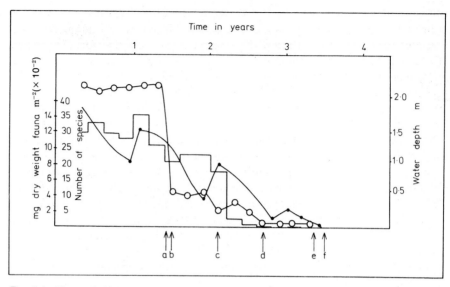

Fig. 9.4 The probable sequence of events in a typical drying phase. The number of species surviving (O) and the total weight of fauna (histograms) are shown in relation to decreasing water depth (●) at the lake centre. (a) swamps become stranded; (b) last record of the midge, *Nilodorum brevipalpis*; (c) last record of live snails, (*Lanistes ovum*); (d) last record of the midge, *Nilodorum brevibucca*; (e) water boatmen dead (*Micronecta scutellaris*); (f) lake dry.

2. Recovery phase

Once re-flooding starts, normally at the beginning of the next rainy season, it proceeds rapidly. Colonization of the filling lake by animals was equally rapid. Mature snails, especially *Lanistes ovum*, were seen in abundance within a week of the first rains, and dense populations of chironomid larvae appeared in the mud at the advancing water margin, even before it had reached the swamp fringe. By the time filling was complete, 40 species were present giving a total dry weight of about 5000 mg of animals per square metre of lake bottom. The fauna at this stage was living largely on dead organic matter from the swamp, rather than on algae, which became important only after filling was complete. The change in food was accompanied by a drop in numbers of animals present (McLachlan 1974a, 1977). At this time the vegetation, now in the water again, became a major substrate, supporting a comparable weight of fauna, but with a substantially larger number of species than the mud. The pattern of invasion, based on records from the lake edge, is summarized in Fig. 9.5.

There appear to be at least two mechanisms of invasion involved. Circumstantial evidence strongly suggests that the insects, mainly midges, fly in from adjacent permanent bodies of water and lay eggs, which are found in large numbers in the filling lake. All the species recorded on Lake Chilwa have also been found in small numbers inhabiting springs, rivers and lakes within a hundred kilometre radius of Chilwa and some as close as 20 kilometres from the south shore. This is a very efficient method of colonization and is well known for specially adapted invaders like *Chironomus transvaalensis* Kieffer, (Fig. 9.1c) (Hutchinson 1953, McLachlan 1970, McLachlan 1974b). However, this method is only effective for highly mobile animals occurring in large numbers and producing great quantities of eggs. It is interesting that some snails, with neither of these properties, were no slower in invading. They are able to survive buried in the dry mud and to revive when the mud was rewetted. Several species of mollusc are able to do this, for example the bilharzia vectors *Bulinus* and *Biomphalaria*(discussed by Jordan & Webb 1969) and the freshwater bivalve *Aspatharia* (Kalk, quoted by Beadle 1974), (Fig. 9.2d, c & a, respectively). No aestivating molluscs were found during the dry phase, but the presence of fully grown snails, as soon as flooding started, indicates that aestivation was indeed the mechanism involved.

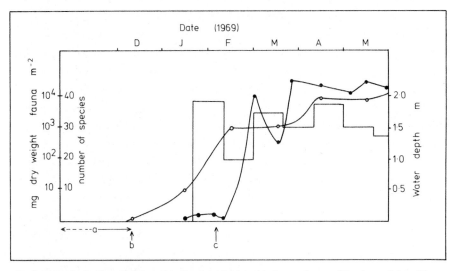

Fig 9.5 Colonization of the re-flooding lake by benthic invertebrates. The dry weight of fauna (histograms) and total number of species (●) present on the available substrata are indicated as the water level (O) rises. (a) lake dry; (b) re-flooding starts; (c) marginal swamp habitats flooded (Redrawn from McLachlan 1974b, McLachlan 1975).

Close examination of the filling phase shows that consideration of 'mud' and 'vegetation' as the two substrata for benthic animals is an over-simplification. There are in fact at least three substrata, each playing a very different role during recovery. Apart from the mud and the *Typha* swamp, there is a habitat found only within the first year or two of the recovery phase. This is a belt of semi-aquatic floodplain vegetation such as normally occurs only landward of the *Typha* swamp. During the dry period, however, conditions on the lake mud which is normally covered with water, apparently provided a suitable habitat for these plants to grow lakeward of the *Typha* belt. The newly flooded lake, therefore, was initially bordered by a double ring of emergent aquatic vegetation: the outer ring of *Typha* and the inner one of floodplain species (see Chapter 7). The latter species are able to survive here only for a limited time after being flooded (see Chapter 7), their presence on the inner edges of the swamp is characteristic of the recovery phase. This habitat is referred to as 'temporary swamp'. The main plant species involved are *Aeschynomene pfundii* or ambatch, a pithy legume about three metres tall (Fig. 9.6a) and *Diplachne fusca*, a floating grass (Fig. 9.6c).

Fig. 9.6 Vegetation of the temporary swamp habitat. (a) live *Aeschynomene pfundii* growing in about a metre of water (photo: C. Schulten-Senden).

Fig. 9.6 (b) dead *Aeschynomene pfundii* floating on the water surface.

Fig. 9.6 (c) dying *Diplanchne fusca*, a floating grass (photos: C. Howard-Williams).

The approximate contribution of each of the substrates in terms of the quantity of the fauna they support appears in Table 9.2. Also shown in the table is the estimated total weight of fauna on each substratum over the whole lake, the latter figure being derived from the total area of each habitat available.

The three substrata are considered separately below.

Table 9.2 Approximate dry weight of fauna associated with the three main benthic habitats during the recovery phase. (Based on data from McLachlan 1974 b, 1975).

	Fauna (mg m^{-2})	Area of habitat (km^2)	Fauna for Whole Lake (kg)
Permanent Swamp	300	600	180 000
Temporary Swamp	4 000	80	320 000
Mud of Lake Bed	3 000	100	300 000
Total	7 300	780	569 400

2.1 The mud habitat

Although densities of fauna in the mud were high from the onset of filling the commonest animals, larvae of the midge *Chironomus transvaalensis*, were confined to a narrow belt just behind the advancing water. This expanding ring of fauna, often only about 100 m wide, eventually enters the peripheral vegetation, leaving over 600 km^2 of lake mud virtually devoid of benthic animals. At this time a ring of fresh water was found; it originated as rain-water run-off and it surrounded a more saline core (Chapter 4). The fauna was so closely associated with the fresh-water ring that at first I thought there was some causal relationship between the salinity of the water and the occurrence of mud dwelling fauna (McLachlan 1969a, 1969b). The relationship was again apparent in subsequent rainy seasons; on each occasion an increase of fauna at the lake edge coincided with the development of the fresh-water ring as shown in Fig. 9.7. However, closer examination of the correlation reveals inconsistencies. For example, the fauna at station 8 persisted at conductivities of 1000–1500 μS cm^{-1}, when it had already disappeared at station 1 where conductivities were only 50–1000 μS cm^{-1}.

Laboratory experiments showed that eggs of *Chironomus transvaalensis* could hatch and produce adult flies in water of conductivities between 4000 and 10 000 μS cm^{-1}, a substantially greater concentration of salts than the maximum of about 2500 μS cm^{-1} recorded in the field at any time during the recovery phase. These experiments were carried out on a substratum of fine sand, the water being made up from distilled water with appropriate proportions of salts added. When lake water was substituted for distilled water plus salts, a thin layer of sediment was found to settle over the sand within a few days. The appearance of this sediment was invariably associated with a significant reduction in the salinity tolerance of the larvae. These results suggest that the sediment is in some way responsible for the disappearance of fauna from the more saline water at the lake centre. Indeed, the development of a conspicuous layer of silt over the lake bottom is very much a feature of the filling phase. The material originates from erosion at the advancing water margin and it is only at the lake edge that wave action keeps the mud free of this sediment; it is on this 'clean' edge that the fauna is to be found. Sediment is known to be of fundamental importance to the development of mud dwelling faunas in newly flooded lakes from the evidence reviewed by McLachlan & McLachlan (1976). Normally sediment is beneficial to the mud dwellers, but Chilwa provides an example of the deleterious effects that can sometimes occur.

As pointed out by McLachlan & McLachlan (1976) chironomid larvae have the ability to change the prevailing particle size in their environment. This is done by converting the fine material taken in the search for food into aggregates many times greater in diameter, by the formation of faecal pellets. These pellets are ideally suited to tube building operations and, being bound together

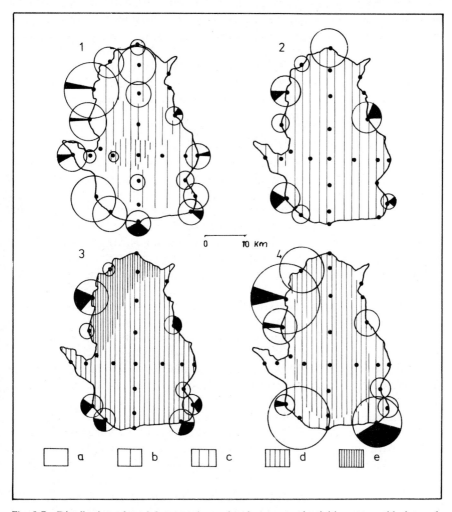

Fig. 9.7 Distribution of mud fauna and associated water conductivities at monthly intervals. 1. end of rains March 1969; 2. middle of dry season June 1969; 3. end of dry season, October 1969; 4. end of rains, February 1970. The area of each circle is proportional to the biomass of the fauna; hatched area of circles, *Chironomus transvaalensis*; open area of circles, remaining species. (a) conductivity less than 500 μS cm^{-1}; (b) 501–1000 μS cm^{-1}; (c) 1001–1500 μS cm^{-1}; (d) 1501–2000 μS cm^{-1}; (e) 2001–2500 μS cm^{-1}. Solid circles indicate sampling sites (After McLachlan 1974b).

155

with silk, persist for a considerable time, perhaps years, before eventually disintegrating. Given time, a larval population can eventually process a sediment, such as that on the Lake Chilwa bed, to create one more acceptable to themselves. It is interesting to speculate that the relatively short 'life' of Lake Chilwa prevents this from happening, each time a dry phase destroys the work already started.

When the advancing lake edge enters the swamp, owing to protection from wind and waves, the mud even here becomes covered with a fine precipitate. Nevertheless, *C. transvaalensis* larvae persisted at the swamp edge (as illustrated in Fig. 9.7). However, the distribution is patchy, larvae being found only in mats of dead swamp plants, which provided colonizable islands in the flocculent mud. In other waters, the bodies of certain living animals may provide a suitable surface. For example, the African fresh-water crab, *Potomonautes*, provides a substratum for larvae of several midge species in the mud of some West African streams (Disney 1975) and for *Simulium* spp. (blackflies) in fast-flowing water (Beadle 1974). Substantial populations of mud dwellers at the swamp edge appear to persist indefinitely. However, *C. transvaalensis*, like most invaders is replaced after a few months by a second species of midge, *Nilodorum brevibucca*.

The lake mud has a high organic content derived from swamp decay, which is the main source of food of chironomids during the filling phase of recovery. At this time the guts of chiromomids in Lake Chilwa contained over 90 per cent organic matter of detrital origin and the dry weight of the fauna was 2967 ± 174 mg m^{-2}. The next year when the lake was full, the diet had changed to about 60 per cent swamp detritus and a greater quantity of algal detrital matter. But this was not enough to maintain the first flush of benthic organisms and the dry weight of fauna fell from 2967 to 1051 mg m^{-2} (McLachlan 1977). The change coincided with the switch from the dominance of *C. transvaalensis* to *Nilodorum brevibucca*, but whether there was a causal connection was not demonstrated.

Precisely what it is about the Chilwa sediment that is inimical is not clear. The most obvious possibility is that the mud is simply too liquid to support the weight of animals. A chironomid larva, for example, attempting to settle on the mud surface in the laboratory simply sinks straight through it. However, there is another possibility. Physical analysis of the mud shows it to have an appreciable content of clay. It is conceivable that the clay creates a barrier to oxygen or even to exchange of ions with the water due to its peculiar exchange properties. Even moderate quantities of clay can have undesirable effects. Harrison and Farina (1965) for example, record the failure of planorbid snail eggs to hatch in a South African river, when covered with a skin of precipitated clay. Indeed, favourite illustrations in text-books on fresh-water biology (for example, Hynes 1972) show structures developed by animals to prevent clay from settling on gill surfaces, which are characteristic of certain stream dwellers, notably the mayflies. The precipitation of clay is especially conspicuous in Lake Chilwa where high ionic concentrations promote the removal of clay from suspension. The endemic fish species, *Sarotherodon shiranus chilwae*, has specially active mucous glands on the gill rakers that keep the respiratory surface free from silt (Cockson 1970). The action of clay on fresh-water animals is an obvious topic for further study, particularly in the tropics and sub-tropics where thunderstorms cause heavy erosion of surface soil which reaches streams and lakes giving a characteristic gross discolouration to the water.

2.2 The temporary swamp habitat

Data given in Table 9.2 show that the temporary swamp habitat is extraordinarily rich in quantity of fauna. The figures are higher than those for permanent swamp or even for the mud at the most productive period, but they require some qualification. It has already been indicated that the plant species in the temporary swamp persist for only a year or two after being flooded; they then die, decompose and disappear from the lake until the next dry phase. However, it is only after their death that they provide a substratum for the benthic fauna. While alive they are of relatively little use in this respect. The death of one of the two dominant plants, *Aeschynomene pfundii*, changes its role and results in increase in dry weight of fauna per square metre of plant surface from 1 to 4000 mg on the death of the plant (McLachlan 1975). Samples of dead *Aeschynomene* tissue, sent to the author by C. Howard-Williams in 1971, showed that the activity of the wood boring mayfly nymph, *Povilla adusta*, (Fig. 9.1a) is largely responsible for the dry weight increase of animals on the vegetation.

Several factors appear to act together to produce the advantages for many animals of dead compared with living wood. The most obvious of these is the fact that borers like *Povilla* do not penetrate the living plant. As illustrated in Fig. 9.6b, there is also a change in orientation of the 'trees' due to the roots losing their purchase in the mud and the plants breaking free and floating horizontally on the water surface. In turbid waters like those of Lake Chilwa, the consequences of the change in orientation are especially obvious. Algal growth is normally confined to a 'splash zone' a few centimetres wide around the circumference of each trunk. When floating horizontally this splash zone is, of course, greatly increased in extent. The animals too occur in the splash zone, presumably because of the food there. The increase in extent of the zone, therefore, has a direct influence on the 'tree' surface inhabitants. Fluctuations in water level no longer cause animals to be stranded or drowned, since the trunks simply float and rise up and down on the water surface with change of lake level. The floating dead wood habitat is therefore also a much more stable one. Floating plant debris from both aquatic and terrestrial environments is (according to Fittkau 1971) an important habitat for benthic fauna in the Amazon River, which like Lake Chilwa is typified, in certain stretches, by frequent fluctuations in water level.

Once dead, the temporary swamp plants on Lake Chilwa last for a few months only, since decomposition is rapid at the high water temperatures. This is especially true because the plants are composed of light spongy material. Their death is therefore followed by a short burst of faunal colonization and reproduction. The fauna thereafter are dependent solely upon the permanent *Typha* swamp and the mud.

The temporary swamp habitat is reminiscent of drowned forests which characterize many of the large reservoirs constructed in recent years in Africa and elsewhere. Here trees, dying when submerged by rising lake waters, are colonized by *Povilla adusta* (or ecologically related animals) in the same way as described for the 'temporary' swamp plants of Lake Chilwa. In some waters these mayfly nymphs are important in the diet of fish. The decomposition of the trees and the consequent disappearance of drowned forests can have important economic implications for the fishing industry as predicted by Petr for the man-made Volta Lake in Ghana (Petr 1970).

2.3 *The permanent swamp*

Although extensive in area the permanent swamp is a poor habitat for benthic animals (Table 9.2). Since the mud, as already described, is largely uninhabitable and the temporary swamp, as the name implies, is ephermeral, the *Typha* plants offer the only long term substratum available to the benthos. It must be remembered, however, that the observations were confined to the inner lake edge of the swamp, an 'ecotone', or interface between two habitats, which is often especially rich. Thompson (1976), for example, notes such a situation at the riverward edge of the Nile swamp. Estimates of total faunal weight given in Table 9.2 may be exaggerated if they are not typical of the whole swamp. (cf Chapter 10). Nevertheless, as the largest area of inhabitable substratum, this is the major benthic habitat on the lake.

As with the temporary swamp vegetation, dead *Typha* leaves, which have fallen into the water, support the bulk of the permanent swamp fauna. Over 97 per cent of the total quantity of 'benthic' animals in the lake was present on the dead leaves at the lake edge of the swamp (McLachlan 1975). It may be recalled that it was these dead *Typha* leaves that provided the only inhabitable surfaces in an otherwise hostile environment. The constant supply of dead leaves to maintain the mud population must be regarded as a key role of the *Typha* swamp, in maintaining the benthic invertebrates and the fish that feed on them. The total biomass of benthic animals in 1970–71 was estimated at 600 000 kg for the whole lake (Table 9.2).

3. Summary

Lake Chilwa is an unstable environment with a cycle of seasonal and long term events dominated by changes in water level. For the benthic animals, concurrent changes in the nature of the substratum appear to be decisive; the dramatic fluctuations in concentration of dissolved salts are of secondary importance.

After a dry period, re-flooding of the lake basin is associated with a dense population of mud dwelling midge larvae. This population follows the advancing shore-line, leaving an unoccupied space of increasing size in the lake centre. Experiments demonstrate that the mud becomes uninhabitable in deeper waters because of the precipitation of a fine sediment derived from erosion of the lake basin at the shore-line.

When the lake is full the mud continues to be inhabited only within the swamp and just off its lakeward edge, the lake bed beneath open water being virtually devoid of benthic fauna. The persistence of animals close to the swamp is apparently due to the accumulation of dead *Typha* leaves providing islands of stable substratum in an otherwise excessively liquid mud environment. The swamp vegetation, therefore, eventually becomes the major habitat for benthos. This is also true of an inner ring of floodplain vegetation which grows on the exposed lake mud during a dry phase and persists for a few years after filling. Towards the end of this period the dead and dying plants become especially attractive as a substratum for a brief period before their decomposition results in the disappearance of this temporary swamp habitat.

Even during a minor recession when the lake does not dry out completely, dry season reduction in water level is often sufficient to cause stranding of the swamp and the drying of the associated mud habitat. The remaining substratum under these conditions, the mud of the open lake, is uninhabitable. Much of the benthos is therefore obliterated at intervals of a few years and would have to re-colonize the swamps afresh from other bodies of water each time this happens.

References

Beadle, L. C. 1974. The Inland Waters of Tropical Africa. An introduction to tropical limnology. Longman, London. 365 pp.

Bowmaker, A. P. 1964. Preliminary observations on the seasonal hydrological changes affecting the ecology of the Luaka Lagoon. Joint Fish. Res. Org. Ann. Report 11:6–20.

Cockson, A. 1970. Polysacharides in the mucous cells in the gills of *Tilapia shirana chilwae* Trewaves (Pisces: Cichlidae). La Cellule 68:207–210.

Darlington, J. P. E. C. 1977. Temporal and spatial variations in the benthic invertebrate fauna of Lake George, Uganda. J. Zool. Lond. 181:95–111.

Dejoux, C. 1969. Les insectes aquatiques du Lac Tchad – aperçu systematique et bio-ecologique. Verh. Int. Verein. Limnol. 17:900–906.

Disney, R. H. L. 1975. Notes on the crab-phoretic Diptera (Chironomidae and Simuliidae) and their hosts in Cameroon. Ent. mon. mag. 111:131–136.

Fittkau, E. J. 1971. Distribution and ecology of Amazonian chironomids. Can. Ent. 103:407–413.

Harrison, A. D. & Farina, T. D. W. 1965. A naturally turbid water with deleterious effects on the egg capsules of planorbid snails. Ann. Trop. Med. Parasit. 59:327–330.

Hutchinson, G. E. 1953. The concept of pattern in ecology. Proc. Acad. nat. Sci. Philad. 105:1–12.

Hynes, H. B. N. 1972. The Ecology of Running Waters. Liverpool University Press, Ontario.

Jordan, P. & Webbe, G. 1969. Human Schistosomiasis. Heinemann. London.

Marlier, G. 1967. Ecological studies on some lakes of the Amazon valley. Amazoniana 1:91–115.

McLachlan, A. J. 1969a. Aspects of the ecology of the bottom fauna of Lake Chilwa. In: N. P. Mwanza & M. Kalk (eds.), Problems of Natural Resources in Malawi, Proc. IBP Symposium, Blantyre, Malawi. 27–29.

McLachlan, A. J. 1969b. Some effects of water level fluctuation on the benthic fauna of two central African lakes. Limnol. Soc. sth Afr. Newsl. 13:58–63.

McLachlan, A. J. 1970. Some effects of annual fluctuations in water level on the larval chironomid communities of Lake Kariba J. Anim. Ecol. 39:79–90.

McLachlan, A. J. 1974a. Development of some lake ecosystems in tropical Africa, with special reference to the invertebrates. Biol. Rev. 49:365–397.

McLachlan, A. J. 1974b. Recovery of the mud substrate and its associated fauna following a dry phase in a tropical lake. Limnol. Oceanogr. 19:74–83.

McLachlan, A. J. 1975. The role of aquatic macrophytes in the recovery of the benthic fauna of a tropical lake after a dry phase. Limnol. Oceanogr. 20:54–63.

McLachlan, A. J. 1977. The changing role of terrestrial and autochthonous organic matter in newly flooded lakes. Hydrobiologia 54:215–217.

McLachlan, A. J. & McLachlan, S. M. 1969. The bottom fauna and sediments in a drying phase of a saline African lake (L. Chilwa, Malawi). Hydrobiologia 34:401–413.

McLachlan, A. J. & McLachlan, S. M. 1975. The physical environment and bottom fauna of a bog lake. Arch. Hydrobiol. 76:198–217.

McLachlan, A. J. & McLachlan, S. M. 1976. Developments of the mud habitat during the filling of two new lakes. Freshwat. Biol. 6:59–67.

Petr. T. 1970. Macro-invertebrates of the flooded trees in the man-made Volta Lake (Ghana) with special reference to the burrowing mayfly, *Povilla adusta* Navas. Hydrobiologia 36:373–398.

Petr, T. 1971. Establishment of chironomids in a large tropical man-made lake. Can. Ent. 103:380–385.

Rzóśka, J. (ed.) 1976. The Nile. Biology of an Ancient River. Monographae Biologicae 29. W. Junk, The Hague, 417 pp.

Thompson, K. 1976. Swamp development in the head waters of the White Nile. In: J. Rzoska (ed.) The Nile. Biology of an Ancient River, W. Junk, The Hague. 177–196.

Williams, W. D. 1972. The uniqueness of salt lake ecostystems. In: Z. Kajak & A. Hillricht-Ilowska, (eds.) Productivity Problems in Freshwaters. P. W. N. Polish Scientific Publishers Warsaw, Poland.

10 Invertebrate communities in the Lake Chilwa swamp in years of high level

M. A. Cantrell

Invertebrate communities in the Lake Chilwa swamp in years of high level

After partial recession in 1973, the lake entered a relatively stable period in 1975 when lake levels were higher than for the twenty five preceding years since records had been kept. The water conductivity reflected the dilution of the lake which never rose above $1600\,\mu$ S cm^{-1} in mid-Kachulu bay, which is typical of the lake as a whole. Sampling invertebrates adequately in dense and often impenetrable swamp vegetation poses many problems, and consequently the swamp margins which are accessible from open water have attracted most attention (see Chapter 9).

Although the bulrush, *Typha domingensis* Pers. covers the greatest area of swamp, there is a mosaic of vegetation types which provides a variety of substrates for invertebrates. These are not always accessible for sampling, but can often be reached along canoe channels cut through the swamp by fishermen. Channels are numerous when viewed from the air, at least on the western and northern shores of the lake, and support their own characteristic fauna and flora. Previously such channels have been used for vegetation studies (Chapter 7).

Swamp invertebrates are known to be important items in the diet of the commercially important catfish, *Clarias gariepinus* (Bourn 1974) and an attempt has been made to investigate faunal abundance and distribution in the area of a canoe channel cut through the swamp 1 km north of Kachulu village (Fig. 10.1). The work started in late 1975 with an identification of the main swamp habitats for invertebrate populations. By this time the temporary swamp, characteristic of the recovery phase, had disappeared. The last remaining stumps of *Aeschynomene pfundii* (Fig. 9.6) for example, were last seen in 1976.

1. The swamp habitats

For my purposes the 'swamp habitats' include marshlands as there are few faunal discontinuities between this habitat and the *Typha* swamp, at least during the periods of high lake levels experienced in this study. The swamp has been divided into a number of major habitats in terms of floristic characteristics. They are adapted from the detailed classification of Howard-Williams & Walker (1974), for a consideration of the swamp invertebrates (Cantrell unpublished data).

The relative position of each habitat is shown in Fig. 10.1 and most can be seen in Fig. 10.2. The swamp margin, habitat (a), investigated previously by McLachlan (1975) is the ecotone where the swamp edge joins the open lake. Here the low density of bulrushes, *Typha*, allows floating plants such as the water lettuce, *Pistia stratiotes* L. and water lilies, *Nymphaea lotus* L., to grow in some areas, together with *Ceratophyllum demersum* L. Floating stems of dead *Typha* often collect, forming rafts which support a higher biomass of invertebrates than other *Typha* substrates. Living, floating vegetation was absent from this habitat during the pre-drying, drying and filling phases (M. Kalk, pers.

comm.) and its presence now may reflect the more stable period of high lake levels with lower salt content experienced between 1975 and 1978.

The second habitat, the *Typha* swamp (b) is the largest and is characterized by dense strands of bulrushes which shade the water surface, preventing the growth of other plants. Large quantities of dead stems accumulate in the water and leaves tend to be trapped in piles above it. Approximately one third of the swamp is said to be burnt annually (though the reason for this is unknown), so

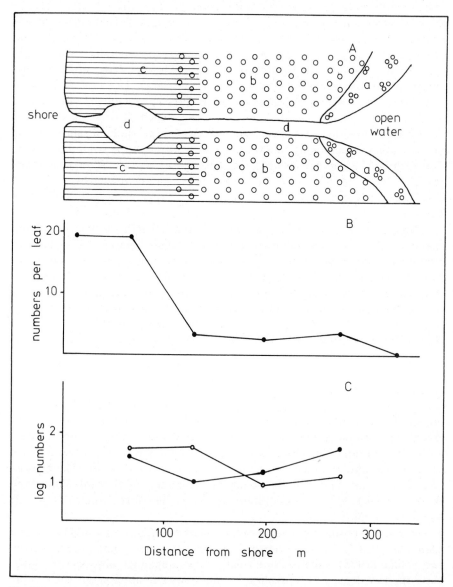

Fig. 10.1 A: major habitats in the swamp – (a) swamp margins; (b) *Typha* swamp: (c) marsh; (d) lagoons and channels. (Marsh vegetation extends some distance into the *Typha* swamp).
B: density of *Bulinus* snails (no. m^{-2}) on lily leaves along the boat channel (n between 2 and 5 for each point).
C: abundance (log. no. m^{-2}) of water boatmen, Corixidae (open circles) and midges, Chironomidae (closed circles) in *Ceratophyllum* beds along the boat channel (n between 2 and 4 for each point).

that fire removes some of this material before it falls into the water. Submerged vertical stems remain after their tips have been burnt, but are less attractive to invertebrates than horizontal mats of dead material. Neither vertical nor horizontal stems appear to support much fauna, for reasons outlined later.

Fig. 10.2 Swamp habitats: a view along the boat channel (d); through flooded marshland (c); towards the *Typha* swamp (b); taken in February 1977 (photo: M. A. Cantrell).

The third habitat, the marsh (c) on the landward side of the swamp includes the marsh vegetation and outer edges of the *Typha* stands. Grasses such as *Leersia hexandra* Swartz and *Vossia cuspidata* Griff form mats which cover the water surface. McLachlan (1975) stressed the importance of floating macrophytes as a substrate for invertebrates compared with rooted vegetation. These mats are a special case where their tangled stems provide a suitable habitat for a wide variety of invertebrates while overlying a column of anoxic swamp water.

Canoe channels and lagoons constitute the final habitat (d). The numerous channels created through the swamp afford the fishermen access to their villages and also to their traps and gill-nets set to catch catfish, *Tilapia* and other fish (Chapter 12). Lagoons, while essentially natural features of the swamp where *Typha* has not become established in the deeper water are often cleared by fishermen to facilitate fishing. These held important fish stocks during the dry phase (Chapter 12). In these open areas of swamp, floating macrophytes such as *Pistia* and *Nymphaea* are common, and often *Ceratophyllum* grows in dense beds. The habitat has therefore similar characteristics to the swamp margins (a) except that *Typha* is absent. Lagoons are little more than expanded canoe channels and have been grouped with them for analysis.

No fauna has been located in the mud underlying the channel through the swamp, not surprisingly in view of the high rates of decomposition there. Even the deposits of *Typha* fragments which overlie the mud in open water, close to the swamp margins, fail to contain chironomid populations of any size, presumably for the same reason. This contrasts with the populations of the midge *Chironomus transvaalensis* which colonized this habitat during the recovery phase (McLachlan 1974). As the mud substrate appears unsuitable for swamp invertebrates it has been excluded from the following discussion.

2. The swamp invertebrates

While McLachlan (1975) worked from open water in the swamp margins, more specialized studies on disease vectors have been based around the shoreline. Snails responsible for transmitting bilharzia were investigated by Morgan (1972) and both snails and malaria-carrying mosquitoes later by M. Magendantz (unpublished data). An investigation of seasonal changes in the distribution of bilharzia snails has been conducted by the present author and is considered separately later.

2.1 *Invertebrates of the swamp margins*

This habitat offers several types of surface for invertebrates and shows much larger faunal standing crops than the *Typha* interior where light penetration is poor (Table 10.1). The larva of the midge, *Nilodorum brevipalpis* (Fig. 9.1) is abundant in the windblown rafts of dead *Typha* which accumulate here. These dead horizontal stems support a much higher faunal biomass per square metre than dead vertical stems. Not only are they less subject to water level fluctuations, but they offer greater surface area for the growth of epiphytic algae on which several invertebrates feed. Live vertical stems of *Typha*, on the other hand, support only a scanty fauna composed of midges and, occasionally, leeches.

The fauna on *Pistia* and *Ceratophyllum* beds shows greater biomass and greater variety than on *Typha* or *Nymphaea* surfaces. Besides midges, several species of dytiscid water beetle are common, particularly *Synchortus simplex* Sharp and *Hydrocanthus* sp. (Fig. 9.1), the latter being associated with several types of floating vegetation. *S. simplex* was recorded in Lake Chilwa by Omer-Cooper (1957) together with *Canthydrus notula* (Eriksen) which is also found. Dytiscid larvae are predaceous and appear to find an abundance of food on both *Ceratophyllum* and *Pistia*. Interestingly, these genera of beetle dominated the same plants in Volta Lake, Ghana (Petr 1968) so that the Chilwa fauna is by no means unique. Water bugs (Hemiptera) are less numerous, though several predaceous Naucorid bugs and the water boatman, *Micronecta* (Fig. 9.1) occur in this habitat. The latter feeds on the epiphyton on the *Typha* stems and though common, occurs at much lower densities than those recorded in the drying phase (see previous section). The snail, *Bulinus globosus* (Fig. 9.2) is only occasionally recorded here. As indicated in other studies such as those on Lake Volta (Petr 1968) and Lake Kariba (McLachlan 1969), each species of macrophyte tends to support a characteristic invertebrate fauna.

2.2 *Invertebrates in the Typha stands*

Although this is the habitat with the largest area, it appears to support a poor invertebrate standing crop. Shading of the water surface prevents the

166

growth of epiphyton communities on the *Typha* stems which were rich in the previous habitat considered (see Chapter 6). Small populations of midges are found here and occasionally snails and leeches (Table 10.1). Sampling has been restricted to vertical stands as areas where leaves have fallen into the water are impenetrable by canoe. The figures in Table 10.1 (based on samples collected within 20 m of the canoe channel) should only be regarded as approximate for the habitat as a whole. The stems must be cut to obtain a sample so that some animals may escape. However, beetles and water bugs often seen in other habitats have never been observed here, which confirms the general absence of insect fauna in these dense *Typha* stands.

2.3   Invertebrates of the marsh

Situated on the landward side of the *Typha* stands, marsh vegetation is an important habitat for invertebrates, especially during the rising lake level in the rainy season. Water beetles, including *Synchortus* and *Hydrovatus*, have been recorded in large numbers, and the standing crop for this habitat is much higher than that of other habitats. The large amphibious snail, *Lanistes ovum* (Fig. 9.2) is also present, but at such low densities that it is rarely collected in samples and has not been included in Table 10.1. This contrasts with the remarkable density of *Lanistes* which were stranded on the exposed mud flats in the drying phase (Fig. 9.3). The snail is eaten by Openbill storks, *Anastomus lamelligerus*, and chipped empty shells are regularly found on *Leersia* mats. In addition, leeches, *Limnatis* sp., evade capture and are more common than the data suggests, as fishermen collect leeches for bait by standing in the marshland, picking them off their legs every few seconds. Levels of oxygen in the water fall rapidly as the water recedes in the dry season and this affects the populations of the snail, *Bulinus globosus*, which are abundant here (Fig. 10.1B), and presumably other invertebrates.

2.4   Invertebrates of channels and lagoons

Although this habitat has similar floristic characteristics to the swamp margins, the faunal standing crop is generally higher as molluscs are more common, especially near the shore. Those plants present support similar numbers and types of invertebrates as other habitats, both midges and beetles dominating the benthos (Table 10.1). With the exception of snails, the biomass of invertebrates depends not so much on the position in which samples are collected along the channel, but on the species of macrophyte concerned. *Ceratophyllum*, for example, supports similar numbers of fauna throughout the channel's length (Fig. 10.1C). This suggests that channels act as corridors for the movement and dispersion of the invertebrate populations. Aerial colonizers such as dragonflies are often seen moving along them, though it remains to be seen whether longer channels such as the one 10 km long through the swamp near Python Island (see Fig. 2.1) are evenly colonized throughout their length. (A provisional list of Odonata collected by J. M. Bereen and M. J. Parr at Lake Chilwa can be found in the checklist in Appendix A).

The numerous large lagoons at the northern end of the swamp have not been investigated due to their inaccessability and no firm conclusions can be drawn

HABITAT

	a: Swamp margins					b: Typha swamp		c: Marsh		d: Lagoons and Channels	
	C	N	P	T vl	T vd	T hd	T	L	C	N	P
Bulinus globosus (snails)	2	+	3	0	2	+	8	8	31	50	7
Chironomidae (midges)	42	26	55	82	71	79	76	15	10	25	55
Coleoptera (Beetles)	17	1	20	3	0	6	0	58	34	4	20

Table 10.1 Benthic animals recorded in the main swamp habitats. Values are the percentage contribution (numbers) by each taxon to the community on the species of plants: C, *Ceratophyllum*; N, *Nymphaea*; P, *Pistia*; T, *Typha*; vl, vertical live; vd, vertical dead; hd, horizontal dead; L, *Leersia*; + = <1%; n=number of samples, which were collected over a period of 12 months, are also shown.

Emphemeroptera (Mayflies)	9	+	6	2	0	5	0	0	8	50	+
Hemiptera (Bugs)	4	0	2	0	0	1	0	1	8	4	5
Hirudinea (Leeches)	3	+	2	5	3	1	13	5	1	0	1
Oligochaeta (Segmented worms)	0	3	6	2	22	5	0	3	+	4	2
Zygoptera (Damsel flies)	18	+	3	3	0	+	0	2	5	7	+
Others	5	69	3	3	2	2	3	7	3	7	10
Mean Total Dry Weight (mg m^{-2})	230	52	283	8	6	99	61	1322	403	446	1239
n	4	8	8	8	6	6	4	7	12	8	10

about their fauna. Recently, large numbers of invertebrates have been discovered at the ends of channels around beached canoes, so that biomass estimates in this habitat may be even higher.

An interesting feature of channel clearance in the swamp is that the improved light penetration results in the development of floating plants, and an increase in invertebrate communities, so enriching the food web. In other words, this traditional method of swamp clearance clearly increases the food of the catfish, *Clarias*, which is caught in the channels in large numbers by gill-net and traps.

3. Invertebrate biomass of the entire swamp

Using data now available on invertebrate standing crops in each habitat, in conjunction with the vegetation map of Howard-Williams & Walker (1974), the total biomass of invertebrates in the swamp has been calculated (Table 10.2). Such an estimate can at best be only a rough approximation, nevertheless the figure of about 600 000 kg is of the same order as an earlier estimate based on the standing crop in the permanent swamp margins and the temporary swamp during the recovery phase (Table 9.2). The later estimate reflects the rich fauna of the marshland which was previously not investigated. This habitat has by far the highest invertebrate biomass, as 90 per cent of the standing crop in the littoral vegetation is found here. The results suggest that the tangled stems of marsh grasses such as *Vossia* and *Leersia* offer the largest surface area for both algae and animal communities, and at the same time maintain them in oxygen-rich conditions near the water surface.

Table 10.2 Dry weight of fauna found in the main swamp habitats. In a and d the estimate takes into account the percentage cover by each species of macrophyte. ^{x}= biomass based on 50% cover of water by macrophytes.

Habitat	Biomass mg m^{-2}	Total area km^2	Fauna for whole swamp (kg)	% Contribution to total
a: Swamp margins	186	40	7 440	1.3
b: *Typha* swamp (vertical stems)	61	600	36 600	6.3
c: Marsh	1 322	390	515 580	88.8
d: Channels & Lagoons	332^x	60	19 920	3.5
	Total biomass		579 540	99.9

4. The distribution of the bilharzia snails

The snail, *Bulinus (Physopsis) globosus* (Morelet) is of medical importance as it transmits *Schistosoma haematobium* which causes the debilitating disease, urinary bilharzia. The dominant occupation of the people of Lake Chilwa is fishing, so that contact with the marginal water occurs daily. This is particularly true in the swamp fishery, where the techniques used, such as gill-netting and trapping often necessitate wading for long periods (see Chapter 12).

Morgan (1972) found *Bulinus* (Fig. 9.2) mainly in affluent streams in 1970–71, although he also located specimens on leaves of water lilies, *Nymphaea*, which were then colonizing the lake/swamp edge north of Kachulu village. None were found to be infected with human schistosomes, a conclusion supported by earlier workers, but bovine schistosomes were common. Snail populations were small in the pre-drying phase, and re-population of the lake was thought to have been effected by migration from the snail reservoir in freshwater streams (Morgan 1972). However, aestivating snails have recently been found in exposed marsh vegetation so that snails probably survived by this mechanism in the swamp during the dry phase. Snail density remained low during the minor recession of 1973–74 but increased remarkably in 1975. M. Magendantz (pers. comm.) found that snails now harbour human schistosomes and that a large number of the male population of Kachulu village was infected (see Chapter 19).

The focal nature of cercarial transmission has been stressed in a number of recent studies in other parts of Africa, for example Lake Volta (Odei in press) and Lake Kariba (Hira 1969). Failure to locate infected snails previously in Lake Chilwa was possibly due to the presence of isolated foci in the swamps and streams in years of recessions. The Volta studies, for instance, have shown that water-contact points (washing areas and the ends of boat channels) are important in cercarial transmission.

Further studies of the ecology of bilharzia snails were started in 1975 to investigate what effect the seasonal changes in water level, which are characteristic of the lake, had on the distribution and abundance of the snail populations, a factor of potential importance in affecting cercarial transmission.

The investigation was largely confined to the shore areas around Kachulu village and only snails of the genus *Bulinus* were found, though *Biomphalaria* has been recorded previously by other workers (see Table 9.1). From this study, some tentative conclusions can be made about the distribution of *Bulinus* in the lake as a whole. Although snails have twice been observed floating in open water in Kachulu Bay (A. J. McLachlan and M. Magendantz, pers. comm.), the most striking feature is the absence of snails in the more open areas of the lake and they are not found on exposed beaches or in beds of the grass, *Paspalidium geminatum* (Forsk.) which extends into open water. Wave action is probably one factor which makes these habitats unsuitable. Snails are numerous in ditches and rice schemes near the lake, and in canoe channels through the swamp, particularly in the marsh at the water's edge where densities of up to 155 m^{-2} have been recorded on the sandy bottom. This substrate allows rapid movement of the snail populations with changing water levels, but in deeper water deoxygenation of the water overlying the bottom may prevent it being inhabited (Chapter 5 and Cantrell unpublished data). The snail populations extend from the water's edge along canoe channels as far as the swamp margins, and snail abundance has been estimated by recording the number of snails on the undersides of lily leaves along such a channel (Fig. 10.1B). The results show a high density of snails at the water's edge which declines rapidly into the swamp. Only small populations exist in the dense

171

Typha swamp, however, as snails prefer horizontal to vertical surfaces on which to attach themselves. For example, in an area where vertical stems were cut to lie horizontally, a three-fold increase in snail density was observed within one week. In addition, the dense *Typha* stands are an unsuitable snail habitat because the water surface is shaded and so prevents the growth of epiphytic algae on which they graze.

By marking individuals with paint it has been found that snails are extremely mobile in the marsh vegetation. Only 7 per cent of marked snails, for example, were recaptured in the same area one week later. The snails actively disperse from the deeper canoe channels into the inundated marsh vegetation as the lake level rises during the rains. Here, more open water was preferred, especially where there are floating macrophytes. Eggs are laid during February and March on leaves of *Nymphaea* which grow in marsh lagoons. The number of eggs laid can be used as an indication of snail abundance. The results again point to the high numbers of snails near to the shore (mean = 10 egg masses per leaf) as compared with the swamp/lake margins (mean = 0.6). Greatest density therefore coincides with areas where fishermen have most contact with the water around their beached canoes, and this undoubtedly has serious consequences for cercarial transmission as infected snails have been found in the same boat channel at certain times of the year (Magendantz, pers. comm.).

By the end of May the lake level begins to fall, and snails gradually disappear from the marsh vegetation, remaining in the lagoons and channels which hold water. The low oxygen concentrations caused by rotting marsh vegetation appear critical in affecting the snail distribution as the water recedes (Cantrell unpublished data). Some snails are stranded on sandy areas and do not survive as their empty shells are regularly found. Dormant snails have been located in organic detritus, and these revived successfully when immersed in an aquarium. This organic substrate may prevent desiccation and even predation, as suggested by Schiff (1964), but further examination of the mechanism is needed.

Bulinus globosus therefore exhibits greatest abundance in areas which are seasonally exposed by a fall in water level in the marsh, and lives at lower densities on vegetation in deeper water. By cutting channels through the vegetation for their canoes and fishing gear, fishermen create suitable habitats for the snails which thrive in more open areas of floating vegetation. On the other hand, there are instances where fishermen provide some degree of snail control. One fishing technique, beach seining, requires the clearance of vegetation along the lakeshore, and the continual movement of the nets keeps these areas free of weeds. Such beaches are without snails and there is therefore no cercarial transmission. The use of these beaches by women and children for water-based activities, in preference to marshy areas where snails are common, could in the future prove important in controlling bilharzia. Biological control seems unlikely as no lake fish are known to eat the snails, though they may be eaten by the large flocks of Openbill storks, which do eat *Lanistes*, and are known to feed almost exclusively on molluscs (Roberts 1940).

5. Summary

The swamp vegetation at Lake Chilwa is an important habitat for invertebrate populations, and a variety of substrates are available. Floating surfaces, such as those offered by the water lettuce, *Pistia*, rafts of dead *Typha* stems, and grasses in the marsh support several types of invertebrate, particularly midges (Chironomidae) and water beetles (Coleoptera). In contrast, the bulrush swamp, although large in area, has only a scanty fauna, and some of the possible reasons for this are discussed.

The snail, *Bulinus globosus*, the intermediate host of *Schistosoma haematobium* causing urinary bilharzia in man, is particularly common at the ends of canoe channels through the swamp around beached canoes, but is not found in the more open waters of the lake. As the level rises during the rains, snails migrate from channels into the inundated marshland to breed. At the end of the rains, the water recedes, confining snails to the deeper channels which have been identified as important foci of cercarial transmission.

References

Bourn, D. 1974. The feeding of three economically important fish species in Lake Chilwa, Malawi. Afr. J. Trop. Hydrobiol. Fish. 3:135–146.

Hira, P. R. 1969. Transmission of Schistosomiasis in Lake Kariba, Zambia. Nature 224:670–672.

Howard-Williams, C. & Walker, B. H. 1974. The vegetation of a tropical African lake: classification and ordination of the vegetation of Lake Chilwa, Malawi. J. Ecol. 62:831–854.

McLachlan, A. J. 1969. The effect of aquatic macrophytes on the variety and abundance of benthic fauna in a newly created lake in the tropics (Lake Kariba). Arch. Hydrobiol. 66:212–231.

McLachlan, A. J. 1974. Recovery of the mud substrate and its associated fauna following a dry phase in a tropical lake. Limnol. Oceanogr. 19:74–83.

McLachlan, A. J. 1975. The role of aquatic macrophytes in the recovery of the benthic fauna of a tropical lake after a dry phase. Limnol. Oceanogr. 20:54–63.

Morgan, P. R. 1972. The effect of natural alkaline waters upon the ova and miracidia of *Schistosoma haematobium*. Central African Journal of Medicine 18:182–186.

Odei, M. A. 1979. Some ecological factors influencing the distribution of the snail hosts of Schistosomiasis in the Volta Lake. In: Kainji Lake and River Basins Development. The Ibadan Conference (in press).

Omer-Cooper, J. 1957. Dytiscidae from Nyasaland and Southern Rhodesia. I. General Introduction. II. Noterinae. J. Ent. Soc. S. Africa 20:353–377.

Petr, T. 1968. Population changes in aquatic invertebrates living on two water plants in a tropical man-made lake. Hydrobiologia 32:449–485.

Roberts, A. 1940. Birds of South Africa. 4th edition revised by L. R. McLachlan & R. Liveridge (1978). The trustees of the John Voelcker Bird Book Fund, Cape Town. 660 pp.

Schiff, C. A. 1964. Studies on *Bulinus (Physopsis) globosus* in Rhodesia. III. Bionomics of a natural population existing in a temporary habitat. Ann. Trop. Med. Parasit. 58:240–255.

Fishes: Distribution and biology in relation to changes

M. T. Furse, R. C. Kirk, P. R. Morgan & D. Tweddle

1 Fishes: Distribution and biology in relation to changes

Most fishes survive changes in their environment as adults, unlike zooplankton, and they must remain within the system, unlike insects and birds (Chapters 9 and 14). A peculiar exception is the little cyprinodont fish, *Nothobranchius*, of temporary tropical pools in Africa which is well known to go into a diapause phase during dryness, as a retarded embryo within the egg membrane (Wourms 1964). These general constraints have imposed a very wide tolerance of changes in temperature, salinity, alkalinity and turbidity on those very few species of fish which live in the open water of Lake Chilwa. The Chilwa streams and swamps, on the other hand, are inhabited by a diverse riverine fauna, some of which are well-adapted to swamps.

There are two ways in which fishes adapt to a new inhospitable environment: firstly, the species with wide distributions and which therefore are more likely to possess regulatory mechanisms or morphological features which endow them with tolerance of the environmental changes encountered, will persist. This type of selection has been described for zooplankton (Chapter 8). Secondly, when fishes have been isolated in a new environment for as little as 4000 years, new species of fishes may evolve (Greenwood 1965). In Lake Chilwa, both ways have occurred.

Several studies on changes are included in this chapter: (1) the geographical affinities of the fish fauna of the Chilwa Basin, which indicate some effects of morphological changes in the lake basin in geologically recent times; (2) the normal seasonal changes in the biology of the fishes of the open lake and their seasonal distributions between swamp and lake; (3) the behaviour of the three lake species in the phases of the decline and recovery of the lake during the major recession; and (4) the evolution of a new sub-species since the lake became endorheic.

1. The zoogeography of the fish fauna of the Lake Chilwa basin *by* D. Tweddle

Kirk (1967 a,b) discussed the zoogeographical affinities of the Lake Chilwa and Lake Chiuta fish fauna within the limits imposed by the restricted amount of information available at that time about the fish fauna of the surrounding rivers and streams. Since then extensive electro-fishing surveys have been conducted of Lakes Chilwa and Chiuta and their affluent streams, and also the majority of other river and lake systems in Malawi. These have revealed the presence of many more species in the Chilwa–Chiuta depression than were known to exist previously. This knowledge has helped to clarify the relationship of the Chilwa–Chiuta fauna with the fauna of neighbouring river and lake systems, and lent support to the interpretation of the origin of Lake Chilwa described in Chapter 2.

Appendix A lists 36 species in the Chilwa and Chiuta basins whose presence has been confirmed by our survey and those of Kirk (1967 a,b) and Tarbit (1972a and unpublished data). Five species listed are either not yet positively identified, or are undescribed. The characters by which they can be distinguished are considered fully elsewhere (Tweedle & Willoughby 1979a). Until recently only one species of *Haplochromis*, *H. callipterus* Gunther, had been recorded from Chilwa–Chiuta. Loiselle (1974) recorded the presence of a second, *H. acuticeps* (Steindachner) in aquarium shipments to the U.S.A. A *Haplochromis* similar to *H. callipterus*, but with a more elongated snout, has been recorded by the author in Lake Chiuta and also in the Likangala River which may be the former species. The status of the Lake Chilwa *Haplochromis* species is currently being investigated by Jackson (in prep.).

1.1 *Comparison of distribution and origin of fish species in Lakes Chilwa and Chiuta basins*

Kirk (1976b) described zoogeographical evidence for a former connection of the Lake Chilwa basin with the Rovuma River system and concluded that much of the present fish fauna has affinities with that of the eastward flowing rivers of Africa. This was supported by Bailey (1969a), who listed the non-cichlid fishes of the eastward flowing rivers of Tanzania. A comparison of Appendix A with Bailey's species list would show that all Chilwa–Chiuta species (excluding small *Barbus* species) have been recorded either in the Rovuma itself or in neighbouring rivers. Small *Barbus* are excluded because of the continuing taxonomic uncertainty of their identity. However, even in the genus *Barbus* a relationship is apparent between Chilwa–Chiuta and east coast rivers. The common minnows *B. kerstenii* Peters, *B. paludinosus* Peters, *B. radiatus* Peters and *B. toppini* Boulenger are all eastern river species and *B. atkinsoni* Bailey, a very common Chilwa–Chiuta species, was described only recently from the Rufiji system (Bailey 1969b). The other *Barbus* species have close relatives in the eastern river fauna (Tweddle unpublished data).

The close relationship between Lakes Chiuta and Chilwa basins is clearly indicated by the presence of ten *Barbus* species known to occur in Lake Chiuta in a 100 m stretch of the Sombani River, the Chilwa stream furthest removed from Lake Chiuta.

The fish fauna of the Chilwa–Chiuta basin is comprised predominantly of the cyprinids described above. Among the six cichlid species found, three of them, *Haplochromis callipterus* Gunther, *Pseudocrenilabrus philander* (Weber) and *Tilapia rendalli* Boulenger, are fairly widespread through Central and Southern Africa. The two sub-species of *Sarotherodon shiranus* are peculiarly distributed. The nominate sub-species *S. s. shiranus* Boulenger is found in Lake Chiuta and in Lake Malawi and its streams, but not in Lake Chilwa. The other sub-species *S. s. chilwae* (Trewavas) appears to be endemic to Lake Chilwa only. Its evolution is discussed in section 11.7.

The possible effects of the isolation of Lake Chilwa from Lake Chiuta

Lancaster (Chapter 2) considers the sandbar between Lake Chiuta and Lake Chilwa to date from about 9000 years B.P. It might reasonably be expected therefore that the two lakes would have similar faunas. Apart from *S. s. chilwae*, which may have evolved in this time, only two species in the Lake Chilwa catchment area, *Marcusenius livingstoni* (Boulenger) and *Barbus* sp. A, have not been found in the Chiuta basin. Both species are rare in Chilwa streams and they may have been overlooked in the Chiuta streams. The former was, in fact, first described from the Rovuma River to which Lake Chiuta is still indirectly connected (Bailey 1969a).

Similarly the cyprinodont, *Nothobranchius kirki* Jubb has not been recorded further north, and this may also be due to incomplete collecting. *N. kirki* lives only in temporary pools of the floodplain during the wet season, although a few have been seen in the swamp and one caught in the lake. Its life-history is very short, since it breeds and then dies in its first year.

On the other hand, there are several fish which occur in Lake Chiuta but not in Lake Chilwa nor in its streams. *Mormyrus longirostris* Peters and *Bagrus orientalis* Boulenger are relatively large species, which are adapted to lake conditions or large rivers. When Lake Chilwa enters a drying phase, the small streams and swamp clearings which provide refuge for other species may well be uninhabitable by these fishes. The same explanation might perhaps suffice for other Chilwa absentees, *Engraulicypris brevianalis* (Boulenger), *Synodontis zambesensis* Peters and *Glossogobius giuris* (Hamilton-Buchanan). But there is an alternative explanation, that these fishes invaded Lake Chiuta from the Rovuma system after the formation of the sandbar cutting off Lake Chilwa from Lake Chiuta, when the rate of river flow must have declined. *Glossogobius giurus* is a euryhaline species, which may have had access to the Rovuma River, but may have penetrated Lake Chiuta more recently than the sandbar formation. Its absence from Lake Chilwa is otherwise difficult to explain as it is able to tolerate extremes of salinity and it is unlikely that it would have been eliminated by the saline conditions sometimes experienced in Lake Chilwa.

1.2 *Migration of Chiuta/Chilwa species of fish to Lake Malawi*

It is now fairly certain that Dixey's theory (1926), questioned by Kirk (1967b), that Lake Malawi at one time was connected to Lake Chilwa, which drained into the Lower Shire via the River Ruo, is incorrect. The fishes of the Upper and Middle Shire and the affluent streams of Lake Malawi are now well known (Tweddle & Willoughby, 1979a; Tweddle, Lewis & Willoughby 1979) and it is clear that the similarities are due to Lake Chilwa–Chiuta species finding their way into the Lake Malawi catchment, rather than the other way round. In addition the fauna of the Upper Ruo River is different from that of Lake Chilwa as will be shown below (Tweddle & Willoughby 1979a).

Bell-Cross (1972) and Fryer & Iles (1972) have discussed the origin of the Lake Malawi fauna. Bell-Cross suggested that the relationship between the Rovuma and Lake Malawi faunas could be explained by watershed exchange between the two systems, whereas Fryer & Iles suggest that the south Rukuru, which enters Lake Malawi in the northwest, was part of the Rovuma River system prior to the formation of the present lake. Evidence suggests that more than one invasion of Lake Malawi by fish species from eastern rivers may have occurred. The presence of two closely-related species of each of the genera, *Marcusenius*, *Labeo*, *Bagrus* and *Mastacembelus* suggests connections with eastern rivers in the past.

Marcusiensis nyasensis and *M. livingstoni*, are very closely related and may be synonomous (Tweddle & Willoughby unpublished data). The two specimens obtained from Lake Chilwa streams are intermediate in form between the 'types' of the two species. They are both found in large rivers as well as in small streams and occur in the Middle Shire River (Tweddle et al. 1979). This is another species which could have invaded Lake Malawi by crossing the divide between the Chiuta streams and tributaries that reached Lake Malawi.

Tarbit (1972a) pointed out that the Lake Malombe–Chiuta watershed is eroding eastwards, cutting into the Chiuta basin (Dawson 1970), and suggested that it is theoretically possible for fish species of captured streams, which originally supplied Lake Chiuta, to have gained access on a number of occasions to the waters of Lake Malombe and the Shire River (Fig. 11.1). He discussed, in particular, the close proximity of the headwaters of the Nyenyesi River, which flows into Lake Amaramba near its outlet into the Lugenda River, and the headwaters of the Masanje River, which is a perennial tributary of Lake Malombe. The Nyenyesi River and the Mandimba River, which it joins near the lake, have now been fished in their upper reaches. The species caught, *Amphilius platychir* Gunther, *Barbus atkinsoni*, *B. kerstenii*, *B. paludinosus*, *B. trimaculatus* Peters, *Chiloglanis neumanni* Boulenger, *Clarius gariepinus* (Burchell) and *Leptoglanis rotondiceps* Hilgendorf, all occur in the Lake Malawi catchment as well as in the Chilwa and Chiuta basins. *B. atkinsoni* is of particular interest as it was known previously only from the Rufiji River, and now it has been found in the Chilwa–Chiuta basins as well as in Lake Malawi. Specimens of this species from Chiuta and Chilwa all have the typical 3 spot coloration described by Bailey (1969b), also typical of Middle Shire specimens. In Lake Malawi specimens the spots are elongated and occasionally broken and often form an almost complete lateral stripe, although the tendency to 3 spots is apparent.

Although there is evidence of faunal links between the Lake Malawi and Lakes Chilwa–Chiuta systems, many Chilwa–Chiuta species are not present in the Lake Malawi catchment, and it is therefore considered highly unlikely that the outflow of Lake Malawi entered Chilwa–Chiuta as suggested by Dixey (1926). *Barbus* cf. *afrohamiltoni* Crass, *B.* cf. *viviparus* Weber, *Barbus* sp. *A*, *Pareutropius longifilis* (Steindachner) and *Haplochromis* sp. do not occur in Lake Malawi, and the closest relative of *Engraulicypris brevianalis* of Lake Chiuta, *E. sardella* of Lake Malawi has clearly been separated for a considerable period.

1.3 *Connections with the Upper Ruo River*

The fauna of the Upper Ruo River, which rises on the east face of Mt. Mulanje and enters the Shire River just before its confluence with the Zambesi River, is different from that of the Chilwa Basin (Tweddle & Willoughby 1978a).

The zoological evidence suggests that Lake Chilwa–Chiuta has never drained through the Ruo River into the Shire River, although limited faunal exchanges have clearly occurred recently. Thus the Tuchila River, a tributary of the Ruo, has a 'Chilwa stream' fauna (Fig 11.1). This stream arises on the Mulanje and flows through the Phalombe plain in close proximity to the Phalombe River which enters Lake Chilwa. It was a former tributary of the Phalombe River. Tributary capture has certainly occurred here and it is possible that even now fish transfer can occur across the plain in exceptionally wet weather. The Tuchila fauna however is quite distinct from the fauna of the rest

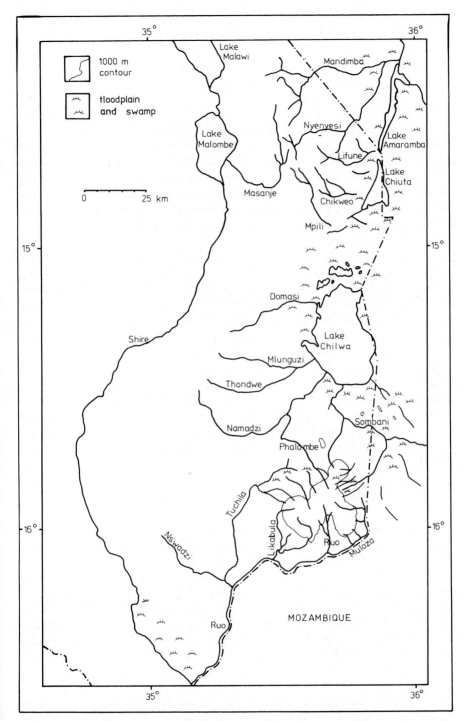

Fig. 11.1 Map of the river systems north of Lake Chilwa to show the indeterminate watershed between rivers to Lake Chiuta and to Lake Malombe in the Rift Valley, and in the south to show the indeterminate drainage between the River Tuchila, tributary of the Upper Ruo and the tributaries of the Phalombe River.

of the Upper Ruo system, and *Barbus* sp. *B*, *B. radiatus*, *B.* cf. *viviparus* and *Hemigrammopetersius barnardi* (Herre), all common in the Tuchila, have not been found anywhere else in the Ruo system. *B. atkinsoni*, *Barbus* sp. *C.* and *Petrocephalus catostoma* (Gunther) are other 'Chilwa' species found in the Tuchila and known elsewhere in the Ruo system from single specimens only.

In the Ruo River itself and in other tributaries six species occur which are not found in the Chilwa–Chiuta basin. Only two of these species are known to occur anywhere else in Malawi. The presence of *B. choloensis*, *Varicorhinus* cf. *nelspruitensis* and *Amphilius natalensis* suggests an affinity with the fauna of eastern rivers south of the River Zambesi, and it seems possible that the Upper Ruo fauna is relict, protected from invasion by Zambesi species by the 60 m high Zoa Falls. The presence in the Ruo of so many species not found in Lake Malawi or Chilwa–Chiuta effectively discredits the theory that Lake Chilwa may at one time have found an outlet to the Zambesi via the River Ruo.

In brief, it seems as though there is a distinct Chilwa–Chiuta fish fauna, which was derived from the eastward flowing rivers to the north, at a time when the ancient proto-Lake Chilwa–Chiuta was much larger (Chapter 2). Where a few of its species are found in the Lake Malawi catchment, the invasions have been from Lake Chiuta via an indeterminate watershed between its affluent streams and those of the Lake Malawi basin. Similarly, there is no faunal similarity with the Upper Ruo River, although one of its tributaries has a 'Chilwa' fauna due to the swampy headwaters shared with the Chilwa streams.

Nevertheless, this common Chilwa–Chiuta fish fauna does not live in Lake Chilwa itself, but in the swamps and streams of the basin. This is unlike the distribution in both lake and streams of Lake Chiuta. The open water of Lake Chilwa lies at the extreme opposite end of the scale from Lake Malawi, which has 245 species of fishes with 93 species endemic to it. Lake Chilwa open water has three main fish species, of which one is an endemic sub-species.

2. Species composition of the fish fauna *by* M. T. Furse

The open water of Lake Chilwa is totally dominated by three species only: the minnow, (cyprinid) *Barbus paludinosus* Peters (known locally with other small fish as *matemba*), the catfish, (clariid) *Clarias gariepinus* (Burchell) (*mlamba*) – the northern form, often known previously as *C. mossambicus*, and the endemic *makumba*, *Sarotherodon shiranus chilwae* (Trewavas), a mouth-brooding cichlid. Their maximum lengths in Lake Chilwa are usually 120 mm, 780 mm and 250 mm respectively. These three species, of which the proportions change with the fluctuations of levels of the lake, provide the basis for the flourishing fishery of the open water described in Chapter 12.

Fishermen in the swamp catch another ten species or so (Table 11.1), which are very common and comprise about 30 per cent of the total fishery of the ecosystem. Only *Haplochromis callipterus* and *Hemigrammopetersius barnardi* are commonly found in the lake in the present years of high level, but they are relatively unimportant in the fishery.

The fauna of the open lake appears to be a relict of the former rich fauna, which was shared with Lake Chiuta before the formation of the sandbar, which divided the ancient Lake Chilwa–Chiuta into an endorheic part and an open freshwater lake. The thirty odd species which are common to both systems today are mostly restricted to the swamp and streams around Lake Chilwa, but live in the open water of Lake Chiuta.

Lake Chiuta has a conductivity of $400\,\mu\mathrm{S}\,\mathrm{cm}^{-1}$, it is deeper and therefore cooler and considerably clearer than Lake Chilwa (Chapter 4). Submerged growths of *Ceratophyllum* occur and large areas are covered by the water lily *Nymphaea* (Furse & Bourn 1971). The Chilwa swamp is much more saline but these plants occur in the channels, lagoons and edges only (Chapter 5) where water is protected by the aerial vegetation and is less turbid and cooler than the open lake. Although the environments of the Chilwa swamp and Lake Chiuta differ notably in conductivity, the swamp has remained hospitable enough to allow much of the former fish fauna to persist.

Table 11.1 Fishes common in the swamps of Lake Chilwa.

Cichlidae:	*Haplochromis callipterus* (Gunther),
	Tilapia rendalli (Boulenger),
	Pseudocrenilabrus philander (Weber);
Cyprinidae:	*Barbus trimaculatus* Peters,
	Labeo cylindricus Peters,
	Alestes imberi Peters,
Characidae:	*Hemigrammopetersius barnardi* (Herre);
Mormyridae:	*Petrocephalus catastoma* (Gunther),
	Marcusiensis (Gnathonemus) macrolepidotus (Peters)

The open water of Lake Chilwa is particularly inhospitable to the palustrine and riverine fishes in the dry season, near the end of which *Clarias* and *Barbus* migrate into the swamp to spawn. During the rainy season this clear division breaks down to a minor degree, but the swamp and river species remain peripheral. They enter the lake with the swollen waters of the affluent rivers in the wet season. Occasionally a temporary incursion may also be brought about by reaction to a wind tide. Trawl catches along the lake-swamp interface south of Chidiamphiri (Python Island) in December 1971 illustrate this movement well. Catches at this place were usually dominated by the three open lake species (Chapter 12). On this date, however, the commonest fish captured in terms of wet weight were first *S. s. chilwae*, and not much less were *B. trimaculatus* and *Tilapia rendalli*. The two invading species were each a heavier proportion of the catch than *B. paludinosus* and *Clarias gariepinus*. In two years of trawling (1971–72) this was only the second time that *T. rendalli* had been caught. It is caught in the lake in the years of very high level (e.g., 1978) (T. Jones pers. comm.), when it occurs in open water except when breeding inshore from January to April. In 1971, the water in which these fish were caught was clearly of swamp origin, being stained dark-brown by humic acid. About 600 m away, only the three lake species were caught in the typically murky waters. The lake water had a conductivity of $2500\,\mu\mathrm{S}\,\mathrm{cm}^{-1}$, pH 9.25 and modified Secchi disc measurement 135 mm; the flushed out swamp water had a conductivity of $3000\,\mu\mathrm{S}\,\mathrm{cm}^{-1}$, pH 8.8 and its Secchi disc reading was 220 mm.

183

The incursion was short-lived and the fishes returned to the swamp. Wind tides occur many times a year, ensuring the mixing of swamp and lake water (Chapters 5 and 13) yet the fishes still return to the swamp.

Experimental gill net catches made in open water in December 1968 showed that *Labeo cylindricus, Alestes imberi, Pareutropius longifilis* and *Marcusenius macrolepidotus* had entered the marginal areas (Morgan 1971), although the conductivity was 2500 μS cm^{-1}. However, only two months later these visiting species had almost completely retreated to the swamp and feeder streams. The reasons for this behaviour from the ecological standpoint are discussed below and experimental data are given in section 11.6.

The swamp species are as tolerant of normal changes of salinity (a factor of 2 or 3 times a year) as the lake species, yet they leave the lake when its conductivity is lowest and return to the swamp. High oxygen demand is more likely in the swamp environment than in the lake because of the respiratory requirements of *Typha*, epiphytic algae and decaying organic material, and available oxygen is variable. Turbidity is certainly less in the swamp, and may be a limiting factor in the lake (Chapter 5). High concentrations of inorganic material in suspension may damage fish gills (Herbert et al. 1961) and there is reason to suppose that Chilwa fish species differ in the degree of protection of their gill epithelia by mucus (Cockson & Morgan unpublished data). The silt load in the lake restricts the penetration of light and there is no submerged vegetation to offer a habitat to the fishes venturing in from the swamp. Thus Lake Chilwa is deprived of this rich food source (Kirk 1967a). Similarly, the benthos in the lake declined dramatically in open water in the post-filling phase (Chapter 9) and so the main food available was zooplankton and decaying plant material. Perhaps the restriction of variety in food excludes those species with different dietary requirements. Much more field and experimental research is needed before the restrictive contributions of the various environmental parameters can be quantified.

3. The diet of the three dominant Lake Chilwa species

Bourn (1972, 1973) analyzed the diets of the three dominant fish species according to 13 recognizable food categories found in their guts: the herbivorous components i.e. higher plant detritus, filamentous and non-filamentous green algae and blue-green algae, and diatoms; zooplankton: crustaceans and rotifers; gastropods, aquatic and terrestrial insects, fish eggs and fish (Tables 11.2, 11.3 and 11.4).

Bourn demonstrated that there are differences in the proportions of food items taken by the fishes dependent on various factors: the maturity of the fishes (the second size in each table indicates mature fish); the proximity to emergent vegetation where the density of plant detritus would be greater; the wet and dry seasons, indicated in the tables by conductivity; whether feeding by night or day. He compared three different capture gears: gill net, trawl and seine net which each automatically selected different ranges of size.

One food item was common in the three species, namely crustacean zooplankton. In the juveniles of all species zooplankton accounted for 48 to 70 per cent of the gut contents. Bourn (1973) showed that the variety of food increased and the proportions of zooplankton decreased as the fishes matured

Table 11.2 Barbus paludinosus: Analysis of fish diet (after Bourn 1973).

Food item as a % of total recognizable contents	Higher plants	Green algae filamentous	Green algae non-filamentous	Blue-green algae filamentous	Blue-green algae non-filamentous	Diatoms	Crustacea	Gastropoda	Aquatic insects	Terrestrial insects	Rotifera	Fish eggs	Fish remains	Number of fish	% empty
Combined Results															
Total gut contents	14	3	19	1	0	1	56	0	3	0	1	1	0	586	6
Length															
1–39 mm	3	4	15	2	0	0	65	0	0	0	6	5	0	44	0
40–79 mm	13	3	19	1	0	1	60	0	2	0	1	0	0	458	7
80–119 mm	31	5	17	2	0	2	33	0	10	0	0	0	0	85	8
Vegetation															
Present	15	4	20	2	0	2	52	0	4	0	1	1	0	470	71
Absent	11	0	13	0	0	0	74	0	0	0	0	0	0	118	4
Conductivity Season															
<1500 μS cm^{-1} wet	14	5	22	2	0	2	48	0	4	0	2	0	0	352	2
>1500 μS cm^{-1} dry	16	1	13	0	0	0	69	0	1	0	0	0	0	201	2

Table 11.3 Clarias gariepinus: Analysis of fish diet (after Bourn 1973).

Food items as a % of total recognizable contents	Higher plants	Green algae filamentous	Green algae non-filamentous	Blue-green algae filamentous	Blue-green algae non-filamentous	Diatoms	Crustacea	Gastropoda	Aquatic insects	Terrestrial insects	Rotifera	Fish eggs	Fish remains	Number of fish	% empty
Combined Results															
Stomach contents	12	2	2	1	0	1	47	1	13	2	0	0	20	474	30
Intestinal contents	17	4	4	3	0	3	37	2	20	4	0	0	8	369	13
Length															
1–149 mm	3	0	2	0	1	0	69	0	12	3	0	0	11	110	23
150–299 mm	14	2	3	1	0	1	46	0	15	1	0	0	17	220	22
300–449 mm	16	3	1	1	0	1	26	2	12	2	0	0	35	136	44
450–599 mm	0	0	0	0	0	0	0	13	0	0	0	0	88	8	75
Vegetation															
Present	14	2	3	1	0	1	39	1	16	2	0	0	22	374	33
Absent	7	1	0	1	0	0	67	0	11	2	0	0	11	98	16
Conductivity Season															
<1500 μS cm^{-1} wet	7	3	3	1	0	1	39	1	16	2	0	0	27	303	38
>1500 μS cm^{-1} dry	21	0	1	1	0	0	56	0	9	1	0	0	11	145	21
Capture method															
Gill net – night	10	4	1	3	0	1	29	2	18	1	0	0	29	196	44
Trawl day	7	1	3	0	0	0	60	0	11	2	0	0	15	258	20
Seine	62	0	0	0	0	1	5	0	18	0	0	0	14	22	5

and grew bigger. *Barbus* and *Sarotherodon* adult diets included far more higher plant detritus, and *Clarias* used more fish as food, which was usually *Barbus*.

Bourn stressed the importance of the diversity of food which these three species consume, although the categories: non-filamentous blue-green algae and fish eggs proved not to be used as food at the time of this study and very few molluscs were consumed. Filamentous blue-green algae and diatoms were selected only by *Sarotherodon* in significant amounts. He suggested that fish with highly specialized diets would be at a disadvantage in Lake Chilwa. 'Each of the three species may be described as non-specialized opportunist feeders, consuming in varying proportion a wide variety of different food items which are available and which they are capable of ingesting' (Bourn 1973).

Table 11.4 Sarotherodon shiranus chilwae: Analysis of fish diet (after Bourn 1973).

Food items as a % of total recognizable contents	Higher plants	Green algae filamentous	Green algae non-filamentous	Blue-green algae filamentous	Blue-green algae non-filamentous	Diatoms	Crustacea	Gastropoda	Aquatic insects	Terrestrial insects	Rotifera	Fish eggs	Fish remains	Number of fish	% empty
Combined Results															
Stomach contents	28	16	7	12	1	10	14	0	0	0	7	0	5	360	40
Intestinal contents	43	16	5	10	1	15	2	0	0	0	9	0	0	267	8
Length															
1–99 mm	9	15	12	7	0	9	46	0	0	0	2	0	0	70	19
100–199 mm	34	17	4	15	1	10	2	0	0	0	10	0	6	267	45
200–299 mm	51	8	6	5	0	18	1	0	0	0	1	0	10	24	50
Vegetation															
Present	31	18	8	14	1	12	1	0	0	0	9	0	6	300	45
Absent	20	10	1	6	1	4	57	0	0	0	1	0	2	59	17
Conductivity Season															
<1500 μS cm^{-1} wet	32	18	7	14	1	11	1	0	0	0	9	0	6	282	47
>1500 μS cm^{-1} dry	43	13	7	12	2	11	5	0	0	0	3	2	2	40	30
Capture method															
Gill net – night	39	18	6	18	1	12	1	0	0	0	3	0	2	217	52
Trawl – day	18	14	7	7	0	8	26	0	0	0	11	1	7	144	24

Tables 11.2, 11.3 and 11.4 show that when the availability of any food item increased in the water, such as macrophyte detritus near emergent vegetation, crustacean zooplankton in the dry season and rotifers and green algae in the wet season (cf. Chapter 8) consumption of that item increased.

Investigation of the digestive enzymes possessed by the fish species (Cockson & Bourn 1972, 1973) confirmed that the fishes are able to digest and utilize a wide spectrum of food. *Barbus* (which has no anatomical stomach) showed amylase activity throughout the length of the intestine, while trypsin was confined to the posterior end. *Clarias* had high pepsin activity in the stomach and a lower activity of a protease in the intestine where amylase activity was also shown. *Sarotherodon* showed amylase activity through the gut and protease activity in the intestine. The pH in the gut of *Sarotherodon* was never lower than 4 in this investigation, which suggested that blue-green algae were not being digested. Since diet has an effect on the physiology of the gut (Cockson & Bourn 1972), this species may accomodate its digestive activity to the source of food at the time when only blue-green algae and diatoms were available, reported by Kirk (1970). In Lake Victoria *S. esculentens* feeds by filtration and ingests many components of the plankton, but digests only diatoms (Fish 1951).

The diet is not only varied, but changes with seasons and with the periodic recessions of the lake which affect the fauna and flora. In the pre-drying period when filamentous blue-green algae reached a high density, the gut of *Sarotherodon* was full of these algae, although they appeared to be undigested (Kirk 1970). *Anabaena* sp. *Oscillatoria* sp. and *Arthrospira* sp. were taken as each became dominant from 1966–67 in the pre-drying and drying periods (see Chapter 6). Evidence from Lake George (Moriarty et al. 1973) has shown that *Sarotherodon niloticus* and *Haplochromis nigripinis* can assimilate *Microcystis*, a non-filamentous blue-green algae. Lysis of the algae was best at pH 1.4–2, and it was possible to tell whether assimilation had occurred because the faeces in fishes in which algae had not been digested were green, and if the algae had been partly assimilated they were brown. The genus *Sarotherodon* is not herbivorous in the sense of browsing on submerged vegetation; in any case, none is available in Chilwa. In 1967 the diatom, *Anomoensis sphaerophora* was ingested, and since this is an epipelic species living on the lake bed, feeding on the bottom was indicated (Kirk 1970).

Clarias gariepinus in Lake Chilwa is neither an obligatory insectivore nor piscivore, it is a generalized feeder which depends on the availability of different foods. In the swamp there is more opportunity to feed on aquatic insects, and fish eating is confined to older individuals. In the open lake it takes zooplankton and fishes. The habits of *Clarias* in Lake Chilwa are similar to those of the same species in Lake Victoria (Corbet 1961) but differ from those in Lake McIlwaine (Munro 1967), where conditions were different.

Curiously, *Sarotherodon* fed more by day than at night, while *Barbus* and *Clarias* fed in both periods. It is probably impossible for fish to select food by sight in such a turbid lake, although cases in other lakes are reported where planktivorous fish select the largest specimens of the largest species of crustacean by sight (Zaret & Kerfoot 1975). All the three species of fish have numerous fine gill-rakers which enable them to filter microscopic animals. These are closer together in juveniles and the selection of zooplankton when the fishes are smaller may be a mechanical one. It was difficult to identify crustacean species in the guts of the fishes, but from evidence of the composition of zooplankton populations and their variation with fish predation, the main species consumed are the calanoid copepod *Tropodiaptomus kraepelini*, and the cladocerans *Diaphanosoma excisum* and *Daphnia barbata* (see Chapter 8).

The flexibility in diet may be considered an advantage for surviving in Lake Chilwa. Fishes which could not live in the lake (and which do live in the swamp) may have a narrower range of feeding habits such as being predominantly herbivorous, bottom feeders, epiphyton feeders or feeders on benthos associated with submerged vegetation. The diets of the swamp fishes, however, have not yet been studied. The trophic adaptability of the three lake species which enables them to utilize the available food supply is an important asset. Survival in the lake also requires a rapid growth rate, early maturity and high reproductive rate, which the three species exhibit in varying degrees, so that in a lake beset by periodic recessions they are capable of a high rate of natural increase.

4. Reproduction and growth in the three open water species of fishes by M. T. Furse

4.1 Barbus paludinosus

The breeding season of *Barbus paludinosus* is an extended one through the months of the rainy season and, in some years, even longer. The major recruitment to the fishery in 1971–72 occurred in March when a cohort (group of the same age) with a modal length of 30–35 mm appeared (Fig. 11.2). In 1971, a small fast-growing cohort of this size first appeared in November, followed by the usual recruitment in March and April. It is however possible that commercial catches give an inaccurate picture of the times when young fish first appear. From the experimental trawling data of 1971–72, the rapid growth of fish through the dry season can be followed (Fig. 11.2). By November, a modal length of 52 mm is obtained by the March recruits, rising to 54 mm by the following February. As maturity in both sexes is usually achieved by 50 mm (and even 46 mm in 1976) most recruits reached maturity in their first year of growth and in time for breeding. After the minor recession in 1973, during the recovery phase of 1975 and 1976, new cohorts seemed to appear in the fishery at intervals of approximately two months, as though there was a compensatory mechanism of increased breeding frequency after the population decline of 1973 (T. Jones pers. comm.).

Females grow more rapidly than males, and there is evidence that they mature at a slightly smaller size (Kirk 1972, Furse unpublished data), so that females outnumber the males in most size categories from 45 mm upwards throughout the breeding season. This is a significant factor in maintaining a high reproductive rate. The relative proportion of females increases further with increased size and few males exceed 75 mm in length. The maximum size recorded in Lake Chilwa is 120 mm for females. This species appeared to be slightly 'dwarfed' in Lake Chilwa compared with the maximum size of 140 mm found elsewhere in east African rivers (Jubb 1967).

Gonads were ripe from November to April in 1971 (Furse 1972), but ripeness has been recorded later, in June 1966 (Kalk 1969), and earlier too in 1976 (Fisheries Dept. 1978). Under more dilute conditions in 1976 maturation began in July and about 70 per cent of females and 50 per cent of males were mature by November, so that each female may spawn several times in the year.

This small species has a very high relative fecundity and therefore meets one criterion for survival, namely a high intrinsic rate of natural increase. A fecundity range of 255 to 801 eggs has been recorded at 50–60 mm, and a maximum value of 2513 eggs has been recorded at 112 mm in length. It is probable that females may live for three years or more (Kalk 1969), but three years after the major recession of 1968 very few of the larger size group were seen.

In 1971 and 1972, both the commercial seine catches and the experimental

188

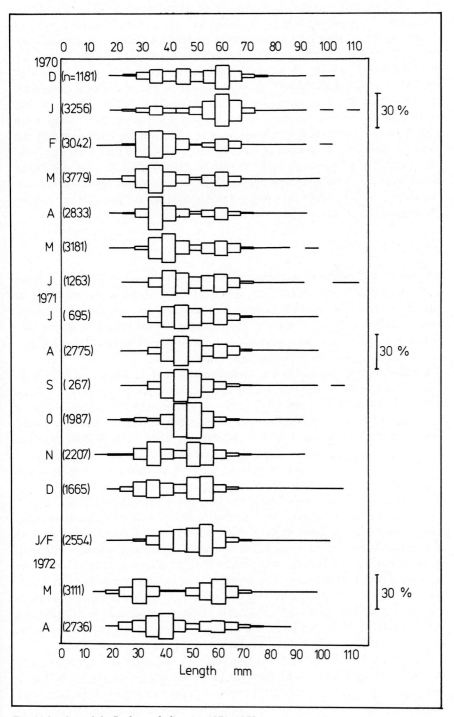

Fig. 11.2 Growth in *Barbus paludinosus*, 1971–1972.

trawl catches fell markedly in the open lake from September to February and very few spent fishes were seen in the open water of the lake. This suggests the possibility of a spawning migration in *Barbus*. At that time the populations of *Barbus* were recovering from the period of dryness and were comparatively low. In 1976, when the lake carried heavier populations of *Barbus* it could be seen that not all individuals leave the lake at the same time, since high catch rates were obtained all through the year. Dispersal from the deeper part of the lake in the southeast sector was indicated by a fall to 25 per cent of its June value from October to December. At peripheral sites near the swamp much higher catches are made and it is assumed that the fish are intercepted by the trawl on their way in and out of the swamp (Chapter 12). Unfortunately no direct evidence of egg-laying in the swamp is available. It is obvious that *Typha* stands would provide a substratum on which the gelatinous egg masses can be laid (Kirk 1967a) and would offer the newly-hatched fry protection from the turbulent and turbid conditions in the lake.

4.2 *Clarias gariepinus*

Juveniles first appeared in commercial catches in January 1971 at a mean modal length of about 70 cm. This was likely to have been inaccurate as an estimate of when the juveniles returned to the lake, since smaller ones would have been missed by the fishing techniques used. By the following November, catfish in their first year had increased in length to 193 mm and risen to 235 mm by the following February (Furse unpublished data).

Maturity is first attained at lengths between 170 and 200 mm for both male and female and therefore the majority of fishes are capable of breeding in their first year. These figures are smaller than those reported by Kirk (1972) for 1966, when the onset of the rains was delayed. Sizes given for maturity in that year were much higher and of later occurrence, whereas in 1976 *Clarias* was ripe in October (Fisheries Dept. 1978). It is well known that if there is insufficient flooding of the land beyond the swamps or river that spawning is postponed and the pre-spawning period is lengthened (Hoar & Randall 1971). Maturation usually begins in September and gonads develop rapidly. In 1971 by November approximately 40 per cent of fishes over 195 mm, both male and female, were sexually mature. Kirk (1972) had found a similar pattern in the years before the lake dried. The breeding season was more discrete than those of *Barbus* and *Sarotherodon* and few ripe individuals were found after January. From January onwards spent fish became common in the catches. It seemed improbable that there is a second period of spawning, as has been observed elsewhere (Jackson 1962). The 1976 trawling data confirmed this general pattern (Fisheries Dept. 1978). Eggs are laid singly and attach themselves by a little adhesive disc to submerged vegetation (Jubb 1967).

Breeding behaviour involves the characteristic migrations from the lake into the surrounding swamps and floodplains, so commonly recorded in Africa (Jackson 1962, Kelley 1968, Holl 1968). *Clarias* has a fairly high estimated fecundity. The smallest ripe female (141 mm) had a fecundity of 169 ripe eggs and in the largest examined (783 mm) it was estimated at over 180,000 mature eggs. This surpasses the fecundity of *Barbus paludinosus*, but is not as high as in some marine fishes (Kyle 1926).

190

Kirk (1970) made an intensive study of *Sarotherodon shiranus chilwae*. The first recruits to the lake population were found in commercial catches in January and February at lengths between 20 and 30 mm. By November, at the onset of the breeding season, the modal length of the new generation was about 100 mm. There may be considerable variation in the timing of first breeding from year to year, apparently depending on the salinity of the water. In ponds, aquaria and streams this species breeds earlier and at a smaller size (100 mm). In the open water of the lake, only a few individuals will breed within their first year, the majority doing so early in their second year, at about 125 mm for females and 150 mm for males. The inability to reach maturity within the first year, in contrast to the two other dominant species, is a disadvantage, compensated for to some extent by a long breeding season from September to May. The principal spawning period is from November to January, but a female may breed two or three times in one year.

The average maximum size achieved by *S. s. chilwae* is usually only 250 mm, so that in comparison with other *Sarotherodon* species, it is dwarfed. The eggs are laid in saucer-shaped nests before being fertilized and taken into the mouth of the female for brooding (Kirk 1967a, 1970). Such a habit requires a fairly firm sandy substratum, which is found in Lake Chilwa on the eastern edge of the lake and in pockets along the west and northwest, as was seen when the lake bed was exposed, and was found in bottom samples in the intensive sampling after the lake recovered from dryness (McLachlan et al. 1972, McLachlan 1974). This breeding distribution would explain the paucity of *Sarotherodon* in the southern sector during the breeding peaks, and the density of juveniles, *kasawale*, which fill the *Barbus* trawls in February in the sandy areas. The spawning habit of *S. s. chilwae* appears similar to that of the endemic species in Lake Rukwa, which does not leave the lake to spawn (Mann 1964) but dissimilar to *S. macrochir* which leaves the lake to do so (Bowmaker 1965).

The eggs of the Lake Chilwa species are fairly small for a mouth-brooding cichlid, measuring 2.5 × 2.0 mm across. They hatch from the egg membrane in the maternal buccal cavity on the sixth day and in 14 days there is considerable reduction of the yolk sac and the larvae make temporary forays into the water and feed on zooplankton. After 3 weeks the young are 10 mm long, and leave their maternal refuge. After 32 days the fry are 14 mm in length, the yolk sac is almost completely absorbed and they are free. After 90 days the fish is between 40 and 50 mm long.

The guarding of newly hatched eggs and fry gives a lower initial mortality rate than would be expected in the unprotected eggs and fry of the other two species of fish in Lake Chilwa. Nevertheless, the brooding efficiency, as measured by the technique of Welcomme (1967) was 42 per cent for 3 mm larvae and as low as 15 per cent at 10 mm (Kirk 1970). Both adults and juveniles fall prey to *Clarias* and to predatory birds.

Estimations of fecundity of *S. s. chilwae* at various sizes over the normal adult range have been derived from Kirk's regression analyses (1970); 100 to 400 eggs are ripe in females 120 to 200 mm long. This is a larger number than for many other mouth-brooding cichlids (Lowe–McConnell 1959).

These reproductive characteristics of *S. s. chilwae* are of considerable advantage to the fish in the fluctuating environment of Lake Chilwa. It is smaller in adult size than *S. s. shiranus* of Lake Malawi and matures earlier; it produces smaller and more numerous eggs and has two main spawning periods in an extended breeding season instead of one period. It also matures more quickly in fresh water (Kirk 1970) so that in times of lake recession the reservoir population of lagoons and streams can repopulate the lakes to some extent even in the first year of recovery.

When Lake Chilwa refilled in 1969, it was restocked with *S. s. chilwae* at the suggestion of Kirk (1968) by a joint effort of the University and the Fisheries Department. Morgan (1971) of the Lake Chilwa Research Project had reared about 300,000 fingerlings in the Fisheries Department culture ponds, which were successfully transferred alive to the open water of the lake. Although some were marked by fin clipping, no trace of them was ever found, and despite this effort the population of *Sarotherodon* did not approach its former density until two years later.

5. Diseases, parasites and predators of Lake Chilwa fishes *by* R. C. Kirk

In the pre-drying year, 1966, fishes of Lake Chilwa were examined by Kirk (1972) for parasites and diseases. Three kinds of parasites were recovered from *Sarotherodon*: nematodes, a trematode and a fungus on the tail. *Clarias* was host chiefly to nematodes and *Barbus* to a large species of tapeworm. These were identified by A. C. Pike of the Commonwealth Bureau of Helminthology.

The monogenean trematode, *Cleiodiscus halli* Price was found on the gills of *S. s. chilwae*. It is common to both sub-species of *S. shiranus*, and has also been found on *S. mossambicus* (Price pers. comm.). Since parasites are usually host specific, this is confirmatory evidence that the three are closely related (see section 11.7).

Between 30 to 50 per cent of the *Sarotherodon* examined in 1966 appeared to be infested by the nematode, *Contracaecum* sp. (Pike pers. comm.) found in the gills and body cavity. There was little variation in the degree of infestation throughout the year. None of this parasite were found in the nominate sub-species, when examined by Lowe (1952) or Kirk (1970). Another *Contracaecum* species infects *Clarias gariepinus* in Lake Chilwa (Pike pers. comm.). The adult stage of the parasite lives in an avian host and in Lake Chilwa this is probably the very abundant African grey-headed gull, *Larus cirrocephalus* L.

The fungal rot affecting the caudal fin of *S. s. chilwae* was found to be particularly prevalent towards the end of the dry season, when the conductivity of the lake rose steeply. It also appeared in similar conditions in laboratory experiments when the fish were kept in very saline and alkaline water (Morgan 1972). On occasion the entire tail fin was rotted away, leaving only the stumps of the fin rays. It is known that stress of any sort is conducive to fungus infections in fish and in Lake Malawi they are more susceptible at low temperature (D. Eccles pers. comm.).

It was at this time that mass mortalities of *Sarotherodon* were observed (discussed in section 11.6). The population had become prone to disease and almost moribund.

The only predatory fish in Lake Chilwa which is of any consequence, is *Clarias gariepinus*. Examination of remains of fish in the guts of almost a hundred specimens revealed 90 per cent of the prey to be *Barbus* and 10 per cent *Sarotherodon*. On one occasion, a *Clarias* stomach contained *Sarotherodon* eggs and on another some *Barbus* eggs (Fisheries Report 1966).

The effect of bird predation is probably considerable. Lake Chilwa is noted for the wide range of fish-eating birds: gulls, terns, cormorants, pelicans and herons were common all over the lake before the major recession. The heaviest predators in 1966 were the pelicans, *Pelecanus onocrotalus* L. and *P. rufescens* Gmel. The former was very common before the lake dried up, but were much rarer in the first few years after filling, but the reed cormorant *Phalacrocorax africanus* L. feeds mainly on *Barbus* in Lake Chilwa (Schulten & Harrison 1975). The fish eagle did not occur very commonly (Chapter 15).

Other animals which prey upon fishes are crocodiles (*Crocodylus niloticus* L.) (Lowe 1952) and monitor lizards (*Varanus niloticus* L.), which are confined to the river mouths. The otter, *Aonyx capensis* Schinz is very active in the feeder streams and often causes considerable damage to fish traps.

6. The effects of the receding and rising lake on the fishes of Lake Chilwa
by **P. R. Morgan**

During the years 1965 to 1968 the lake was transformed from a large body of open water into a dry lake basin, and this drastic change led to a dramatic fall in the fish landings. The decline of the populations of the three species of fishes and their recovery (1969–1972) are reflected in Table 11.5 which gives the estimated total fish landings in the lake and swamp and the percentages of each species in these years (Fisheries Report 1973) and in 1976, a year of high level three years after a minor recession, for comparison (Fisheries Report 1977). Of course, fish landings are, at best, an approximation to the actual populations and relative proportions of species in the lake, since different fishing gear is used for different species (Chapter 12). However, since fishermen choose the method of fishing which is most profitable at any time or place, their catches are some reflection of the species composition in changing conditions in the lake and swamp. The examination of percentages overcomes, to some degree, changes in recording systems.

The total landings decreased dramatically from 1965 to 1968 and then gradually increased to 1972, but 1973 saw a minor recession of the lake level and fishing declined. By 1976, however, the fisheries had recovered and the tonnage recorded about doubled that of 1965. Even allowing for a very great improvement in gathering statistics, one can have no hesitation in concluding that the fisheries had recovered from the catastrophe of 1966–68 and the decline of 1973–74 (Fisheries Report 1977).

The three species of fish responded to the worsening situation differently. *Sarotherodon shiranus chilwae* was most sensitive to change, and their numbers declined three years before dryness. *Barbus paludinosus* actually increased in numbers and proportion in 1966 – possibly due to the more intensive use of nylon seines nets (see Chapter 12) – but then in 1967 and 1968 declined in number as drastically as did the cichlid. *Clarias* fishing was maintained at its high level in 1967 (largely in the swamp) and failed only in the year of complete dryness of the lake bed in 1968.

The stages of recovery also differed between the species. The catch of *Clarias* reached its former weight in the first year of recovery. *Barbus* has been steadily increasing since its recovery in the second year after dryness, so its 'failure' was of three years duration. *Sarotherodon* reached 25 per cent of the total only three years after dryness, thus the *Sarotherodon* fishing failed for 5 years. The fishery for this species was also affected by the minor recession of 1973, since in 1976 it had not reached the tonnage or the proportions shown before the crisis began, whereas *Clarias* and *Barbus* fisheries had completely recovered.

Kirk (1968, 1970) made observations during the pre-drying and drying phases and Morgan (1972) attempted to study the responses of the fishes to various critical environmental parameters in laboratory and field experiments. Morgan & Kalk (1970) had made a study of the chemical composition of the water during the critical phases in the history of the lake and these data were used in later simulation experiments.

6.1 *Sarotherodon shiranus chilwae*

The most critical period in the decline of *Sarotherodon* took place during the second half of 1966 and the first half of 1967, beginning two years before the lake dried up (Table 11.5). Mass mortality of *Sarotherodon* was recorded after two days of very strong wind on the 17th and 18th October 1966 and more prolonged and more drastic mortalities were recorded during December 1966 and January 1967 in very calm hot weather, when the lake had reached its lowest level for seventeen years; mortality continued until all the fishes in the open water were dead (Table 11.6). Large numbers of *Barbus* were also killed, but *Clarias* was unaffected.

The stress factors: elevated levels of silt, inorganic salts and temperature and decreased levels of oxygen were chosen for investigation of the responses of *Sarotherodon* fingerlings. In certain experiments comparisons were made be-

Table 11.5 Total fish landings in metric tonnes from Lake Chilwa and swamps and percentage relative proportion of the three species, *Sarotherodon shiranus chilwae*, *Barbus paludinosus* and *Clarias gariepinus* during normal, predrying, drying, filling, post-filling phases and a high level year.

Date	Total	*Sarotherodon*	*Barbus*	*Clarias*	Min. depth m
1965	8820	48	26	26	1.08
1966	7200	19	39	42	0.43
1967	3139	7	5	88	0.35 (dry at edge)
1968	97	4	7	89	0 (lake bed dry)
1969	3326	0.4	2	97.6	1.18
1970	4166	8	35	57	1.05
1971	3595	25	35	40	1.36
1972	5246	25	42	33	0.90
1973	1903	n.a.	n.a.	n.a.	0.02 (dry at edge)
1974	3171	n.a.	n.a.	n.a.	2.21
1975	2809	n.a.	n.a.	n.a.	1.55
1976	19 746	14	52	34	2.00

n.a. not available.

194

Table 11.6 Chemical records for L. Chilwa during critical period for *Sarotherodon*. (Data of A. Morgan 1968), with monthly rainfall figures and fish deaths. The amount of mortality is indicated by the number of crosses.

Date	Year	Conduct. μS cm^{-1}	Alk. meq l^{-1}	Temp. °C	Oxygen mg l^{-1}	pH	Depth m	Rain cm	Mortality
Sept. 5	1966	5050	27.4	25.0	8.3	–	1.07	–	–
Sept. 19	1966	5800	28.6	25.2	7.0	8.8	0.97	0.63	–
Oct. 3	1966	6000	32.0	22.0	6.4	8.6	–	–	–
Oct. 17/18	1966	–	–	–	–	–	(high winds)	–	++++
Oct. 28	1966	5600	34.2	26.4	5.7	8.6	–	–	–
Nov. 14	1966	7000	42.8	29.0	9.0	8.8	–	2.64	–
Nov. 26	1966	7100	44.0	23.0	7.4	8.8	0.63	–	–
Dec. 10	1966	12000	61.6	28.7	3.7	9.3	0.50	11.76	++++
Dec. 30	1966	6000	34.0	26.0	10.2	9.5	0.43	–	++++
Jan. 11	1967	5550	27.0	35.2	16.1	9.6	0.46	15.49	++++
Feb. 1	1967	6800	41.6	37.4	15.7	9.6	0.52	11.68	+++
Feb. 24	1967	6480	35.6	27.0	4.9	9.5	0.48	–	++
March 11	1967	1630	9.8	30.0	–	8.4	0.51	29.92	+
March 29	1967	276	1.3	25.4	7.7	8.0	0.84	–	+
April 19	1967	2630	15.4	33.4	8.0	8.4	1.01	3.30	+
May 2	1967	4090	22.8	30.0	–	8.4	1.07	–	+

tween the endemic *S. s. chilwae* and *T. rendalli* (Boulenger), a cichlid of more widespread distribution, but one that is common in the Chilwa swamp and which also enters the open lake in small numbers at certain times when the lake level is high.

6.1.1 Oxygen stress

Lethal levels of oxygen were first determined by enclosing 20 *S. s. chilwae* within 2.5 litre sealed vessels which had been filled with tap water. Water samples were drawn off at intervals and tested for oxygen, while nitrogen gas was used to displace the water withdrawn. Fish behaviour and times of death were recorded. Confirmatory tests were performed by enclosing fish within containers for periods between 15 and 105 minutes. Oxygen levels and the number of deaths were recorded when the containers were re-opened in series.

S. s. chilwae were able to tolerate oxygen content as low as 0.6 mg l^{-1} for brief periods, but sustained levels between 0.3 mg l^{-1} and 0.4 mg l^{-1} were found to be lethal, if endured for more than a few minutes. The fish became very active as the oxygen level fell, followed by a quiescent phase with a loss of stability and finally death.

6.1.2 Silt stress

The adverse effects of released mud were studied by placing two alloy tubes of 60 cm diameter in the lake and transferring 20 *Sarotherodon* into baskets and lowering them into the tubes. Mud at the bottom of the experimental tube was stirred up into suspension, whilst the other tube was left alone as a control. The baskets were inspected regularly for fish deaths, and water samples in both tubes were analyzed at regular intervals for dissolved oxygen. Histological examinations were also made of the gills of fish exposed to high silt loads and compared with normal gills.

Mud released into the water by stirring in the experimental tubes (simulating the high winds of October 1968) caused a sudden depletion of oxygen, and subsequently all the fish died. The suspended solid content of stirred water was raised from 1400 mg l^{-1} to 18 000 mg l^{-1} within a few minutes, and oxygen levels fell to zero within two minutes of stirring. Normally day time dissolved oxygen in Lake Chilwa ranges from 5.4 mg l^{-1} to 8.2 mg l^{-1} in different seasons. Fish came to the surface gulping for air within a minute of stirring and continued this for 15 minutes. After this period their movements weakened and they fell to the bottom; all were dead within 50 minutes. Conditions appeared normal in the control tube: the fish did not gulp and were in excellent condition at the end of the experiment.

Laboratory experiments confirmed that suspended mud had a high affinity for oxygen, and that it was capable of deoxygenating sixteen times its own volume of aerated water. Total deoxygenation had occurred when sufficient mud had been added to previously aerated water to raise the level of suspended solids to about 12,860 mg l^{-1}. No deaths occurred when fish were subjected to high levels of mud in suspension if the water was artificially aerated without stirring.

The gills showed a high degree of mucus secretion in which particles of mud

196

were trapped. Histological examination showed hyperplasia of the mucus cells of the gill epithelium which could be interpreted either as a response to mechanical irritation or as an excessive secretion of the acid mucopolysaccharide of the mucus cells 'which serve as an electrolyte carrier and ion exchanger' according to the experimental electron microscopic studies of Philpot & Copeland (1963).

The first mortalities

One may now return to the details of the actual circumstances of mortality of *S. s. chilwae* in Lake Chilwa and interpret the sequence of events. During October 1966 when the first deaths occurred on a big scale, there were no prolonged chemical changes in the lake at the time when very strong winds (*mwera*) occurred. But a layer of liquid mud 0.3 m deep persisted in about 1.0 m of water for two days after the winds had subsided (A. P. Mzumara pers. comm.) strongly suggesting that turbulence had stirred up the bottom sediments to at least the lethal level of 12 000 mg l^{-1}, that had been found experimentally to cause oxygen depletion and death of *Sarotherodon*. Most deaths occurred in the north and west, where the lake was shallowest and the fetch of the southeast winds was greatest. Similar mortalities took place in 1955 and 1960, when the lake was low and the latter date is on record as having had extraordinarily heavy winds (Mzumara 1967). Gauge heights at Kachulu on the west, for the lowest levels in 1955, 1960 and 1966 were 1.06 m, 0.83 m and 0.43 m respectively, and the gauge height in October 1966 was 0.83 m. It is possible that a gauge height below 1.0 m may be a critical level, below which bottom erosion is likely to occur from wave action caused by high winds.

Mud with a high oxygen demand is often found in highly productive eutrophic water, where a considerable amount of organic matter falls to the lake bed. Cichlid mortalities attributed to the overturn of the mud or deoxygenated water have been reported from several shallow lakes in tropical Africa. These include Lake George in 1957 (Hickling 1961) and in 1969 (Burgis et al. 1973); the Nampongwe River (Zambia) in 1964 (Tait 1965); Lake Victoria (Fish 1955), Mwera-wa-Ntipa (Zambia) (Bowmaker 1965) and Lake Chad (Benech et al. 1975). Usually only catfish and lung fish with accessory breathing organs survive. In some cases the flushing out of stagnant mud by excessive inflow of rain water can cause deoxygenation, but this could not be the case in Lake Chilwa, because the months of fish mortality and inflow of water do not coincide.

Many *Sarotherodon* in Lake Chilwa actually survived this first catastrophe however, and fishing continued on a small scale. Possibly they had found shelter near or within the *Typha* swamp where the wind and turbulence is much less (Chapter 5). In common with other fish, *S. s. chilwae* becomes oxykinetic (very active at low levels of oxygen) and this behaviour enables them to escape into more oxygenated water, if it is close by. Thompson (1925) noted that fish have the ability to avoid water deficient in oxygen, and are able to recover quickly from anoxia if they are forced into deoxygenated water temporarily. This behavioural reaction would have considerable survival value if the areas of oxygen depletion were localized or if the fish found itself on the edge of a stress zone. Winds and rough weather build up quickly on Lake Chilwa however, and large areas of the lake bed could be raised into suspension within a short space of time, even overtaking fishes fleeing to the swamp.

The second mortalities

The very dense growths of blue-green algae, which flourished during the second type of mass mortality in January and February 1967 (Table 11.6) caused supersaturation of oxygen in the water during the day and conversely very low and lethal levels of oxygen at night (Chapter 4). The nocturnal respiration of algae causes deoxygenation of water and even the decomposition of algae during the day can have catastrophic effects on fish (Olsen 1932, Hutchinson 1936). At this time the algae of the pre-drying phase (*Oscillatoria planctonica* and *Anabaena torulosa*, were dying and being succeeded by *Arthrospira platensis*, and *Spirulina major*, characteristic of the drying phase (Chapter 6). Coe (1966) reported extensive mortalities of *Sarotherodon grahami* in Lake Magadi in 1960 following periods of excessive rainfall when the water was filled with *Arthrospira*. Similar events occurred in Lake Natron, another alkaline East African lake. Coe (1966) ascribes both events to deoxygenation of the water following algal flushes. During the Sahelian drought (1972–74), in Lake Chad mass mortalities were attributed to anoxia due to turbulence in the north basin and to macrophyte decomposition in the east basin (Benech et al. 1976).

6.1.3   Alkalinity stress

The tolerance of freshwater fishes to low oxygen concentrations is generally decreased in more saline water. For example in *Sarotherodon niloticus* (L) approximately 29 per cent of the total oxygen consumption is required for osmoregulation in water of salinity equivalent to 30 per cent of that of sea water (Farmer & Beamish 1969). In the drying period, Lake Chilwa reached 12 000 μS cm^{-1} in conductivity, which was roughly equivalent to 12 g l^{-1}, or 30 per cent seawater. Others have found that respiratory distress in fish is aggravated by high alkalinity, pH and temperature (Erichson-Jones 1964).

In order to test whether waters of high salinity and alkalinity adversely affected *Sarotherodon* in Lake Chilwa, it was necessary to prepare water with the required concentration in large volumes. The ions Na$^+$, Cl$^-$, CO$_3^{2-}$, and HCO$_3^-$ account for about 97 per cent of the total ions in Lake Chilwa water, and the chloride and bicarbonate ions are in similar proportions with the alkalinity increasing at higher conductivities (see Chapter 4). The addition of sodium chloride and sodium carbonate in calculated proportions to the dilute lake water in the recovery phase could thus simulate the conditions of the pre-drying phase. These solutions were tested for conductivity, alkalinity and pH over a wide range of concentrations and they compared well with the recorded parameters of the lake at different times in the pre-drying and drying phases (cf. Table 11.7).

A series of large tanks were prepared and the endemic *Sarotherodon* and *T. rendalli* were added; daily measurements of conductivity, alkalinity and pH were made and fish deaths were noted (Table 11.7). Similar determination of lethal concentrations were made using sodium chloride only as an additive to make equivalent conductivities, in order to separate the effects of alkalinity and chlorinity (Table 11.8).

Both cichlid species proved to be clearly euryhaline as expected (i.e. they tolerated a wide range of salinity), but the endemic species was consistently more resistant to high ionic stress. The degree of stress that could be tolerated depended to a certain extent on the rate of acclimation of the fishes, as so often demonstrated (Hoar & Randall 1971). They defined acclimation as 'bringing an animal to a steady state by setting one or more conditions to which it is to be exposed at a certain level for an appropriate time before a given test.' When the rate of increase of conductivity in artificial conditions was similar to the rate of increase during the natural critical phase of the lake itself, namely, 35 μS cm^{-1} per day, the median lethal conductivity was 12 800 μS cm^{-1} for the endemic *Sarotherodon* and 10 800 μS cm^{-1} for *T. rendalli*. Median lethal alkalinities were 61 meq l^{-1} and 53 meq l^{-1} respectively when the pH was 9.4 (Table 11.7). When *Sarotherodon* were subjected to sodium chloride additive alone, without alkali, they withstood much higher concentrations (Table 11.8).

The widespread occurrence and the extended duration of the December and January mortality of *Sarotherodon* suggests that ionic concentrations had

Table 11.7 Tolerance of *S. s. chilwae* and *T. rendalli* to artificial lake water. Duration of experiment = 43 days; mean daily increase in conductivity = 340 μS cm^{-1} per day; temperature = 22.5°C; 3 tanks + control, with 14 fishes in each.

	Tank no.	T. rendalli 14 + 14 + 14	S. s. chilwae 14 + 14 + 14
Conduct. (Alk.) (pH) at 1st mortality:	1	9 000 (44.8 (8.4)	5 200 (26.7) (9.2)
	2	8 475 (44.4) (9.4)	2 550 (13.8) (9.0)
	3	10 500 (50.0) (9.4)	4 800 (24.5) (9.2)
Conduct. (Alk.) (pH) at 50% morality:	1	11 000 (55.0) (9.4)	12 600 (63.0) (9.4)
	2	10 500 (50.0) (9.4)	12 000 (61.0) (9.4)
	3	11 000 (56.0) (9.4)	12 000 (61.0) (9.4)
Conduct. (Alk.) (pH) at 100% morality:	1	11 500 (58.0) (9.4)	15 000 (75.0) (9.4)
	2	11 000 (56.0) (9.4)	14 000 (71.0) (9.4)
	3	12 000 (61.0) (9.4)	15 000 (71.0) (9.4)

Conductivity in μS cm^{-1}; alkalinity in meq l^{-1}.

Table 11.8 Tolerance of *S. s. chilwae* to NaCl solutions. Duration of experiment = 29 days; mean daily increase in conductivity = 1 790 μS cm^{-1} per day; temperature = 22°C.

	Conductivity μS cm^{-1} and g l^{-1} NaCl on solution		
	1st mortality	50% mortality	100% mortality
Tank 1	25 000 (12.5)	38 500 (20.2)	51 000 (27.2)
Tank 2	27 000 (13.7)	37 500 (19.8)	52 000 (27.8)
Tank 3	38 000 (20.0)	44 000 (23.0)	47 000 (25.0)

reached lethal levels. The experimental results had shown that 50 per cent mortality had been reached at 12 000 μS cm^{-1}.

The incidence of corneal opacity in alkali-stressed fish in natural conditions and in the laboratory experiments is indicative of similar physiological responses to stress in the two environments and it may have been caused by deposition of calcium carbonate. (No parasites could be found in the chambers of the eyes). Conditions in 1967 completely annihilated *Sarotherodon* in the lake by June, but remnants of the population which had escaped to refuges in the swamp in October 1966, while there was continuity between the lake water and the swamp, survived there.

Although *S. s. chilwae* is moderately tolerant of high ionic concentrations, it is much less resistant than *S. grahami* (Boulenger) of Lake Magadi, where alkalinity rises above 1000 meq l^{-1} and *T. alcalica* (Hilgend) of Lake Natron where it exceeds 2000 meq l^{-1} (Coe 1966). It is evident that the tolerance levels of species are closely allied to their histories of physiological stress over the long term, and that the endemic sub-species of *Sarotherodon* in Lake Chilwa is undergoing a similar selection process.

6.1.4   Heat stress

The mean point of heat coma and the upper tolerance limit of temperature were tested by placing fishes in tanks in which water temperature was raised with aquarium heaters. *Sarotherodon* was able to survive relatively high water temperatures. The median lethal temperatures for *S. s. chilwae* acclimated to 24°C was 42°C and for *T. rendalli*, 41°C. When subjected to 40°C for extended periods, *S. s. chilwae* was again the more resistant. The mean time of survival for *S. s. chilwae* was about 420 minutes, whereas it was only 166 minutes for the less resistant *T. rendalli*.

The endemic *Sarotherodon* of Lake Chilwa is remarkably tolerant of high temperature, like the cichlids in hot soda lakes, and it seems that this factor alone would not have caused natural mortality. The highest temperature recorded in the surface water of the lake was 40°C in calm hot weather. But such conditions are rare. Although heat coma may not have caused deaths in Lake Chilwa, elevated temperatures may have decreased the fishes' resistance to other stress factors. Temperature, like salinity, has a considerable effect on the oxygen consumption of the fish: both the rates of blood circulation and gill irrigation increase. In addition, the saturation value of oxygen in water is lower at elevated temperatures and chemical and biological consumption are increased in rates, so the oxygen concentration falls. Thus the elevated temperature during January and February may have placed considerable stress on both respiratory and osmotic mechanisms and caused mortality, although the ionic content of the water was lower than the lethal limits in the laboratory at a lower temperature.

6.1.5   Breeding restriction

Yet one more factor may limit the populations of *Sarotherodon*, more particularly after the minor recessions, which occur every six years or so. Restricted opportunity for breeding during the late dry season, when the sandy areas on the periphery of the lake are exposed in the north and west, may prevent nesting. The sandy areas on the east remain covered because the lake bed there is more steeply sloping (Chapter 2). Since the normal breeding season is a protracted one over eight months, limited later breeding will still occur when the next wet season covers the exposed lake bed.

Recovery

Those fish which had earlier fled to the swamp, witnessed swimming in large numbers up channels by R. G. Kirk (pers. comm.) could not return to the lake when there was no longer any connection between permanent streams and lake as the level fell. These fish bred and grew in suitable sandy areas in lagoons and streams in the marshes. Subsistence fisheries seriously reduced their numbers, but since they become mature in freshwater earlier than in the lake (see above, section 11.4), the total increased during the year that the lake bed was dry.

When the lake filled in 1969, they returned to the 'new' lake, but it took three years for the proportion to build up again to that of the pre-drying year. The reasons for this have been discussed under reproduction in section 11.4.

6.2 Barbus paludinosus

Mortality was observed among *Barbus* during the high winds in October, described above, and later on a much reduced scale. *Barbus* was able to escape the harsh conditions of December 1966–February 1967, to which it is not any better adapted physiologically than *Sarotherodon*, by a behavioural adaptation. Fishes usually migrate from the centre of the lake, where they spread in the cooler months, to the periphery and then enter the swamp to breed (see section 11.4). They could not return until two years later because there was no connection between the swamp lagoons and the lake. The lake populations of *Barbus* recovered their former importance in two years after dryness.

6.3 Clarias gariepinus

This catfish has no scales (like an eel) and its skin is tough and leathery, fairly impermeable to water loss in humid atmosphere. By various means *Clarias* is able to keep the osmotic concentration of its blood constant in lake water with a conductivity as high as $10\,000\,\mu S\,cm^{-1}$ (M. Kalk pers. comm.). It can aestivate in mud and would have avoided the ionic stress of the drying period by its annual migrations to the swamp for spawning, which commence in September. Even if some returned to the lake in the wet season of 1967, they must have left again travelling over mud with their spiny pectoral fins, before the water level fell too low. This species was recorded in the lake up to June 1967 in small numbers.

The respiratory stress described above for *Sarotherodon* would not have occurred in *Clarias* since it has an accessory respiratory organ in the branchial cavity. These are a pair of stiff, vascular, bushy projections on the second and fourth gill arches in each enlarged branchial cavity. They are supported by a cartilaginous skeleton to which is attached a small muscle capable of erecting the trees in air (Hoar & Randall 1977). The epithelial surface is much folded into very small multicellular villi, similar to those of gills, but unlike the latter in that the acid mucopolysaccharide secreting cells are scattered over the surface, one to each villus. The gills cannot separate when out of water for the mucus secreting cells are confined to the tips of the secondary lamellae and not between them. The acid mucopolysaccharide (staining only with Alcian Blue) is considered to form a barrier to water loss in air (Cockson 1972) and to be an ionic regulator (Philpot & Copeland 1963). Since *Clarias* can withstand oxygen depletion in water (Alexander 1965), it was unaffected by possible nocturnal periods of anoxia.

Clarias even more than *Barbus*, is particularly well adapted to a fluctuating lake level in that its breeding migrations normally occur when the lake level is lowest and it can breed later in the shallow water over the floodplain instead of being confined to streams or swamps. Their fecundity ensured a speedy repopulation of the lake during the subsquent 'filling' year. The stimuli which bring about the migrations to and from the swamp have not been studied at Lake Chilwa. Reaction to rising salinity and alkalinity are probably not the triggers since the centre of the swamp is more saline and alkaline than the lake at the time they enter it (Chapter 5 and section 11.2). Although the swamp waters may be cooler, the shallow water over the floodplain is much warmer than the lake in the wet season.

6.4   *Other species of fishes*

The drying of the lake and margins of the swamp progressed until only deeper 'lagoons' and enlarged mouths of the larger rivers in the swamps remained as 'oases of survival'. There was a concentration of all species of fish there and intensive fishing. The swamp species were probably in competition with the lake species and must have afforded excellent food for *Clarias*, which becomes carnivorous when it grows to a large size.

When the lake refilled, several swamp species entered the lake during the rains, but remained around its periphery near the swamp. These were found in experimental gill nets, for the first two months only and then returned to the swamp (section 11.2).

Nothobranchius kirki, a small fish, has perhaps the most highly evolved characteristic of any fish in its complete adaptation to the presence and absence of water. During the annual rains, depressions, ditches and even footpaths become full of water. Here the eggs of this fish, buried in the mud during the long dry period, quickly imbibe water and continue their organogenesis to hatch into small fry. They mature, breed and die within four months leaving their partly developed eggs in the mud. Many were caught in the waters around Lake Chilwa but attempts to complete the life cycle in the laboratory did not succeed (D. M. Bourn pers. comm.).

7. Speciation of *Sarotherodon shiranus chilwae* by R. G. Kirk

In African freshwaters, cichlids exhibit adaptive radiation to an intense degree (Fryer 1967). The opportunity for divergence from a common stock in the ancient Chilwa–Chiuta basin had been provided by the formation of the sandbar which cut off Lake Chilwa from Lake Chiuta and the northeastern streams about ten thousand years ago (Chapter 2). A brief consideration of the taxonomy of the related species is necessary to appreciate the identity of the new sub-species and the degree of adaptation of *S. s. chilwae*.

Many well known cichlid species were formerly assigned to the widespread genus *Tilapia*. More recent research has led to the recognition that this broadly defined genus includes two distinct groups separated on both morphological and behavioural grounds and each meriting generic status; these are the substratum-spawning *Tilapia* and the mouth-brooding *Sarotherodon* (Trewavas 1973).

Seven species of *Sarotherodon*, unlike the majority of them, have four spiny rays to the anal fin and inhabit the eastward flowing rivers of Kenya, Mozambique, Tanzania and Rhodesia, and the males of at least four of them show an elongation of the jaw (Trewavas 1966). The latter feature is also shown by the three-spined *S. mossambicus* which shares other features with *S. shiranus, S. placidus* and *S. rovumae* and so form a southern sub-group. The distinguishing characteristics of these species and the two sub-species of *S. shiranus* are mainly varying degrees of coarseness of the pharyngeal teeth (Trewavas 1966).

The two sub-species of *S. shiranus* are in general similar in colour pattern (Fig. 11.3.i) but differ conspicuously in the breeding livery of the males, which is so essential for mate recognition (Kirk 1970). *S. s. chilwae* seems to have a fairly labile general coloration, as it can appear much more like *S. s. shiranus* if kept in garden ponds or aquaria, or even when they are found in feeder streams (Fig. 11.3). Both sub-species are silvery, greenish dorsally with a black mid lateral band. The dorsal and caudal fins are edged with vermilion in *S. s. chilwae* and with yellow in *S. s. shiranus*. In freshwater *S. s. chilwae* the red fades to rust.

The main differences in the male breeding colours are striking (Fig. 11.3.ii). Although both sub-species are darker than usual and have the same background pattern of dark vertical bars, the *S. s. chilwae* males have iridescent white spots on the dorsum during courtship and *S. s. shiranus* males acquires a greenish iridescence they did not possess before. The Chilwa sub-species has a very dark belly with contrasting white genital papilla, whereas the other shades to white on the belly. The throat, the anal and pelvic fins of the former are black and of the latter yellow. The edges of the dorsal and caudal fins are usually vermilion in '*chilwae*' and yellow in '*shiranus*'. These differences were considered by Trewavas (in combination with other morphological differences) to suggest that the fishes were sub-species evolved from a common ancestor, *S. shiranus* which previously existed in northeastern rivers, such as the River Rovuma into which the ancient Lake Chilwa–Chiuta drained to the Indian Ocean, via the River Lujenda.

The present distribution of the two sub-species differs curiously. *S. s. shiranus* inhabits the streams entering Lake Malawi and Lake Malombe, and the lakes themselves as well as the Upper Shire River and its tributaries. It also occurs in Lake Chiuta (E. Trewavas, personal communication to D. Tweddle). Its arrival in Lake Malawi appears to be recent, since it differs considerably from the species flocks in Lake Malawi. Very possibly its spread occurred after the formation of the sandbar between Lake Chilwa and Lake Chiuta.

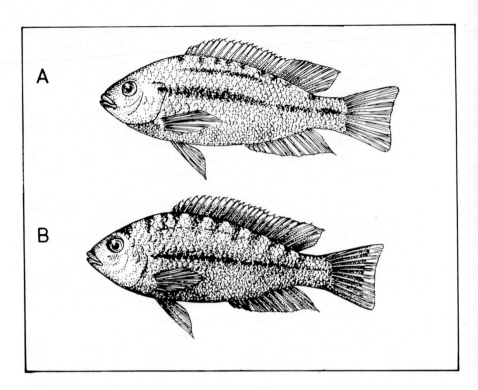

Fig. 11.3 Two sub-species of *Sarotherodon shiranus*. A. *S. s. shiranus*; B. *S. s. chilwae*. (i) Males in non-breeding condition: (ii) Males in breeding condition showing differences in male breeding patterns (drawings by D. Lewis).

Confirmation of the morphological distinctions have been given by the electrophoretic studies of haemoglobin by Tarbit (1972b) who demonstrated that these two kinds of fish, described as sub-species by Trewavas (1966) are similar enough to be considered related at a sub-species level and different enough from other species in the southern sub-group, mentioned above, to be considered specifically different from them. Unfortunately Tarbit was unable to obtain blood from *S. rovumae*, which on morphological grounds appeared to be the most nearly related species to the *S. shiranus* sub-species.

The two sub-species under consideration differ not only in distribution, blood proteins and male breeding livery, but also in many aspects of their breeding biology: in the smaller size at maturity and in the larger number and smaller size of eggs, and the length of the breeding period (Kirk 1970). Dwarfing in cichlid fishes has been considered an adaptive mechanism involving growth and breeding characteristics which enable populations to withstand high mortality rates (Iles 1973). Compared with *S. shiranus*, the Lake Chilwa sub-species is stunted. In becoming mature when smaller, the fish can breed earlier, and perhaps sooner after some unfavourable change has occurred in the environment. As is described in more detail in the previous section of this chapter, *S. s. chilwae* suffered high mortality as early as two years before dryness during a time of high winds which churned up the shallow lake and in the year before dryness when the water became anoxic at night (Chapter 4).

The prolonged breeding season of *S. s. chilwae*, the length of which varies from year to year, ensures that smaller individuals living in the shelter of the swamp–lake fringe may persist when adults suffer mass mortality (section 11.6). In 1965–66 the breeding season started at a later date than the previous season, but breeding activity continued over a longer period. There are probably only a few months of the year when only a small proportion of mature fish might be expected not to be in breeding condition. By the end of the first year, females averaging 97 mm in length were beginning to enter the spawning stock. The equivalent spawning in the following year took place in January 1966, but growth was more rapid in that year (Kirk 1970). The rate of population regeneration is therefore variable and seems adapted to climatic fluctuations in some degree.

It would seem therefore that this fish is sensitive to some environmental parameters, at present not known. Although it does not normally migrate annually to the swamp or up rivers to breed as do so many other cichlids in other lakes, it did seek refuge in the permanent streams and lagoons in the marshes during the major recession. Acclerated maturity while in 'conservation' areas of the ecosystem is useful in times of recession; but it was three years before its numbers were high in the lake. The 'opposite number' of this species, namely *Sarotherodon rukwaensis* in the variable Lake Rukwa, is a distinct species which has evolved still further along the line of adaptation to a fluctuating environment. In contrast to *S. s. chilwae*, *S. rukwaensis* generally becomes sexually mature in its first year of growth, at a size of 120–140 mm when the lake level is high (Ricardo 1939).

Exceptionally large individuals of *S. s. chilwae* have been recorded in Lake Chilwa; one specimen over 390 mm in length and two over 320 mm in length (Kirk 1970). It is not known how old these fish were, since growth rings on scales or otoliths are not visible in this species. They could possibly have survived the minor recession (1960) before the major catastrophe of 1967–68.

The comparative incompleteness of the adaptation of *Sarotherodon* in Lake Chilwa is suggested by their being more prone to disease than the other two species and by the longer time they take to build up populations after a recession. Their sandy nesting grounds on the periphery of the lake are at great risk since they are exposed even in a minor recession. The reduction in the number of eggs produces fewer fish, in spite of the maternal care of mouth-brooding.

Eggs and fish fry are more sensitive to changes in salinity than juveniles and adults since they have not yet developed osmoregulating apparatus, but mouth-brooding does not exclude the young from contact with the water since the maternal buccal cavity is irrigated for respiration. This is another dis-advantage.

One might therefore conclude that the existence of the sub-species endemic to Lake Chilwa demonstrates a species in the making rather than a completely successful clear-cut case of final speciation.

References

Alexander, R. M. 1965. Structure and function in catfish. J. Zool. Lond. 148:88–152.

Bailey, R. G. 1969a. The non-cichlid fishes of the eastward flowing rivers of Tanzania. Rev. Zool. Bot. Afr. 83:170–199.

Bailey, R. G. 1969b. Notes on certain small *Barbus* (Cyprinidae) from Tanzania, East Africa, with a description of a new species. Rev. Zool. Bot. Afr. 83:88–96.

Bell-Cross, G. 1972. The fish fauna of the Zambezi River System. Arnoldia, 5(29):1–19.

Benech, V., Lemoalle, J. & Quensiere, J. 1976. Mortalités de poissons et conditions de milieu dans le Lac Tchad au cours d'une periode de secheresse. Cah. O.R.S.T.O.M., ser. Hydrobiol. 10:119–130.

Bourn, D. M. 1972. The feeding, diet and ecological relations of the three species of fish of economic importance in Southern Malawi. M.Sc. thesis, University of Edinburgh.

Bourn, D. M. 1973. The feeding of three commercially important fish species in Lake Chilwa, Malawi. Afr. J. Trop. Hydrobiol. Fish. 3:135–145.

Bowmaker, A. P. 1965. The fisheries of Zambia, Mweru-wa-Ntipa. 60 pp. Natural Resources Handbook, Ministry of Lands and Natural Resources, Zambia.

Burgis, M. J. Darlington, J. P. E. C., Dunn, I. G., Ganf, G. G., Gwahaba, J. J. & McGowan, L. M. 1973. The biomass and distribution of organisms in Lake George, Uganda. Proc. Roy. Soc. London, B. 184: 271–298.

Cockson, A. 1972. Notes on the anatomy, histology and histochemistry of the respiratory tree of *Clarias mossambicus (gariepinus)*. Zool. Beiträge. 18:101–109.

Cockson, A. & Bourn, D.M. 1972. Enzymes in the digestive tract of two species of euryhaline fish. Comp. Biochem. Physiol. 41:715–718.

Cockson, A. & Bourn, D. M. 1973. Protease and amylase in the digestive tract of *Barbus paludinosus*. Hydrobiologia 43:357–363.

Coe, M. J. 1966. The biology of *Tilapia grahami* in Lake Magadi, Kenya. Acta Tropica, 23:146–177.

Corbet, P. S. 1961. The food of non-cichlid fishes in the Lake Victoria basin, with remarks on their evolution and adaptation to lacustrine conditions. Proc. Zool. Soc. London 136:1–101.

Dawson, A. L. 1970. The geology of the Lake Chiuta area. Geol. Surv. Bull. 34. Government Printer, Zomba, Malawi. 31 pp.

Dixey, F. 1926. The Nyasa section of the Great Rift Valley. Geogr. J. 68:131–133.

Erichsen-Jones, J. R. 1964. Fish and River Pollution. Butterworth. London.

Farmer, G. J. & Beamish, F. W. H. 1969. Oxygen consumption of *Tilapia nilotica* in relation to swimming speed and salinity. J. Fish. Res. Canada, 26:280–282.

Fish, G. R. 1955. Some aspects of the respiration of six species of fish from Uganda. J. exp. Biol. 33:186–195.

Fisheries Department Reports. 1966, 1972, 1973, 1977. Ministry of Agriculture and Natural Resources, Zomba, Malawi.

Fisheries Department. 1978. Data from experimental trawling in Lake Chilwa 1975–1977. Fisheries Dept. Lilongwe Malawi (unpublished).

Fryer, G. 1967. Speciation and adaptive radiation in the African Lakes. Verh. int. Ver. Limnol. 17:303–313.

Fryer, G. and Iles, T. D. 1972. The cichlid fishes of the great lakes of Africa: their biology and evolution. Oliver & Boyd, London. 641 pp.

Furse, M. T. 1972. In: Howard-Williams, C., Furse, M. T., Schulten-Senden, C. M., Bourn, D. M. & Lenton, G. Lake Chilwa, Malawi: studies of a tropical ecosystem. Report to IBP/UNESCO Symposium, Reading, England. 28–34.

Furse, M. T. & Bourn, D. M. 1971. Lake Chiuta trip. July 1971. Cyclostyled Report, University of Malawi (unpublished).

Greenwood, P. H. 1965. The cichlid fishes of Lake Nabugabo, Uganda. Bull. Brit. Mus. (nat. Hist.) Zool. 12:315–357.

Herbert, D. W. M., Alabaster, J. S., Dart, M. C. & Lloyd, R. 1961. The effect of china-clay wastes on trout streams. Int. J. Air, Wat. Pollut. 5:56–74.

Hickling, C. F. 1961. Tropical Inland Fisheries. Longmans, London.

Hoar, W. S. & Randall, D. J. 1971. Fish Physiology. VI, Academic Press, New York, 559 pp.

Holl, E. A. 1968. Notes on the spawning behaviour of the barbel, *Clarias gariepinus* Burchell in Rhodesia. Zool. Africana. 3:185–188.

Hutchinson, G. E. 1936. Alkali deficiency and fish mortality. Science 84:18.

Iles, T. D. 1973. Dwarfing and stunting in the genus *Tilapia* (Cichlidae); a possibly unique recruitment mechanism. Rapp. P-v. Reun. Cons. perm. int. Explor. Mer, 164:247–254.

Jackson, P. B. N. 1962. Ecological factors affecting the distribution of freshwater fishes in tropical Southern Africa. Ann. Cape Prov. Mus. 2:223–228.

Jubb, R. A. 1967. The Freshwater Fishes of Southern Africa. Balkema, Cape Town. 248 pp.

Kalk, M. 1969. Lake Chilwa and the plankton-eating fish matemba, *Barbus paludinosus* Peters. Proc. IBP Sci. Conf. 21–23, University of Malawi.

Kelley, D. W. 1968. Fishery development in the Central Barotse Plain. FAO Rep. TA 2554:1–8.

Kirk, R. G. 1967a. The fishes of Lake Chilwa. J. Soc. Malawi. 20:1–14.

Kirk, R. G. 1976b. The zoogeographical affinities of the fishes of the Chilwa–Chiuta depression in Malawi Rev. Zool. Afr. 76:295–311.

Kirk, R. G. 1968. Restocking Lake Chilwa after the major recession. Fisheries Department Cyclostyled Report.

Kirk, R. G. 1970. A study of *Tilapia (Sarotherodon) shirana chilwae* Trewavas in Lake Chilwa, Malawi. Ph.D. Thesis, University of London (unpublished).

Kirk, R. G. 1972. Economic fishes of Lake Chilwa. Fisheries Bulletin, 4:1–13. Ministry of Agriculture and Natural Resources, Zomba, Malawi.

Kyle, H. M. 1926. The Biology of Fishes. Sidgwick & Jackson, London. 396 pp.

Loiselle, P. V. 1974. The cichlid fishes of the Chilwa–Chiuta depression Part 2. Buntbarsche Bull. 45:12–18.

Lowe, R. H. 1952. Report on the *Tilapia* and other fish and fisheries of Lake Nyasa, Colon. Off. Fish Publ. H.M.S.O. London. 126 pp.

Lowe-McConnell, R. H. 1959. Breeding behaviour patterns and ecological differences between *Tilapia* species and their significance for evolution within the genus *Tilapia* (Pisces: Cichlidae). Proc. Zool. Soc. London. 132:1–30.

Mann, M. J. 1964. The fisheries of Lake Rukwa, Tanganyika. EAFFRO Occ. Pap. 6:58 pp.

McLachlan, A. J., Morgan, P. R., Howard-Williams, C., McLachlan, S. M. & Bourn, D. M. 1972. Aspects of the recovery of a saline African lake following a dry period. Arch. Hydrobiol. 70:325–340.

McLachlan, A. J. 1974. Recovery of the substrate and its associated fauna following a dry phase in a tropical lake. Limnol. Oceanogr. 20:54–63.

Morgan, A. 1968. A physico-chemical study of Lake Chilwa. M.Sc. thesis, University of Malawi.

Morgan, A. & Kalk, M. 1970. Seasonal changes in the waters of Lake Chilwa in a drying phase, 1966–68. Hydrobiologia 36:1–23.

Morgan, P. R. 1971. Breeding of *Tilapia (Sarotherodon) shirana* in fish ponds in Malawi. Soc. Malawi J. 24:74–79.

Morgan, P. R. 1972. Causes of mortality in the endemic *Tilapia (Sarotherodon)* of Lake Chilwa, Malawi. Hydrobiologia 40:101–119.

Moriarty, D. J. W., Darlington, J. P. E. C., Dunn, I. G., Darlington, C. W. & Tevlin, M. P. 1973. Feeding and grazing in Lake George, Uganda. Proc. Roy. Soc. Lond. B. 184:299–319.

Munro, J. L. 1967. The food of a community of East African freshwater fishes. J. Zool. Lond. 151:389–415.

Mzumara, A. J. P. 1967. Mortality of fishes in Lake Chilwa. Fisheries Department, Malawi (unpublished).

Olsen, T. A. 1932. Some observations on the inter-relationships of sunlight, aquatic plant life and fishes. Trans. Amer. Fish. Soc. 62:278–289.

Philpot, C. & Copeland. D. 1963. Fine structure of chloride cells from three species of *Fundulus*. J. Cell. Biol. 18:389–404.

Ricardo, C. K. 1939. The fishes of Lake Rukwa. J. Linn. Soc. (Zool.) 40:275–625.

Schulten, G. M. & Harrison, G. 1975. An annotated checklist of birds of the Lake Chilwa area, Malawi. Soc. Malawi J. 28:1–30.

Tait, C. 1965. Mass fish mortalities. Fish. Res. Bull. Dept. Game and Fisheries, Zambia 3:28–30.

Tarbit, J. 1972a. Preliminary report on a short fisheries survey in Lake Chiuta. August 1972. Fisheries Department, Malawi (unpublished).

Tarbit, J. 1972b. Protein taxonomy of the *Tilapia* of Malawi. Limnol. Soc. sth. Afr. Newsl. 1972. 12 (abstract).

Thompson, D. H. 1925. Some observations on the oxygen requirements of fish in the Illinois River. Bull. Ill. Lab. Nat. Hist. 15:423–433.

Trewavas, E. 1966. Fishes of the genus *Tilapia* with four anal spines in Malawi, Rhodesia, Mozambique and southern Tanzania. Rev. Zool. Bot. Afr. 74:50–62.

Trewavas, E. 1973. On the cichlid fishes of the Genus *Pelmatochromis* with a proposal of a new genus for *P. congicus*; on the relationship between *Pelmatochromis* and *Tilapia* and the recognition of *Sarotherodon* as a distinct genus. Bull. Brit. Mus. nat. Hist. (Zool.) 25:1–26.

Tweddle, D. & Willoughby, N. G. 1979a. The fish fauna of the Chilwa–Chiuta depression: a key, annotated checklist and discussion on geographical affinities. Bull. J. L. B. Smith Inst. Ichthyology (in press).

Tweddle, D. & Willoughby, N. G. 1979b. Report in an electro-fishing survey of the Viphya-Chinteche area of Malawi. FAO working paper (in press).

Tweddle, D., Lewis, D. S. C. & Willoughby, N. G. 1979. The nature of the barrier separating the Lake Malawi and Zambesi fish faunas. Bull. J. L. B. Smith Inst. Ichthyol. (in press).

Welcomme, R. L. 1967. The relationship between fecundity and fertility in the mouth-brooding cichlid fish *Tilapia leucosticta*. J. Zool. London. 151:453–468.

Wourms, J. P. 1964. Comparative studies on the early embryology of *Notobranchius taeniopygus* and *Aplocheilichthyes pumilus* with special reference to the problem of naturally occurring diapause in teleost fishes. Ann. Rep. E. Afr. Freshw. Fish. Inst. 68–69.

Zaret, T. M. & Kerfoot, W. C. 1975. Fish predation on *Bosminia longirostris*: Body size versus visibility selection. Ecology 56:232–237.

2 The fisheries of Lake Chilwa

M. T. Furse, P. R. Morgan & M. Kalk

The fisheries of Lake Chilwa

The importance of Lake Chilwa as a commercial fishery was first recognized by Hornby (1963). He 'conservatively' estimated that 9000 tonnes wet weight of fish were being cropped annually from Lake Chilwa. He based this estimate partly on net sales from his Blantyre factory and partly on a short survey of fish leaving the jetty at Kachulu. As early as 1942 Hickling had recommended that information be gathered on the fish and fisheries of Lake Chilwa and this recommendation was repeated by Lowe (1952), because the fish yields were inconsistent and dependent on the condition of the lake. Catches of fish were reduced at times of very low water when fishing became difficult and the water notably alkaline and unsuitable for many fish. It was not until 1961 that the colonial administration appointed a Fish Ranger to supplement the Fishery Assistant who had been periodically based at Kachulu since 1952. The first official estimate of fish landings was for 1963, the year that Hornby circulated his private findings. The government figure was 3260 tonnes. This figure rose to 8820 tonnes by 1965 and the value of the fishery was established beyond doubt. The first Fisheries Research Officer was appointed by the Malawi Government in 1964.

It was well known that a catastrophic recession similar to the one of 1968 occurred in 1914–15 and records for reduced fish catches or fish mortalities were available for 1931–33, 1955, 1960–61 as well as for 1914–15 and 1966–68. According to several local reports the fishery usually took 3 or 4 years to recover from recessions. It was widely known amongst the fishermen that the *makumba* (*Sarotherodon*) always succumbed first in a recession, followed by *matemba* (mainly *barbus*) and then *mlamba* (*Clarias*). *Mlamba* was always established first after a recession and *makumba* last (Morgan 1971).

1. Fishing techniques

The fishing methods in common usage in 1963 were those described by Mzumara (1967). These were basically the traditional techniques long practised by the local fishermen. By this time fishing gear was showing signs of twentieth century sophistication, at least in materials.

The earliest fishing craft were believed to be rafts of marsh grass. These were replaced by bark platforms, strengthened by thwarts, curved upwards at the sides, and pegged and caulked bow and stern. These craft were already essentially canoe-shaped. With improvement in available iron tools, these bark craft were gradually replaced by canoes shaped out of local trees such as *mbawa* (*Khaya nyassica* Stapf.), *chonya* (*Adina microcephala* Del. Hiern.), *mtondo* (*Cordyla africana* Lour.) and others (Chirwa et al. 1966). In 1976 canoes were still the most commonly used fishing craft but about 10 per cent were western style plank boats made by the local Fisheries Institute (Figs. 12.1 & 12.2).

In a survey of the fisheries in 1976, there were 951 wooden canoes, 84 plank boats without engines and one with an engine, operating from 36 beaches, and 37 bark canoes in the northern swamps (Fisheries Report 1977). The bark canoes reappeared in 1973 and are increasing in number.

211

Fig. 12.1   Canoes at Kachulu jetty in 1966 (photo: N. Lancaster).

Fig. 12.2   'Modern' plank boats and ferry in 1978 (photo: A. J. McLachlan).

212

The six fishing methods still in common use in 1976, were gill nets, scoop nets, seine nets, fish traps, long lines and spears. The original netting material used on the lake was fibre of the local shrub *bwazi, Securidaca longepedunculata* Freson and *Secamone* sp. (Mzumara 1967) woven around the four fingers making a large mesh. This traditional material was replaced in turn by threads removed from pneumatic tyres, sewing cotton and more recently synthetic nylon netting.

Gill nets, used from 1963, had nylon head and foot ropes held taut by vertical bamboo spacers at approximately 5 metre intervals. The spacer canes are forced into the mud substrate but stouter bamboo stakes at intervals of about 20–30 m fix the whole net firmly in position. The nets function by trapping fish around the gill region in their meshes, usually 70 mm mesh size. *Barbus paludinosus* may also be trapped by entangling their serrated dorsal spine in the netting.

In early years the gill nets were only 50 cm deep and did not fish the whole water column. Gill nets, because of their mode of operation, are very selective, being efficient only for a narrow size range for each species and mesh size.

Scoop nets are composed of a conical net bag mounted on two bamboo supports, crossed near their base, which gives the net a triangular mouth (Fig. 12.3a). Cross members, near the junction of the bamboo supports, form handles at the apex of the triangle (Mzumara 1967). In operation one fisherman holds the net in the water at the prow of the canoe whilst a second fisherman propels the boat. The net acts like a small trawl, efficiently funnelling the fish it encounters into the belly of the net (Fig. 12.3b).

A similar principle is employed by fishermen along the eastern shoreline who fashion basket scoops out of reeds *Phragmites mauritianus* and split bamboos. The scoop is roughly conical, open at one end only. It is supported at the front by a cross piece that keeps the mouth open. The closed end is held to the fisherman's belly and he propels the scoop by wading in the water (Mzumara 1967). The baskets are large, up to 3 m long, and cumbersome and the operators are vulnerable to attacks by crocodiles.

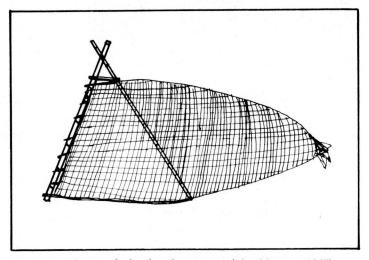

*Fig. 12.3*a   Diagram of a hand-made scoop net (after Mzumara 1967).

*Fig. 12.3*b   Scoop net in use from a canoe (photo: D. Bourn).

Seine nets are beach operated and as such are restricted to relatively few suitable shoreline areas (Fig. 12.4). Although designs vary from beach to beach the method of operation is similar to that used traditionally in Europe and elsewhere. The net is set from a boat, being laid in a semi-circle some distance off the beach to which it is then hauled by means of ropes attached to the ends of the net. A team of men handles each rope (Fig. 12.4). The fish are trapped in the belly of the net. The head line of both seine and gill nets may be supported by floats, often made traditionally of the trunk segments of the local marshland tree, *Aeschynomene pfundii* (Mzumara 1967). More recent seine nets are constructed from machine-made fibres, whose impact on catches became particularly apparent in 1965. The seine nets are by far the most efficient means of fishing. In 1976 the catch per unit effort was 33 kg per pull, being much less selective than other gear (Fisheries Report 1977).

Fig. 12.4 Hauling in a shore seine on a flat beach (photo: N. Lancaster).

Fish traps, like basket traps are made of split bamboo canes (Fig. 12.5a). The baskets are generally tapered with a valve at the broad end. A plug at the tapered end can be removed to empty the trap (Mzumara 1967). Traps may be either baited or unbaited. They function according to the principle of western lobster pots. Fish may enter through the valved opening, which is itself tapered, but find it difficult to find their way out again. Traps may be used singly in the open lake or more commonly in series along canoe channels in the swamps (Fig. 12.5b) or in a circle around clearings in the swamp. Gill nets and long lines may also be used in similar situations in the swamp. These clearings are often created by fishermen in an area where they may construct huts upon mounds of cut *Typha* stems, called *zimbowera* (Fig. 17.2a). These huts may form temporary homes for fishermen for long periods of the year, but women do not normally live in them.

Fish traps may also be set in gaps along weirs and dams floating below the surface of the water, attached to bamboos stuck into the muddy substratum (Fig. 12.5b). Traps are normally short, stout and rather barrel-shaped but in irrigation channels of rice fields on the floodplain long, narrow, cylindrical traps have occasionally been seen set in crude mud dams. Migrating fish are thus forced into the trap when passing along the channel.

Long lines are a series of hooked traces hanging from a horizontal cord, which runs horizontally just below the lake surface in radial directions from swamp to open water. This line is supported periodically by long bamboo stakes sunk into the mud and protruding above the lake surface. Hooks are baited with various lures including snails, worms, fishes and even Lever

Brothers' 'blue' soap. On the same principle single hooks may be suspended from floats in small clearings in the swamps (Mzumara 1967).

Finally spears are of simple metal blades, plain or barbed, lashed to or inserted in the ends of bamboo canes of about 2 m length. These are principally used on the floodplain but may also be employed in clearings or lagoons in the swamps and even in the open lake by the experienced (Mzumara 1967).

Barbus paludinosus, haplochromids and other small juvenile fishes (collectively termed *matemba*) are principally caught by seine and scoop nets and in fish traps. Gill nets are particularly useful in catching *Sarotherodon shiranus chilwae* (*makumba*) although these may also be caught by other methods.

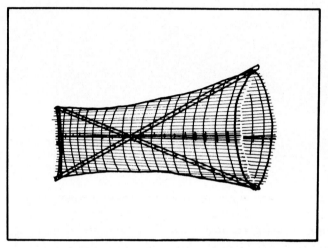

Fig. 12.5   (a) Diagram of a fish trap (after Mzumara 1967).

Fig. 12.5   (b) Fish traps in use across a weir in a swamp channel (photo: N. Lancaster).

216

Clarias gariepinus (*mlamba*) may also be caught by gill nets and more particularly by long lines. Spearing is traditionally associated with the migration of catfish, often very large, on to the floodplains to spawn.

Once caught, the fish are normally preserved by one of two traditional methods. The simplest method is direct drying in the sun, and is usual mainly for *Barbus*. The fishes are laid out on grass mats, without cutting off the head or taking out the viscera, and left for several days until quite dry. *Clarias* and *Sarotherodon* are gutted and dried or smoked over wood fires. In 1970 preservation of fish in brine was being introduced. After preservation, the fish are either consumed by the fishermen and their families, sold locally along the lakeshore villages or sold to traders who transport the fish by bicycle, lorry or bus. The principal outlets for fish from the lake are from Kachulu in the west, Swang'oma in the southeast and Mposa in the northwest. Jetties have been constructed at the western and southeastern outlets.

2. The fisheries during the major recession of the lake *by* P. R. Morgan

2.1 *The rise of the fishery 1960–65*

The moderately severe recession of 1960–61 caused mass mortalities of *Sarotherodon*, and local reports confirm that yields of this fish were relatively poor in the early 1960s. However, approximately 1.7 million metres of gill netting were sold to the Lake Chilwa fishermen between 1959 and 1962 which seems to indicate that fishing was prosperous. The bulk of the catch may have consisted of *Clarias* at this time. A census taken in 1962–63 indicated that the fishing intensity was high. In one survey, 981 fishermen owned a total of 1082 canoes or boats, 1206 fish traps and scoop nets, 14,000 hooks on long lines and 719,000 metres of gill netting (Hornby 1963).

Improved methods of assessing fish landings were initiated by the Fisheries Department in 1963. These data show that while the landings for *Clarias* remained relatively constant during the period 1963–65, those for *Sarotherodon* were rising rapidly to reach a peak in 1965 of 4263 tonnes, nearly twice that of *Clarias* (see Table 11.5). The increases of fish landings over the period 1962–65, although partly due to rising lake levels, largely reflect increasing fishing activity and also increasing availability of locally made gill nets. The sharp increase in the number of *Barbus* in 1965 was the result of the introduction of small mesh nylon seine nets. Towards the end of 1964, the Fisheries Department introduced an education campaign in an attempt to persuade the fishermen to use nets of larger mesh size for gill nets. This exercise was successful and had the effect of increasing the mean size of *Sarotherodon* caught (Mzumara 1967).

2.2 *The decline of the fishery, 1966–68*

Landings of *Sarotherodon* fell drastically after 1965 and this trend was followed by *Barbus* after 1966 and by *Clarias* after 1967. In 1965, *Sarotherodon* accounted for approximately half the total catch; in 1966 it was less than a quarter and by 1967 less than one per cent. The gill net fishery for *Sarotherodon* was virtually abandoned after June 1967 when fishing became difficult and

217

very unprofitable. At this time most of the fishing was diverted to the streams and lagoons on the margins of the lake. Extensive mudflats had to be crossed to reach open water which had become too shallow for the effective use of gill nets. The final decline in *Sarotherodon* catches may have resulted from the cessation of fishing rather than from the absence of fish (Kirk 1970). Remarkably, some *Sarotherodon* remained in the open lake during June 1967, but these were totally annihilated by the end of 1967. A full account of this catastrophe is described in Chapter 11. By 1967, the effects of the recession were beginning to grip and only *Clarias gariepinus* was significant in the catch. The fishery was totally eliminated in the lake in 1968, though entrepreneural fishermen were taking full advantage of fishes trapped in lagoons around the south of the lake until October 1968 (Morgan 1971).

A certain proportion of the *Sarotherodon* did not succumb because they remained either in the upper courses of affluent streams or in freshwater lagoons surrounding the southern perimeter of the lake. These areas did not dry out in 1968, and according to local reports, one lagoon in the Limbe Marsh harboured a relatively large number of *Sarotherodon* during the recession of the main body of water. The fishermen who transferred their effort to this lagoon were able to operate until October 1968.

After this time fishing became unprofitable because fish stocks were depleted due to overfishing. Undoubtedly the fish that remained acted as a nucleus from which new stocks built up when the lake became reflooded. *Barbus* came into the lake from the streams, for the stock that survived the major dry phase remained in the upper courses of the streams with the *Sarotherodon*. The silt laden marshes in the north, west and south of the lake are extensive breeding grounds for *Clarias*, and a large reserve of these catfish remained in these areas and their feeder streams when the lake bed was completely dry.

Many details of the recovery of the fishery have already been touched upon (Chapter 11).

2.3 *The recovery of the fishery, 1969–70*

Clarias appeared in fish catches only one month after the onset of the rains in December, 1968. *Barbus* was recorded for the first time in March 1969. *Sarotherodon* made a much slower recovery and only small numbers were caught in both experimental and commercial gill netting throughout 1969. The initial lack of *Sarotherodon* caused much concern, especially amongst the fishermen, who recalled similar events earlier in the century. *Clarias gariepinus* was best adapted to withstand the drought and dominated the fishery in 1969. Already catches of this fish had reached their highest recorded level, but this was partly because the whole fishing effort was directed towards them. *Barbus paludinosus* was the next fish to recover sufficiently to figure prominently in catches. In the years between 1969 and 1972 catches of this species (or more precisely *matemba*) rose from 89 to 2000 tonnes per annum. This latter figure approached the catches in the best years preceding the drought.

For the reasons outlined earlier (Chapter 11), the recovery of the fishery for *S. s. chilwae* was much slower despite the stocking programme. The Fisheries Department had taken measures before the lake dried out to breed stocks of

S. s. chilwae in fish ponds. During the early months of 1969 about 300,000 fingerling *Sarotherodon* were transferred from these ponds and other dams to the Kachulu area of Lake Chilwa. Certain fisheries regulations were imposed to protect the new stock which included restrictions on seine netting and a fixed minimum of 70 mm for the stretched mesh size of gill nets. The seine net restriction was rescinded in May 1970 which led to improved catches of *Barbus*. The gill net restriction was maintained until later in the year.

Experimental fishing was undertaken with graded fleets of gill netting during December 1969 and in the first two months of 1970 provided evidence that both adult and fingerling *Sarotherodon* were appearing in several marginal areas of the lake together with *Clarias*, *Barbus* and a few riverine species.

Experimental trawling carried out in 1970 and early 1971 revealed that *Sarotherodon* fry and fingerlings were appearing in very great numbers in several regions of the lake, lending support to the predictions of the local fishermen that makumba would take 3 or 4 years to make a recovery. The lengthy recovery period of the *Sarotherodon* fishery undoubtedly reflects the exceptional severity of the 1966–68 drought and the heavy marginal fishing of stocks that would have contributed to its recovery when the lake refilled. In 1972 catches were still less than a third of the peak year of 1965.

In 1973 the lake entered another recession phase and catches declined again, but to a lesser extent. Recovery took three years. A seven year period of decline, recovery and further decline was completed (see Tables 11.5 and 18.2).

3. Fisheries in years of high level *by* M. Kalk

The record catches of over 20 000 tonnes of fish were obtained from Lake Chilwa in 1976, three years after the minor recession in 1973 (Fisheries Report 1977). There appeared to be many reasons for this. A better recording system on which to base estimates was used, which exaggerated the disparity between this year and previous ones to some extent, but it gives a better indication of the sizes of the fish catches. There was a greater fishing effort and in 1977 the number of fishermen in the lake and the swamp increased even more. It was quite probable that the populations of fishes were larger because more time had elapsed since the lake had refilled and breeding had continued unhindered. No data are available on any change of sizes of the fishes caught commercially.

The northern swamp featured more prominently in the statistics, than hitherto, with 34 per cent of the total weight of fish being landed. The swamp is now crossed by very many channels, cut by the fishermen; many clearings have been made and the natural lagoons are, of course, also intensively fished, using traps and gill nets. 45 per cent of the fish was *Clarias*, 11 per cent *Barbus* and 11 per cent *Sarotherodon*; but one-third of the catch comprised species other than these dominant ones (Tables 12.1 and 12.2).

The most productive fishing areas, which produced 43 per cent of the catch, were adjacent to the swamp centred at Kachulu in the northwest and Chinguma in the northeast. *Barbus* contributed 72 per cent of the total at these sites, since there are many suitable beaches for seine nets (Table 12.2).

Swang'oma in the south had the least number of beaches, but the highest number of fishing craft, which implies the greater use of gill nets, but the catch

Table 12.1 Monthly fish landings in tonnes at three distribution centres at Lake Chilwa in 1976. (From Fisheries Report, 1977).

Area	Jan.	Feb.	Mar.	April	May	June	July	Aug.	Sept.	Oct.	Nov.	Dec.	Total	Percentage
N. Marsh Kachulu/	500	450	300	1738	200	1460	500	1190	233	188	246	120	7125	34
Chinguma	1078	1363	1093	923	287	453	501	456	439	438	1164	1164	9359	43
Swang'oma	716	298	163	242	95	120	363	262	1120	408	721	350	4858	23
Total	2294	2111	1556	2903	582	2033	1364	1908	1792	1034	2131	1634	21342	100
Percentage	11	10	7	13	3	9	6	9	8	6	10	8	100	

Table 12.2 Fish catch at Lake Chilwa in three centres (1976) in tonnes showing species, area and month. (From Fisheries Report, 1977).

Species	Sarotherodon			Barbus			Clarias			Others		
Month	Swang'oma	Kachulu	N. Marsh	Swang'oma	Kachulu	N. Marsh	Swang'oma	Kachulu	N. Marsh	Swang'oma	Kachulu	N. Marsh
January	35	132	225	2	791	–	679	154	275	–	–	–
February	73	117	–	–	1146	–	225	100	–	–	–	–
March	62	53	–	12	862	–	88	179	–	–	–	–
April	29	93	126	–	714	152	213	116	862	–	–	597
May	24	47	–	5	347	–	63	93	–	3	–	–
June	25	65	24	18	289	213	77	100	687	–	–	536
July	27	75	–	191	313	–	107	113	–	38	–	–
August	32	49	60	122	369	189	92	37	373	16	–	568
September	57	47	40	876	365	16	187	27	106	–	–	71
October	39	169	39	254	539	22	101	100	113	14	9	14
November	156	95	65	367	281	43	197	61	123	–	2	16
December	76	265	52	208	650	13	49	249	56	17	–	–
Total	635	1207	631	2055	6666	648	2078	1329	2595	88	11	1802
		2473			9369			6002			1901	
Percentage		12.5%			47.5%			30.5%			9.5%	

per unit effort for all kinds of gear was lower. Fishermen were reported to have temporarily migrated to Kachulu and Chinguma where fishing was better. It will be noted that the width of the swamp is much less in the southern than in the northern sector giving a smaller area for breeding. A Cotton Development Project has been started on the Phalombe Plain in the south which may influence the success of fishing in the future by attracting former fishermen as participants. The southern sector contributed only 23 per cent of the catch. The overall composition of the total catch analyzed in 1976 was 48 per cent *Barbus*, 30 per cent *Clarias*, 12 per cent *Sarotherodon* and 10 per cent other species (see Tables 11.5 and 12.2 which are, however, incomplete).

Table 12.2 shows that the highest catches of *Barbus* were recorded from December to April, with a peak in February in the northern lake, but in September in the southern sector. *Clarias* catches were higher in December in the northwest and northeast but in January in the south, while in April and June, at the highest seasonal lake level, many more were caught in the northern swamp. These variations confirm that there are migrations of fishes and that fishing is especially good when the fishes are intercepted in the breeding migration routes.

Commercial fishing in the cooler months, May–August, was not so good on the whole; catches were about one third those of the warmer months. The reasons for the poorer fishing are complex: the cooler weather and strong winds prevent fishermen from venturing out in canoes or from staying out for long or on many days; the domestic calls of crop harvesting are urgent; these are also the months when, traditionally, nets are repaired, boats and houses are built. Different reasons relate to the accessibility of the fishes, which do not seem to be concentrated around the periphery where traditional fishing takes place, but have probably dispersed into open water or into large lagoons in the northern swamp.

The types of fishing gear the fishermen used, which were described above (section 12.1) are closely adapted to the fishing grounds, i.e. the habitats of the fish, and different methods predominate at different times of the year to take advantage of the habits of the fishes. In 1976, fish traps were more productive in the swamp, catching on the average 3.6 kg per trap. Long lines were used for *Clarias*. Gill nets set at night from canoes were by far the most commonly used gear. Although the proportion of gill nets was high, they landed only 20 per cent of the total catch. The depth of the net is now being increased so as to fish the whole water column (T. Jones pers. comm.). Seines for *Barbus* were the most efficient method, averaging 33.3 kg per pull (Fisheries Report 1977).

According to the Fisheries Report (1977) about 17 per cent of the fish landed were sold at the larger urban markets within 100 km radius from Kachulu harbour, the largest of which was Mulanje in the south where the workers on the extensive tea plantations live. A small proportion was exported to Zambia. The rest went to the very numerous smaller village markets and to the local people (see Chapter 18).

Records for the first six months of 1977 showed a similar, high success in fishing (Fisheries Report 1977). The catastrophic changes in depth, salinity and alkalinity in the lake eight years before, and the lesser changes of the minor recessions three years previously, had been surmounted by the fish which had taken refuge in the swamp, streams and lagoons. Now by their enormous reproductive capacity they have increased to a density which is of the same order as a stable lake in tropical Africa like Lake George. *Sarotherodon shiranus chilwae*, the mouth-brooding cichlid which produces far fewer eggs, still lags behind, most probably because some of the breeding sites suffered exposure during the recent recession. The 1976 catch of *Sarotherodon*, however, did equal that of the previous record year of 1965.

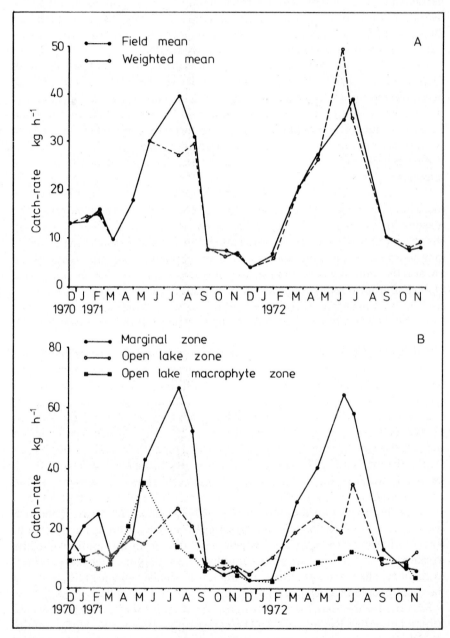

Fig. 12.6 Experimental trawling: catch rate of *Barbus paludinosus* caught in monthly sampling programme, 1970–1972. (A) total catch; (B) catch rate in marginal zones, open water and near vegetation.

222

4. Experimental trawling, 1970–72 *by* M. T. Furse

This period of recovery of the lake depth and fisheries was also marked by the innovations in extension work of the incumbent Fish Ranger, Cliff Ratcliffe. He was largely responsible for the introduction of two new mechanized fishing methods and for the encouragement of the production of a series of relatively cheap and easy-to-construct plank boats. Probably the more important of the new fishing techniques was trawling and in particular paired craft trawling.

The idea of small craft shallow water trawling originated in Britain with the designs of Buckingham (1968). Ratcliffe modified these designs and made them suitable for shallow Lake Chilwa. Details of the designs and methodology of trawling are given by Ratcliffe (1971, 1975) and will be summarized here. Unlike gill nets, the trawl nets were constructed to fish the entire water column. The approximate spread of the net in operation was 10 m with a bosom opening of 3 m. Mesh size of the side panels and bosom varied but was of the order of 5 cm tapering to a 1.5 cm cod end. The warp of the trawls was of variable length, dependent upon lake levels. In operation the trawl was towed by two 5 m flat bottomed planked boats, each powered by a 4–5 h.p. outboard engine. Like seine nets, trawls are relatively unselective, there being a minimum but not a maximum size of fish caught.

The second mechanized technique introduced by Ratcliffe was motorized scoop-netting. The method was not completely new to enterprising local fishermen. Where Ratcliffe refined the technique was in the design of the net. He tapered the panels in such a manner that not only did the net hold a better shape in operation but it also required less netting to make. Catches of up to 30 kg per hour were possible with this technique at certain times of the year.

An interesting development related to the introduction of trawling techniques was the implementation of 'manpower trawling' (Ratcliffe 1972). The boats of paired craft trawling were replaced by two or more wading, towing fishermen. The tow started at the swamp edge and a semi-circular haul moved into the lake and then back to the swamp fringe to empty the catch. The catch could be anything up to 25 kg a haul (Ratcliffe 1971).

The introduction of paired craft trawling offered not only the means for an economically viable new fishing industry but also provided a powerful tool for the study of abundance, distribution and ecology of the fishes of the lake. Ratcliffe (1975) costed the trawling operation and determined that catch rates of 23 kg per hour would make trawling commercially viable. In conjunction with the Department of Fisheries, Government of Malawi, Furse (1972 and unpublished data) introduced a monthly series of experimental trawls at 13 sites over the lake. The purpose of this programme was twofold. First to provide material for ecological studies; many of the results of this facet have been presented (Chapter 11); and secondly, to determine in what areas and in what seasons catches of 23 kg per hour were possible.

Furse (op. cit.) divided his fishing sites into three zones: marginal areas within 50 m of the lakeward edge of the swamp, open lake areas free of macrophyte growth and open lake areas in the region of macrophyte stands. Strong seasonal and zonal patterns could be distinguished, and can be compared with 1976 catches (Fig. 12.6 and section 12.5).

Alternate periods of high and low catch rates were slightly out of phase with the climatic seasons. The best catches occurred between April and August i.e. in the cooler months (Fig. 12.6A). Between August and September, catches fell sharply and it is this decline which was related to spawning movements. During the April to August period catches could be maintained at or around the 23 kg per hour over a wide range of sites, but the August decline led to values below 10 kg per hour – clearly uneconomic. These low levels were maintained through most of the rainy season but began rising again in February and March as fish, including juveniles, entered the open lakes from the swamps (Fig. 12.7).

Fig. 12.7 Part of a catch of *Barbus* (matemba); (photo: A. J. McLachlan).

Imposed upon this strong seasonal pattern were equally distinct spatial variations. These variations were however time dependent. During the April to August period of high catches, the marginal zone yielded consistently higher catches than either of the open lake zones (Fig. 12.6B). In this period each of the lake zones produced catch rates in excess of 23 kg per hour but those in the marginal zones reached and often exceeded twice this value (Furse op. cit.). The highest recorded catch rate was of 103 kg per hour at a marginal site near Nchisi Island in the south. These data show that, with careful selection of fishing zone, catch rates far in excess of subsistence levels can be made in the appropriate season. At the end of the season of high returns, the distinction between catches in marginal areas and open lake areas broke down completely. The catch rate of all zones fell but catches in marginal areas fell more steeply (Fig. 12.6B). As a consequence catch rates in all three zones were approximately equal. No zone allowed worthwhile catches to be made and under no circumstances was trawling an economic proposition immediately before or during the rainy season, in the years 1971–72; but later the situation changed as lake stocks continued to recover (see section 12.5).

Barbus paludinosus comprised the major portion of catches throughout the two year study period (1971–72) and trawling clearly enables this species to be exploited in areas where it was previously underfished in the dry months. This particularly includes the open lake. Two other features also arose from this study. Firstly the catch rate of *Sarotherodon shiranus chilwae* increased markedly in 1972 over 1971 showing the recovery of stocks of this species. Secondly haplochromids, and particularly *Haplochromis callipterus*, were caught at rates approximately 10 kg per hour and were clearly shown to be economically important.

The higher catches of fish in marginal areas in the dry season may also be attributed to the greater food supply available there. This factor has already been discussed in some detail in relation to overall distribution patterns in the Chilwa Basin (Chapter 11). Burgis et al. (1973) stated that similar patterns of greater weight of fish in marginal zones occurred in Lake George, Uganda, also because more food is available there. In the U.S.A. Clark (1974) assessed the relative importance of environmental parameters in controlling the catch rates of fishes in trawls from the Everglades National Park. Three factors appeared to account for most of the variation: vegetation density, substrate conditions and salinity. These factors were often inter-dependent, as in Lake Chilwa, but Clark suggested that differences in vegetation density were primarily responsible for the variations. The effect of salinity appeared to be relatively unimportant in his study.

Trawling was abandoned, and other new and traditional methods were of little avail, when the lake entered a minor recession again in 1973–74. Fish landings dropped to under 2000 tonnes. The subsequent recovery showed many of the hallmarks of the previous decline and recovery, except that after three years, the repopulation was even more spectacularly successful. Estimated landings in 1976 reached 20,000 tonnes (Table 12.1) over twice that recorded in 1965. *Sarotherodon shiranus chilwae* represented only 12 per cent of the commercial catch. Experimental trawling was re-introduced and was more successful than ever previously. Maximum trawl catches of 198 kg per hour along the northwest swamp fringe were twice that of the pre-1973 recession (Fisheries Dept. 1978). The Lake Chilwa ecosystem was once again exhibiting its resilience.

5. Experimental trawling in 1975–76 *by* M Kalk

An experimental trawling programme was again undertaken in the years of high level, 1975–76, by the Fisheries Department (Fisheries Dept. 1978). This time the trawl was adapted to a single plank boat with outboard engine. This survey went further afield than that of Furse (section 12.4), to include the northeast sector and the deep water in the southeast, as well as the west and south. The fishing (mainly for *Barbus*, because of the 13 mm mesh size) yielded more than the economic limit of 23 kg per hour suggested by Ratcliffe (1975), at every site throughout the year, with the exception of the Phalombe river mouth in October, April and May, and of the open water in October, November and in June 1976 (although in the previous June, a higher than average sample had been fished in open water).

The mean values for trawl catches throughout the year were higher than in

225

1972, and most probably related to the denser zooplankton (Chapter 8): approximately 34 kg per hour in the southern sector (including the deepest water), 47 kg per hour in the northeast sector, and 36 kg per hour in the northwest sector. The northeast sector, as was also the case for zooplankton, had the highest average peak (4 sites) of 114 kg per hour in April, the northwest (3 sites) 58 kg per hour in May and the southern sector (4 sites) 51 kg per hour in May (Fisheries Dept. 1978). The higher values in open water were in May while the northeast sites were below average, and the contrary towards the end of the year. This suggests that *Barbus* migrates into the centre of the lake in the cooler months and disperses. Although the data have not yet been completely analyzed, there is a strong indication from experimental trawling that the rich peripheral trawling in the warm months is related to the breeding migration to the edge of the swamp as it was in 1971–72.

Commercial fishing for *Barbus* in the swamp was very low, but this depended on the lack of suitable beaches for seine nets and did not necessarily reflect the presence or absence of the fish species.

Data from the Fisheries Department (1978) on the experimental trawling, shed further light on the reproductive behaviour of fishes. In February it was noted that the trawl was full of the juvenile *Sarotherodon* (*kasawale*) only in the northeast sites, but by March they had disappeared. Examination of gonads showed an extended breeding season for *Barbus*, confirming the findings of Kirk and Furse (Chapter 11), and very few spent fish were recorded in the lake itself offshore. On the other hand, the breeding period of *Clarias* was restricted to October to February and again few spent fish were seen in the lake. *Haplochromis callipterus* (a mouth-brooding cichlid, similar in size to *Barbus*) occurred in significant numbers on the northeast periphery and its breeding season was from March to October. *Sarotherodon*, on the other hand had a longer breeding season than *Haplochromis*.

The new conclusions which can be drawn from the later trawling survey, which was conducted during years of comparatively high lake level are the following: trawling may be more efficient than seine netting at times, but it depends on the density of the fish. There are times of uneconomic density of *Barbus* in the southern sector in the deeper water and the Phalombe River mouth, but, in general, the centre of the lake is worth fishing. The cost of motorized fishing has not yet been compared with the labour intensive seine fishing. When the same, fairly efficient standard trawling method of fishing was used, the catch was very much higher per unit effort at the best times of the year in years of higher lake level than in the years of recovery after dryness. The cost of petrol, engines and plank boats have risen in the last few years and only a few fishermen have, as yet, availed themselves of this seasonal source of fish. The economic minimum profitable catch of 23 kg per hour may now be higher.

6. The fish productivity of Lake Chilwa *by* M. T. Furse

The only indication of the production potential of Lake Chilwa that is available is the annual yield of the lake and swamps, with various fishing techniques. Comparison with other lakes is therefore not fully meaningful, if nets of different mesh, motorized gear or other techniques different from those used on Lake Chilwa are employed. However, it is generally known (Morgan 1971) that shallow tropical lakes with fairly high alkalinity give generally high yields. Fryer & Iles (1972) listed estimated yields from various countries. Table 12.3 quotes the African lakes listed by Fryer and Iles and adds the highest yield so far in Lakes Chilwa and Malombe (a shallow lake just south of Lake Malawi (Fig. 2.1), as well as their average yields over some years. The average yield for Lake Chilwa over a ten year period includes one of the most severe recessions in its history. It will be seen that this yield is similar to that of Lake Rukwa in 1963, when its level was high. The highest yield recorded is near to that listed for Lake George, a more stable equatorial lake. The comparison with most of the other similar African lakes is complicated by the fact that their levels also change over periods, and the state of the fisheries is changing over the years with the introduction of new techniques.

Table 12.3 Yields per unit area of selected lakes in Africa (after Fryer & Iles 1972).

Lake	Year	Mean Annual Yield (tonnes)	Area (km^2)	Yield (kg ha^{-1} yr^{-1})
George (Uganda)	1959–1965	4700	269	174
Chilwa (Malawi)	1976	20 000	1256	159
Mariut (Egypt)	1920–1929	3428	243	141
Malombe (Malawi)	1976	5248	406	131
Kioga (Uganda)	1963–1965	17 700	2280	80
Malombe (Malawi)	1961–1963	1500	400	37
Rukwa (Tanzania)	1963	c.7000–12 000	3302	36
Chilwa (Malawi)	1963–1972	4448	1256	33
Mweru-wa-N'tipa (Zambia)	1967	2840	1540	19
Naivasha (Kenya)	1961–1963	200	140	14
Bangweulu (Zambia)	1952–1964	7000	7777	9
Mweru-wa-N'tipa (Zambia)	1961–1967	c.1350	1540	9

The area used in the calculation of yield per hectare per year includes the swamp of Lake Chilwa, which contributes over 30 per cent of the production. At its best, Lake Chilwa is probably a highly productive lake, and the yield may even now be increased in future. Over an extended period, when it suffers minor and major recessions, it still compares reasonably well with other lakes of similar depth and size.

Within Malawi, Lake Chilwa contributed just under a third of the national total fish landing in 1976. In a country where available protein is limited, the importance of the Chilwa fishery cannot be overstated.

7. The strategy of exploitation *by* M. T. Furse

The question is how best to manage such a fishery. Fryer (1972, 1973) gives many pertinent warnings to those who would seek to exploit African lakes. In essence he urges sound conservational policy. In theoretical terms most fishery administrators would approve of the general concern for sound management policy. Normally they attempt to put theory into practice by implementing controls over numbers of fishermen, of boats or fishing gear licensed, or over the design and type of fishing gear. Fryer's fears were that these controls are ill-chosen or are not adequately implemented and the fisheries are either mismanaged or unmanaged, and he warns against the catastrophic effects of over-fishing.

In Chilwa during the recovery, limits were imposed upon the mesh of gill nets that could be used (a fixed minimum of 70 mm). This offered some protection to breeding stocks of *Sarotherodon* without seriously affecting the catch of other species. In future it would also exploit larger fish which could probably evade a trawl. Seine nets were also banned for 1969 to protect *Barbus* during the first years of recovery (Morgan 1971).

On a more theoretical level several possible schemes have been propounded for stabilizing lake levels in Lake Chilwa, including re-opening links with Lake Chiuta (Shroder & Nicholson 1967), and thus stabilizing annual yields. Morgan (1971) also explored the possibility of developing a series of fish ponds around Lake Chilwa and on affluent streams in which the nucleus of fish for re-stocking could be reared in the case of further recessions. He also proposed that, during recession, bans ought to be imposed on fishing in the swamps and lagoons around the lake.

It can, however, be argued that any practical regulations or theoretical suggestions for such fishery protection ignore the reality of the situation on three counts. The first is that many of the measures advocated seem based upon the concept that it is the *Sarotherodon* stocks which must be conserved and protected. *Sarotherodon* and *Tilapia* form the basis of many important, natural and artificial fisheries. In Chilwa, *Sarotherodon shiranus chilwae* is slow to recover from recessions and is not specifically favoured by the local people. The second objection to conservation measures is that they are very difficult to impose in practice. Many parts of the lake are so remote, particularly in the north and in most of the surrounding swamps, that such paper restrictions would be impossible to implement.

These two arguments alone may not be sufficient grounds for abandoning fishery regulations, but it may be contended that the third and principal argument is the need to forego fishery regulations because of the realities of an unstable lake system. The lake is perhaps never going to have the opportunity to allow a stable community to develop for long. Schemes to provide a stable lake water level are extremely expensive, of unproven efficacy and most unlikely to be implemented. With intervals between recessions of only perhaps five to ten years any imposition of fishing regulations merely denies the local fishermen the chance of cropping the fishes before the lake itself does so through critical physico-chemical demands and mass mortalities. The *Barbus paludinosus* and *Clarias gariepinus* populations are capable of very rapid recovery in newly-inundated conditions. Any attempts to protect *S. s. chilwae* stocks merely denies the fishing industry the opportunity to crop the other species whilst they are there to be cropped unless control can be carried out selectively. Fryer's (1972, 1973) comments are valid and laudable for permanent water bodies, but for Lake Chilwa the fish should be caught whilst stocks exist. Stocks that can recover in two or three years from virtual annihilation by drought are never likely to be overfished to the point of disappearance. Lake Chilwa itself has its own reservoirs in the swamps which serve as the great conservator of fishes!

Discussion of the management of the fishery should not be terminated without one grave warning. Overfishing may not destroy the fishery but accumulation of toxic chemicals could one day do so. The dangers of the accumulation of fertilizers or of directly toxic substances have been clearly demonstrated in open basin lakes in Europe and North America (Fryer 1973). This danger is far more acute in closed basin lakes, like Lake Chilwa (Chapter 18). Lake Chilwa, its fishes and its fishery may have the resilience to overcome the physico-chemical stresses the climate places upon it. It must be protected against the untimely ravages of man.

References

Buckingham, H. 1968. A Buckingham universal trawl – and a mini trawl. World Fishing 17:39–41.

Burgis, M. J., Darlington, J. P. E. C., Dunn, I. G., Ganf, G. G., Gwahaba, J. J. & McGowan, L. M. 1973. The biomass and distribution of organisms in Lake George, Uganda. Proc. Roy. Soc. London. B. 184:271–298.

Chirwa, G., Kanjo, C. & Munthali, M. A. 1966. The dug-out canoes of Lake Chilwa. Soc. Malawi J. 19:58–61.

Clark, S. H. 1974. A study of variation in trawl data collected in Everglades National Park, Florida. Trans. Amer. Fish. Soc. 103:777–785.

Fisheries Report 1977. Statistics of Lakes Chiuta and Chilwa 1976. Fisheries Dept. Lilongwe, Malawi. 24 pp.

Fisheries Dept. 1978. i. Data from experimental trawling in Lake Chilwa 1975–1977. ii. Data from commercial fishing. Fisheries Dept. Lilongwe, Malawi (Unpublished).

Fryer, G. 1972. Conservation of the Great Lakes of East Africa: A lesson and a warning. Biological Conservation 4:256–262.

Fryer, G. 1973. The Lake Victoria fisheries: Some facts and fallacies. Biological Conservation 5:305–308.

Fryer, G. & Iles, T. D. 1972. The cichlid fishes of the Great Lakes of Africa. Their biology and evolution. Edinburgh. Oliver and Boyd. 641 pp.

Furse, M. T. 1972. Fish. In: Howard-Williams, C., Furse, M. T., Schulten-Senden, C. M., Bourn, D. M. & Lenton, G. Lake Chilwa, Malawi. Studies on a tropical freshwater ecosystem. Report to IBP/UNESCO Symposium, Reading, England. 28–34.

Hornby, A. M. 1963. Lake Chilwa fisheries. Unpublished report. Blantyre Netting Company.

Kirk, R. G. 1970. A study of Tilapia shirana chilwae Trewavas in Lake Chilwa, Malawi. Ph.D. thesis, University of London.

Lowe, R. H. 1952. Report on the Tilapia and other fish and fisheries of Lake Nyasa. Colon. Off. Fish. Publ. 1. H.M.S.O. London. 126 pp.

Morgan, P. R. 1971. The Lake Chilwa Tilapia and its fishery. Afr. J. Trop. Hydrobiol. Fish. 1:51–58.

Mzumara, A. J. P. 1967. The Lake Chilwa fisheries. Soc. Malawi J. 20:58–68.

Ratcliffe, C. 1971. Experimental small craft pair trawling. Lake Chilwa 1970–71, Malawi. Fisheries Bulletin, Ministry of Agriculture, Zomba, Malawi 1:1–31.

Ratcliffe, C. 1972. Trawling – Central African style. Fishg. News Internat. 11(9):22–27.

Ratcliffe, C. 1975. Commercial small craft pair trawling trials, Lake Chilwa, Malawi, 1971. Afr. J. Trop. Hydrobiol. Fish. 3:61–78.

Shroder, J. F. Jnr. & Nicholson, F. H. 1967. Lake Chilwa reclamation project. A preliminary investigation into the feasibility of beginning a program of water management for Lake Chilwa. Cyclostyled report. University of Malawi.

Interactions between swamp and lake

Clive Howard-Williams

231

Interactions between swamp and lake

The shallow peripheral area of a lake which is occupied by aquatic macrophytes is usually termed the littoral, and as it forms the interface between land and lake it often has a vital influence on the quantity and quality of material entering the lake from the catchment area. In lakes where the ratio of littoral area to pelagic area is large, such as in Lake Chilwa and many other shallow African lakes, the influence of the littoral on the ecology of the whole lake can be considerable. In Lake Chilwa, almost the entire littoral is occupied by a *Typha* swamp. The very turbid water (Chapter 4) precludes any development of submerged plants in the lake, and the windswept nature of the lake margins usually prevents the growth of water lilies beyond the fringe of the *Typha*. In some very sheltered bays, however, the water lilies *Nymphaea lotus* and *N. caerulea* are sometimes found in patches (Fig. 13.1).

Fig. 13.1   In sheltered bays along the swamp-lake edge, water lilies (*Nymphaea* spp.) grow. Bird life in such areas is particularly abundant (photo: C. M. Schulten-Senden).

Wetzel & Hough (1973), Gaudet (1974) and Howard-Williams & Lenton (1975) have reviewed the role of the littoral region and of aquatic macrophytes in freshwater ecosystems in general, and it appears that the littoral zone of lakes has the following major functions:

i. It provides a diverse habitat for animals and plants, thus increasing species richness in the lake system. It consequently provides a feeding ground for species of the pelagic zone.

233

ii. It acts as a trap and a sieve for particulate and dissolved, organic and inorganic materials.

iii. It can influence the nutrient status of a lake by either speeding up nutrient release from the sediments to lake waters via the plants themselves – the so-called 'nutrient pump' effect (Odum 1971), or on the other hand, by absorbing the nutrients from lake waters and binding these up in the organic material of the littoral vegetation. Thus if the littoral zone is large, the aquatic plants can act as a significant nutrient reservoir.

iv. It can contribute a major portion of the autotrophic production of a lake, and the macrophytes of the littoral may form the basis of the food chain in many shallow lakes.

v. Dissolved organic compounds originating in the littoral zone can markedly influence planktonic production in the open lake. This influence is either in the form of chelators for trace metals in particulate form or as growth factors.

Wetzel & Allen (1972) and Wetzel (1975) have thus claimed that in many lakes the littoral flora, consisting of a complex of macrophytes and epibenthic algae and bacteria, has a considerable regulating influence on the metabolism of the whole lake. It is clear, however, that the extent of this regulatory influence will depend largely on the area occupied by the littoral zone, the volume of the lake, and the nutrient status and particulate load of the inflowing waters. We might therefore expect at the outset that the extensive swamps surrounding Lake Chilwa have a considerable influence on the ecology of the lake. To examine this effect in detail, however, it is necessary to consider the processes occurring within the swamp and the interactions between the swamp and the lake. It is convenient to do this under the headings of the littoral functions listed above, then finally to consider the utilization of swamp production.

1. The swamp as a diverse habitat

The only means of access to the interior of the swamps of Lake Chilwa is by boat or canoe, via the narrow channels cut by the fishermen through the swamps. These channels, some of which are 11 km long, lead from fishing villages on the shores straight through the swamp to the lake. When one first travels down one of these channels, the experience is somewhat claustrophobic, with never-ending walls of tall green *Typha* shoots (over 4 m high) leading down into still dark waters which give the impression not of diversity, but of monotony. And yet close inspection of the system reveals that this is not at all the case.

For instance, consideration of Fig. 5.3 in Chapter 5 shows that there is considerable variation in the physical and chemical factors in the soil and water along a transect from land to the lake edges of the swamp. A statistical comparison of some physical and chemical conditions in 24 stands in the littoral zone and 19 stands in the open lake, scattered as widely as possible, was carried out by Howard-Williams & Lenton (1975). They found that the coefficients of variation in nearly all the parameters measured were significantly greater in the swamps than in the open lake. If we consider dissolved oxygen, we find parts of the swamp which are completely deoxygenated for much of the year, while other areas nearer the lake receive oxygenated water throughout the year. Areas in the central swamp may receive pulses of oxygenated lake water following onshore winds (Fig. 5.5). The waters of lagoons, bays and man-made clearings in the swamps are often highly oxygenated in the daytime due to photosynthesis by submerged macrophytes and epiphytic algae. Here it is worth noting again the considerable temporal variations in the environmental factors in the swamps which we discussed in Chapter 5. During the rainy season when quantities of fresh river water are entering Lake Chilwa, the dissolved oxygen regime of the entire swamp alters as the stagnant swamp waters are replaced by fresh river water. Many animals not found in the swamps during the dry season become temporary inhabitants of the area during the rainy season, and Beadle (1974) has pointed out that although little is known of life histories of many of the inhabitants of shaded swamps, it seems that most reproduction occurs during the rains when the level of oxygen is raised in peripheral swamp waters.

Thus considerable spatial and temporal variations occur in the swamps which are not matched in the open lake during years when the lake is full. Temporal variation is particularly important because organisms which can survive a certain environment in the adult stage may not be able to breed there unless the environment changes suitably at some stage in the year.

Of particular importance to animals which have a benthic stage in their life cycle, is the nature of the substrate of the area in which they live. McLachlan (1974 and Chapter 9), for instance, points out that the bottom muds of the open lake are unsuitable for colonization by invertebrate benthos. Within the swamps, however, there exists a variety of substrates which changes through the swamps both vertically and horizontally: live macrophyte stems, large dead floating *Typha* shoots, submerged detritus, a fairly compact mud with macrophyte roots, and all sediment types, ranging from sand to dense layers of the fine 'mud blanket' referred to in Chapter 4. With this habitat diversity in the swamps it is not surprising that the rich invertebrate community of the swamps (McLachlan 1970) exceeds the open lake mud fauna both in numbers of species and biomass (Chapters 9 and 10). The diversity of the fish species is considerably greater in the swamps than in the open lake, again reflecting the habitat diversity here. In addition, the swamp provides shelter and a substrate for the spawning of *Clarias gariepinus* and *Barbus paludinosus*, two of the most important fish species of the open lake. A spawning migration of both these species into the swamp occurs each year (Chapter 11). Rzoska (1974) pointed out how tropical swamps harbour a unique and very diverse fauna, and Beadle (1974) discusses the importance of tropical swamps as a habitat for a large variety of terrestrial animals. The work on the birds of Lake Chilwa by Schulten & Harrison (1975) and Schulten (1974) (Chapter 14) emphasizes the importance of the swamps as a habitat and feeding area for the very rich bird life of the lake basin. *Homo sapiens*, an important predator in the lake system, has also utilized the seemingly impenetrable swamps as a habitat. Man has adapted to the swamp conditions by the use of dugout canoes which are very long and extremely heavy, and can move easily through the *Typha* on the lake edges. He has made canoe channels through the central swamps to land, and even lives in the swamps by cutting large areas of *Typha*, heaping the cut material into a mound above the water level, and weaving a temporary reed hut on this support.

The Lake Chilwa swamps enrich, to a considerable extent, the flora and fauna of the lake ecosystem.

2. The swamp as a sieve and a trap

Everything entering the lake from the rivers of the catchment area of Lake Chilwa must first pass through the swamp. As the first effect of aquatic vegetation is to decrease the speed of water flow, heavy particles in suspension would settle out on the periphery of the swamp, thus enriching the silt and particularly the clay content of the open lake. This has a number of important effects. Firstly the fine-grained clays which pass more easily through the swamps have a cation reserve and exchange capacity than the sands (McLachlan & McLachlan 1969). This therefore influences the water chemistry of the open lake because of the continuous interchange of elements between the sediments and the waters of the lake (Chapter 4). Secondly, these fine clays are a poor substrate for benthic invertebrates (Chapter 9) and thus influence the benthic fauna of the open lake.

In the reverse direction, the swamp also traps suspended sediments from the open lake waters. Fig. 5.5, shows, for instance, the increase in turbidity of the water in the swamp following an onshore wind as turbid lake water is blown into the swamp. Much of the fine-grained sediments in suspension settle out in the swamp and form a particularly thick mud blanket in some areas, especially on shores facing the prevailing southeast winds. This soft mud blanket in the northern swamps is 50 cm thick in some areas. I have suggested (Howard-Williams 1973) that this is the reason why the most extensive encroachment of the *Typha* into the lake in recent years has occurred from the northwest edge of the lake. Recently Weisser (1978) using data from Lake Neusiedlersee, a temperate lake very similar in size and shape to Lake Chilwa, has produced a quantitative model of a siltation system in shallow lakes with littoral vegetation. His work also emphasizes the importance of the littoral in the trapping of fine-grained sediments from the open water of shallow lakes. The littoral swamps of Lake Chilwa play an important part in regulating the composition of the particulate material in the open lake.

3. The influence of the swamp on the nutrient status of Lake Chilwa

Swamps can have an influence on the dissolved substances in the waters passing through them. This influence can be of two types, enrichment or removal. The removal of nutrients from flowing water by swamps is a subject of increasing interest. Talling (1957) suggested that papyrus swamps act in some ways as giant filters or exchange systems. Gaudet (1976) found that nutrients entering papyrus swamps are, through various pathways, eventually trapped as an organic sludge. This can occasionally be washed out into the lake. He suggests that swamps should be looked at not as filters, but as large 'holding tanks'. However, in spite of many similarities, papyrus swamps have very different mechanisms of nutrient cycling from those dominated by *Typha*, such as in Lake Chilwa. Papyrus generally occurs as a buoyant mat, floating in shallow water. Nutrient uptake by the swamp community is thus directly from the water, while in a *Typha* swamp nutrients are absorbed from the sediments, and on death and decay of the above ground portions, they are released into the swamp water.

Let us first consider the situation in the swamps towards the end of the dry season (August to October). At this time of year the above ground biomass of the *Typha* is maximum for the year and the water level in the lake and swamp is at its lowest (Howard-Williams 1972). The support offered to the tall, heavy *Typha* shoots is minimal at a time when it is most needed. As the water level drops, the heavy shoots, unable to support themselves, collapse over large areas of the swamps. This collapsed material together with all the dead leaves which have fallen during the year forms a dense tangle of dead material just above the water. At the start of the rains the water level rises rapidly and I estimate that most of the dead material is covered by inflowing and rising waters within the first month of the rainy season. The question must now be asked: what happens to all the nutrients bound up in the dead plant material as it is covered by the inflowing waters? Samples of the dry mass of dead shoots collected from 15 localities in the swamps at the end of the dry season weighed 1250 g m^{-2} (Howard-Williams & Howard-Williams 1978). The amounts of nutrients bound up in this dead material are given in Table 13.1, which also shows the amounts in swamp water as a comparison.

Table 13.1 Standing stock of nutrients bound up in *Typha* in the Lake Chilwa swamps in live and dead shoots. Standing stocks in the swamp waters are also given. All figures are g m^{-2}.

Element	Live shoots	Dead shoots[+]		Swamp water*
		1	2	
Sodium	30.3	45.7	51.6	180
Potassium	13.9	4.5	5.1	6
Calcium	13.7	14.6	20.0	10
Magnesium	9.5	8.6	9.7	4
Nitrogen	22.7	4.5	5.1	0.1
Phosphorus	2.1	2.5	2.8	0.6

[+] Dead shoots 1 – Nutrients in dead shoot biomass collected at the end of the dry season (Howard-Williams & Howard-Williams, 1978). Dead shoots 2 – Nutrients in total annual production after death (Howard-Williams & Lenton 1975). This represents the amounts of nutrients available for recycling.
* From Howard-Williams & Lenton (1975).

It is clear that considerable quantities of nutrients are bound up in the live shoots, and when one considers the enormous area of the Lake Chilwa swamps, the vegetation here obviously acts as an important nutrient reservoir, particularly for nitrogen and phosphorus. In addition, large quantities of nutrients are present in the dead material when compared with the amounts in swamp water. These are now available for recycling into the water on the decay of the plant material. The decay process by which these nutrients are released from the decomposing plant material occurs in two stages, an initial rapid leaching which is mostly a physical process, and a slower, biologically mediated decomposition process.

Howard-Williams & Howard-Williams (1978) measured, in laboratory experiments, the rates of leaching of various nutrients and organic matter from dead *Typha* shoots. Some of the results are shown in Fig. 13.2.

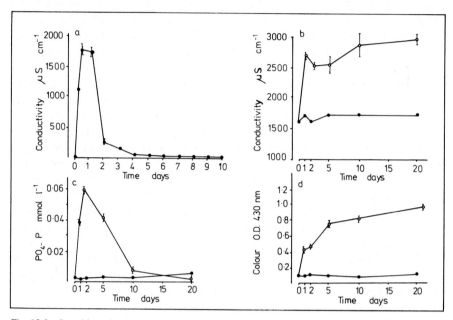

Fig. 13.2 Leaching of substances from dead *Typha* shoots from the Chilwa swamps. (a) Changes in conductivity in continually renewed distilled water in contact with dead *Typha* shoots. (b) Changes in conductivity of swamp water to which dead *Typha* shoots have been added. Open circles = *Typha* present, solid circles = Control, no *Typha*. The drop in conductivities after day 1 is not statistically significant. (c) Changes in PO_4-P in swamp water to which dead *Typha* shoots have been added. Symbols as for b. (d) Changes in the concentration of dissolved organic matter (reflected by the degree of brown colour in the water) in swamp water to which dead *Typha* shoots have been added (after Howard-Williams & Howard-Williams, 1978). Symbols as for b.

Fig. 13.2a shows the changes in conductivity in a laboratory experiment with dead *Typha* shoots in distilled water (in the same weight to volume ratio as found in the field) which was renewed daily and shows clearly that the leaching process of ions from the plant material is very rapid, most of the soluble ions being released within the first day of contact with water. This is reflected in the conductivity rise from 3.5 to 1800 μS cm^{-1} in the first day. Subsequent release was less and the conductivity dropped slowly. In flowing water the leaching of elements from dead *Typha* shoots is thus a pulse-like process lasting 1–3 days. However, much of the Chilwa swamps are not directly in the path of inflowing rivers, particularly the northern swamps. Here the dead *Typha* is subjected to rising but not rapidly flowing water during the rains.

Fig 13.2b shows the changes in conductivity in a laboratory experiment with dead *Typha* shoots in swamp water which we hoped would simulate the field conditions in the northern swamps. Again there is a rapid rise in conductivity on the first day, most of the soluble ions contributing to conductivity being released during this period. However, some elements, particularly phosphorus (Fig. 13.2c) decreased sharply after the initial release in stagnant swamp water. This decrease could be due to adsorption on to the plant surfaces and to uptake

238

by micro-organisms associated with the decomposing *Typha* shoots. The amount of dissolved organic matter released (Fig. 13.2d) shown by increasing brown colour in the water, was also very marked on the first day of contact with the swamp water.

If we relate these experiments to the field, it is possible to calculate the amounts of nutrients bound up in the dead vegetation which are rapidly lost to the water by leaching alone. These results are shown in Table 13.2.

Table 13.2 Percentage of the total standing stock of nutrients lost by leaching during flooding in the Chilwa swamps.

Element	% of nutrient stock lost
Sodium	26
Potassium	7.5
Calcium	9.2
Magnesium	11.5
Phosphorus*	1.5
Nitrogen*	0.3

* Underestimates total N and P lost because only PO_4–P and NO_3–N were determined in the water.

It would seem, therefore, that when waters are flowing through the swamps, considerable amounts of nutrients and dissolved organic matter are leached from the dead swamp vegetation and flow into the lake. The figure for nitrogen is probably increased when the NO_3–N released from the floodplain soils at the beginning of the rainy season is taken into account (see Chapter 5).

This process of lake enrichment during the rainy season from the peripheral swamps is, however, offset by the wind-generated water movement from the lake into the swamp. As described earlier, nutrient-rich suspended clays in the lake water settle out in the swamps. In addition and particularly important, Figure 13.2c shows how in a system which is not flowing, the initial release of phosphorus was reversed and this element was reduced to levels lower than those in the original water. Thus during periods when there is no water flow, PO_4–P may be retained in the swamp. In conclusion therefore, the available evidence indicates that the waters flowing into Lake Chilwa from the catchment have been modified both in their suspended and dissolved loads by the peripheral swamps before they reach the lake.

4. Production and decomposition in the Chilwa swamps

The source of primary production in the littoral zone of lakes is threefold; planktonic algae, epiphytic and benthic algae, and aquatic macrophytes. Few data are available on the algal producers of swamp waters, but due to the low light levels reaching the water surface, algal primary production is probably very low. Production is further hampered by the brown colour of most swamp water which considerably reduces light penetration (Dokulil 1973). Imhof (1973) reporting on the *Phragmites* reed swamp of the Neusiedlersee estimated that planktonic and epiphytic algae account for 3.3 per cent and 6.6 per cent respectively of total swamp production. Fig. 13.3 shows the contrast between chlorophyll *a* levels in the waters of the swamp and open lake of Chilwa in 1972. Chlorophyll *a* (and hence algal biomass) is often undetectable in swamp waters, while it is normally present in higher levels in the lake. Photosynthetic oxygen production showed a similar trend (Howard-Williams & Lenton 1975). The contribution to littoral production by algae in the swamp waters of Chilwa and, I suggest, most other herbaceous swamps, is unlikely to be more than 10 per cent of total net swamp production. An exception may be the narrow zone along the swamp-lake edge and in channels and lagoons, where more light reaches the water surface, and where dense growths of epiphytes are found on the *Typha* shoots.

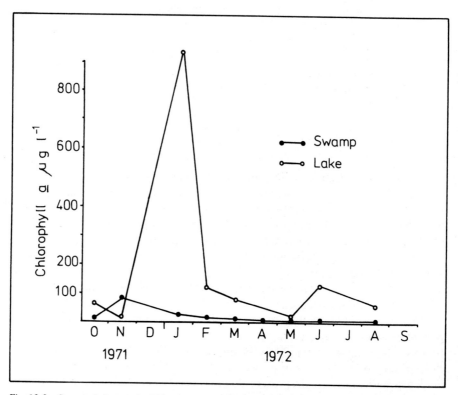

Fig. 13.3 Seasonal changes in Chlorophyll *a* in the swamp and open lake of Lake Chilwa showing the very low values in the swamp (after Howard-Williams & Lenton 1975).

Macrophyte production in the Chilwa swamps is confined almost entirely to *Typha*. There are no submerged macrophytes in the open lake (Chapter 7) and *Typha* makes up virtually all the biomass of the swamp vegetation. Table 13.3 shows biomass values and estimates of production of *Typha* in Lake Chilwa.

The biomass of material below ground is almost as high as that above ground. It is of interest that even during the period of minimum shoot biomass, more than 1 kg dry weight per square metre of live material still stands in the swamps. In other words, the swamps never die back completely, remaining green throughout the year.

The estimated value for above-ground production in dry weight is $1580 \, g \, m^{-2}$ per year. This is likely to be an underestimate rather than an overestimate because of the method of calculation of leaves shed in Table 13.3. The organic matter content is approximately 92 per cent of the dry weight. The amount of organic matter produced in the whole of the Lake Chilwa swamps each year is in the region of 0.8×10^6 metric tons.

What happens to this enormous quantity of organic matter produced by the littoral swamps? Lake Chilwa has no outflow, so the produced material must either accumulate as peat or other kinds of sediment, or decompose in the lake basin itself.

Table 13.3 Biomass values and estimates of production of *Typha* in the Lake Chilwa littoral. (Mostly from Howard-Williams & Lenton 1975).

	$g\ m^{-2}\ yr^{-1}$ dry weight
1. Maximum shoot biomass	2537
2. Below ground biomass (December)	2296
3. Minimum shoot biomass	1122
4. Estimation of leaves shed[+]	165
5. Estimation of net shoot production (1 − 3) + 4	1580
6. Total littoral swamp production in Lake Chilwa	0.8×10^6 tonnes yr^{-1} (dry weight)

[+] From Kvet (1971) for *Typha angustifolia*. This may be a serious underestimate for tropical conditions.

In many swamps particularly those with acid waters, there is considerable accumulation of organic material. Many of the papyrus swamps of Uganda for instance accumulate a deep peat layer (Carter 1955, Beadle 1958). In the valley swamps of western Uganda, papyrus peat reaches depths of 20–30 m (Beadle 1974) although in the region of Lake Victoria the peat is only 2–3 m deep. One somewhat surprising feature of the Lake Chilwa swamps is that there is almost no accumulation of dead organic matter on the bottom. Even in the central swamps undecomposed material seldom forms a layer more than 10 cm thick. This means that virtually all the 800 000 tons of organic matter produced each year is oxidized in the Chilwa basin itself.

Now, in tropical wetland areas there are two main pathways by which organic matter can be oxidized: through consumption by animals, bacteria and fungi and final dissipation as heat, or by fire.

During the dry season in the Chilwa basin, fires are started all round the lake by the inhabitants. Fires move very rapidly through the swamps, and in strong winds extensive areas of the *Typha* swamp are burnt, even those standing in water. Ground and aerial surveys conducted between 1970 and 1972 indicate that as much as one-third of the swamps are burnt each year. In this way a considerable proportion of the produced organic matter is oxidized, thus making *Homo sapiens* an important decomposer. The remaining swamp production must be decomposed by the aquatic organisms in the lake.

The rate of decomposition of *Typha* shoots in the Chilwa swamps has been reported by Howard-Williams & Howard-Williams (1978), and Fig. 13.4 shows the rate at which a known amount of *Typha* disappears from nylon mesh bags suspended in the swamp waters. The decomposition rate is twice that of *Typha latifolia* in the southern USA (Boyd 1970) and over three times as fast as that of *Typha angustifolia* in England (Mason and Bryant 1975). This rapid decomposition of *Typha* in Lake Chilwa is probably due to the relatively high temperatures (Chapter 4) throughout the year, and to the alkaline conditions in the lake waters. An important point, as discussed earlier is that in parts of the central swamps and all around the swamp-lake edge, oxygenated lake water enters the swamp. Thus the factor which limits much aquatic decomposition in other swamps, namely dissolved oxygen, is not limiting in large areas of the Chilwa swamps. Of importance also are the periods when the lake dries out. During the dry phases, especially after rain has moistened the soil in the lake basin, the breakdown of organic matter must be considerable. W. J. Junk (personal communication) suggests that the annual dry phase in many Central Amazonian floodplain lakes plays a major role in the decomposition of aquatic plant material, as during this period it is attacked and rapidly broken down by terrestrial arthropods.

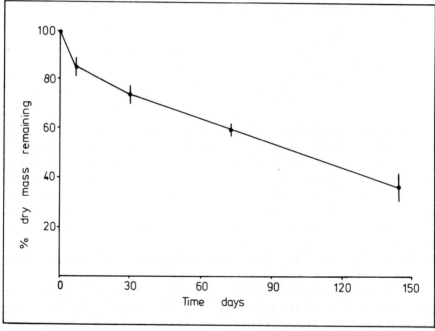

Fig. 13.4 Decomposition curve for *Typha domingensis* in the Lake Chilwa swamps. Vertical bars = 2 Standard Errors (after Howard-Williams & Howard-Williams 1978).

5. The utilization of littoral production

Having concluded that the biological community of Lake Chilwa is responsible for the breakdown of most of the organic matter produced in the littoral it is necessary to examine this in more detail. Man's influence has already been mentioned.

Firstly, the swamps of Chilwa (and virtually all herbaceous swamps) support very few herbivores (see Chapter 10). Few aquatic organisms actually graze live *Typha* shoots. Boyd & Goodyear (1971) specifically point out that *Typha latifolia* is seldom grazed because it has a high fraction of indigestible material. Gaevskaya (1969) in an extensive review of the role of aquatic plants in the nutrition of animals in Europe points out that *Typha* is eaten by geese, some fish species and aquatic rodents, as well as 59 species of invertebrates. However, few of the latter group were truly aquatic species. In addition, actual losses of live *Typha* material by grazing are very low, and in fact almost all of the publications dealing with net production in emergent aquatic plant communities infer that grazing losses of live plant material are either negligible or very low (Imhof & Burian 1972, Kvet et al. 1973, Boyd & Goodyear 1971, Imhof 1973).

Typha shoots only become of real value to the aquatic community once they have died and collapsed into the water. The bulk of the produced material in a *Typha* swamp is thus broken down via a food web based on detritus. This generally applies to other swamps as well (Imhof & Burian 1972, Kvet et al. 1973). The detritus produced by *Typha* during its decomposition in Lake Chilwa appears to play a major role in the ecology of the open lake, and much of the lake's fauna is concentrated on the periphery near the swamp margins especially in years after a recession. A feature which is typical of many African lakes with fringing swamps is a concentration of fish species along the swamp edge (Ratcliffe 1972, Harding 1960, Welcomme 1972).

Detritus from the Chilwa swamps forms an essential substrate for the mud-dwelling invertebrates of the open lake, which again are concentrated on the periphery (Chapter 9, McLachlan 1974). During the filling and post-filling phases of Chilwa, there was evidence that the largest numbers of zooplankton occurred near the periphery of the open lake (Chapter 8). It is likely that particulate organic matter from the swamps, with associated bacteria, forms a large part of the diet of zooplankton here. This is discussed by Kalk & Schulten-Senden (1977) and in Chapter 8.

Zooplankton and macrophyte detritus are items of great importance in the diet of the three major fish species of the lake, *Clarias gariepinus*, *Barbus paludinosus* and *Sarotherodon shiranus chilwae* (Bourn 1972, Chapter 11). Aquatic insects and small fish were important also for *C. gariepinus*, while filamentous algae were a common food item for *Sarotherodon*. All these components of the fish diet are associated with the swamp or swamp–lake interface. As the principal food items for the major fish species are found in or near the swamp, one would expect the major proportion of the lake's fishery to be centred around the periphery of the open lake also, which is indeed the case (Chapter 11), more especially in the years after a recession.

Schulten (1974) showed the overwhelming importance of aquatic plants from the swamps in the diet of the major duck and geese species of Lake Chilwa.

The Lake Chilwa swamp thus provides a diverse habitat for the lake's fauna, and shelter and spawning ground for many fish. The swamp influences allochthonous inputs (particulate matter from the streams and floodplain) into the lake, modifies inflowing waters both from the land and the lake, provides a substrate for open lake mud-dwelling invertebrates, and also provides a major source of food for the food web of the open lake. Overall, therefore, the ecology of Lake Chilwa is very largely influenced by biotic and abiotic interactions between swamp and open lake.

References

Beadle, L. C. 1958. Hydrobiological investigations on tropical swamps. Verh. internat. Ver. Limnol. 13:855–857.

Beadle, L. C. 1974. The inland waters of tropical Africa. Longmans, London. 365 pp.

Bourn, D. M. 1972. The feeding, diet and ecological relations of three species of fish of economic importance in southern Malawi. M.Sc. thesis, University of Edinburgh.

Boyd, C. E. 1970. Losses of mineral nutrients during decomposition of *Typha latifolia*. Arch. Hydrobiol. 66:511–517.

Boyd, C. E. & Goodyear, C. P. 1971. Nutritive quality of food in ecological systems. Arch. Hydrobiol. 69:256–270.

Carter, G. S. 1955. The papyrus swamps of Uganda. W. Heffer & Sons, Cambridge. 25 pp.

Dokulil, M. 1973. Planktonic primary production within the *Phragmites* community of Lake Neusiedlersee (Austria). Pol. Arch. Hydrobiol. 20:175–180.

Gaevskaya, N. S. 1969. The role of higher aquatic plants in the nutrition of the animals of freshwater basins. Vol. I–III. (English translation published by National Lending Library for Science and Technology, Boston Spa, England). 629 pp.

Gaudet, J. J. 1974. The normal role of vegetation in water. In: D. S. Mitchell (ed.) Aquatic vegetation and its use and control. UNESCO, Paris, pp. 24–37.

Gaudet, J. J. 1976. Nutrient relationships in the detritus of a tropical swamp. Arch Hydrobiol. 78:213–239.

Harding, D. 1960. Lake Kariba: The hydrology and development of the fisheries. In: R. H. McConnell (ed.). Man-made Lakes. Academic Press, London, pp. 7–18.

Howard-Williams, C. 1972. Limnological studies in an African swamp: seasonal and spatial changes in the swamps of Lake Chilwa, Malawi. Arch. Hydrobiol. 70:379–391.

Howard-Williams, C. 1973. Vegetation and environment in the marginal areas of a tropical African lake (L. Chilwa, Malawi). Ph.D. thesis, University of London.

Howard-Williams, C. & Lenton, G. M. 1975. The role of the littoral zone in the functioning of a shallow tropical lake ecosystem. Freshwater Biology 5:445–459.

Howard-Williams, C. & Howard-Williams, W. 1978. Nutrient leaching from the swamp vegetation of Lake Chilwa, a shallow African lake. Aquatic Botany.

Imhof, G. 1973. Aspects of energy flow by different food chains in a reed bed. A review. Pol. Arch. Hydrobiol. 20:165–168.

Imhof, G. & Burian, K. 1972. Energy flow studies in a wetland ecosystem. Special publication of the Austrian Academy of Sciences for the I.B.P. Vienna. 15 pp.

Kalk, M & Schulten-Senden, C. M. 1977. Zooplankton in a tropical endorheic lake (Lake Chilwa, Malawi) during drying and recovery phases. J. Limnol. Soc. sth. Afr. 3:1–7.

Kvet, J. 1971. Growth analysis approach to the production ecology of reedswamp plant communities. Hidrobiologia (Bucur.) 12:15–40.

Kvet, J., Ulehlova, B. & Pelikan, J. 1973. Structure of the reed belt ecosystem of the Nesyt fishpond. Pol. Arch. Hydrobiol. 20:147–150.

Mason, F. & Bryant, R. J. 1975. Production, mineral content and decomposition of *Phragmites communis* Trin. and *Typha angustifolia* L. J. Ecol. 63:71–95.

McLachlan, A. J. 1970. The bottom fauna. In: M. Kalk (ed.) Decline and recovery of a lake. Government Printer, Zomba, pp. 34–35.

McLachlan, A. J. 1974. Recovery of the mud substrate and its associated fauna following a dry phase in a tropical lake. Limnol & Oceanogr. 19:74–83.

McLachlan, A. J. & McLachlan, S. M. 1969. The bottom fauna and sediments in the drying phase of a saline African lake (L. Chilwa, Malawi). Hydrobiologia 34:401–413.

Odum, E. P. 1971. Fundamentals of Ecology. 3rd Edition. W. B. Saunders Company, Philadelphia. 574 pp.

Ratcliffe, C. 1972. The fishery of the Lower Shire River area. Malawi Fisheries Bulletin No. 3. Fisheries Dept., Malawi.

Rzóśka, J. 1974. The upper Nile swamps, a tropical wetland study. Freshwater Biology, 4:1–30.

Schulten, G. G. M. 1974. The food of some duck species occurring at Lake Chilwa, Malawi. Ostrich 45:224–226.

Schulten, G. G. M. & Harrison, G. 1975. An annotated list of birds recorded at Lake Chilwa (Malawi, Central Africa). Soc. Malawi J. 28, 2:6–30.

Talling, J. F. 1957. The longitudinal succession of water characteristics in the White Nile. Hydrobiologia 11:73–89.

Weisser, P. 1978. A conceptual model of a siltation system in shallow lakes with littoral vegetation. J. Limnol. Soc. sth. Afr.

Welcomme, R. L. 1972. A brief review of the floodplain fisheries of Africa. FAO Committee for Inland Fisheries of Africa. Report CIFA/72/S.17.

Wetzel, R. G. 1975. Limnology. W. B. Saunders Company. Philadelphia. 743 pp.

Wetzel, R. G. & Allen, H. L. 1972. Functions and interactions of dissolved organic matter and the littoral zone in lake eutrophication and metabolism. In: Z. Kajak & A. Hillbricht-Ilkowska (eds.) Productivity problems of fresh waters. P.W.N. Polish Scientific Publishers, Warsaw, Poland, pp. 333–347.

Wetzel, R. G. & Hough, R. A. 1973. Productivity and role of aquatic macrophytes in lakes. An assessment. Pol. Arch. Hydrobiol. 20:9–19.

Amphibians, reptiles, mammals and birds of Chilwa

C. O. Dudley, D. E. Stead & G. G. M. Schulten

Amphibians, reptiles, mammals and birds of Chilwa

The amphibian and reptilian fauna of south-eastern Malawi has recently been reviewed by Stevens (1974), who recorded 159 species. His area included great diversity of physiographic regions and an overlap of tropical, transition and temperate faunal elements. Thirteen species of amphibians have been reported to occur on the low plains and waters of the Chilwa area and are listed in the check list (Appendix A).

1. Amphibians: distribution in relation to water level *by* C. O. Dudley

A range from complete dependence on water for their life-histories, through adaptation to low sporadic rainfall on the Chilwa plain, to complete independence of water, will be illustrated by eight species of amphibians observed in the last few years by the author. They occur along a gradient from the lake's edge to the dry edge of the floodplain.

(1) Mueller's clawed frog, *Xenopus muelleri* (Peters) (Figs. 14.1F and 14.2A) belongs to the most aquatic genus of frogs. It is common in the larger temporary pools formed in floodplain depressions and is adapted to aquatic life with neuromast organs (similar to the lateral line organs of fish) which detect pressure differences and can be used to find its prey in murky water (Russell 1971). It lives entirely below water and rises to the surface at intervals of 20 minutes or so, to expel air and refill its lungs. When the water evaporates at the end of the rainy season, this frog aestivates in the soft damp mud, retained under the hard dry caked crust, crowded together in the last area to dry. *Xenopus* also lives in permanent artificial ponds, such as the Domasi fish culture ponds nearby and does not aestivate, but exists actively in the water throughout the year. *X. muelleri* is large (58–72 mm), smooth and slippery, its skin being covered with mucus glands. The skin is highly vascularized for accessory respiration while submerged in water or aestivating in wet mud. The frog has a long larval history (six to eight weeks). It is absent from the lake itself, since its skin is permeable and it cannot be exposed to saline water without dehydration.

(2) The commonest frog of the shallow lake edge is the small puddle frog, *Phrynobatrachus acridoides* (Cope) (Stewart 1967) (Figs. 14.1D and 14.2C). It moves along with the very shallow almost fresh water as it advances over the floodplain during the heavy rains. It is a small (23–30 mm) well-camouflaged frog, being grey-brown with green and tan markings. Like *Xenopus*, the puddle frog's hind feet are completely webbed. In addition its toes end in bulbous swellings, which may enable it to move over soft mud. This frog does not remain under water and feeds primarily on terrestrial and semi-aquatic shoreline insects. The very tiny, black eggs are deposited in a jelly which floats under the water surface, and the tadpoles frequent the very shallow edges of the lake.

249

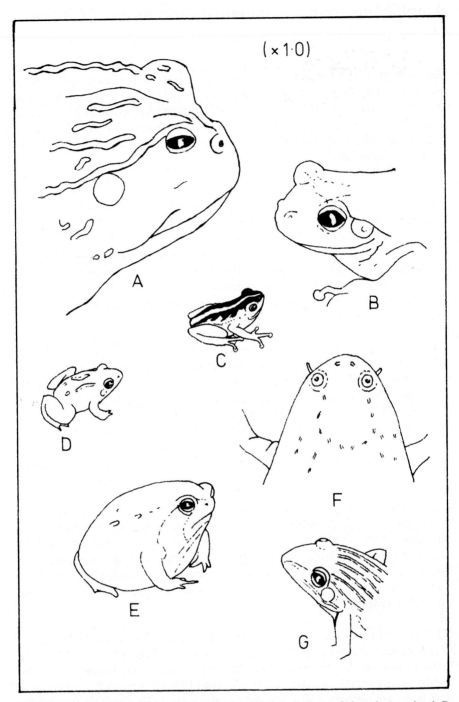

Fig. 14.1 Amphibia of the Lake Chilwa area: natural size. A. *Pyxicephalus adspersus* head; B. *Chiromantis xeramphelina*; C. *Hyperolius parallelus albofasciatus*; D. *Phrynobatrachus acridoides*; E. *Breviceps poweri*; F. *Xenopus muelleri*; G. *Ptychodaena mascareniensis* (after Stewart 1967; Pienaar et al. 1976).

The puddle frog survives the dry season along perennial stream edges under stones and in crevices.

(3) The mascarene ridged-frog, *Ptychadena mascareniensis* (Dumeril and Bibron) (Figs. 14.1G and 14.2B), is the most abundant frog around the mouths of permanent streams (Stevens, 1974). Although somewhat tolerant of brackish water it lays its black and white eggs along the relatively fresh waters of the marsh edge during the rains. *P. mascareniensis* is a large (33–48 mm) active frog, leaping away speedily when harassed – most often away from the water. Because of its colour pattern it is as difficult to locate in the floodplain grasses as it is in the water. Its diet consists principally of terrestrial and semi-aquatic insects and small amphibians. When its habitat dries up the mascarene frog digs into the ground with the assistance of small tubercles on its hind feet. Here it aestivates until the rains.

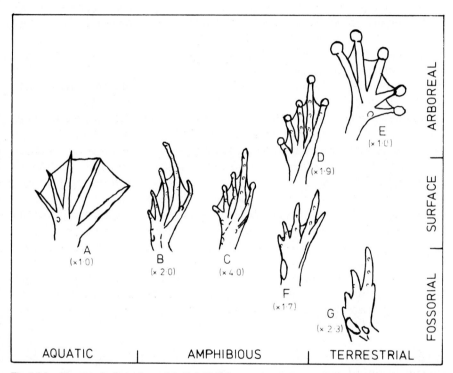

Fig. 14.2   Structural adaptations of the left hind foot to the gradient of habitats from water to land. A. *Xenopus muelleri*; B. *Ptychodaena mascareniensis*; C. *Phrynobatrachus acridoides*; D. *Hyperolius parallelus albofasciatus*; E. *Chiromantis xeramphelina*; F. *Leptopelis bocagei*; G. *Breviceps poweri* (after Rose 1950; Stewart 1967). (Original diagrams reduced to two thirds.)

(4) A sedge frog, *Hyperolius parallelus albofasciatus* (Hoffman) (Figs 14.1C and 14.2D), occurs along the swamp edges (Blake 1965). Its feet are extensively webbed and it has large toe disks. In the rainy season it rests on the vertical stems of bulrushes above the water, but enters the water to breed. Thin hygroscopic skin occurs on the area of the inner flanks of the thighs which are bright red in colour. During the day these flanks are folded away as the frog clings vertically to reeds. At dusk this sedge frog becomes active, feeding on

insects and calling throughout the night. During these activity periods the flanks are exposed and it is expected that water is absorbed (Moore 1964). In late October this small (20–35 mm) frog may be found clinging to the branches of leafless *Acacia* spp. shrubs on the termitaria of the floodplain more than 800 metres from the wet marsh of the lake edge. This suggests that it may aestivate in termitaria.

(5) The largest frog found in Southern Africa is the African bullfrog, *Pyxicephalus adspersus* (Tschudi) (Fig. 14.1A), the males occasionally being more than 120 mm long. The bullfrog is a fossorial species, highly adapted to a short variable rainy season. Adults emerge from hibernation to breed and congregate around very shallow temporary pools on the floodplains. The males, which are larger than the females, are described by Balinsky & Balinsky (1954) as aggressively defending territorial areas within these pools. During mating, the female, while standing in water two to three centimetres deep, smears her eggs on the thighs of the male. Fertilization takes place above the water ('dry fertilization') and the eggs then slide down into the water. The male 'guards' the eggs until they hatch into tadpoles which swarm in a mass around the male. They move with him through the pond and his size and aggression protect them from tadpole-eating frogs such as *Xenopus*. This 'paternal care' may compensate for the hazardous environment of temporary rainpools. Other authors, Loveridge (1950), Wager (1965), Grobler (1972) and Pienaar, Passmore & Carruthers (1976) suggest that this habit might involve cannibalism, although this is a feature of the immature frogs. The most relevant feature of the bullfrog is the very short larval life before metamorphosis, which can be as little as 18–30 days. Balinsky & Balinsky (1954) observed the breeding places of bullfrogs over many years and found that only on about 10 per cent of the occasions were froglets successfully reared before the habitat dried up. Between these short breeding seasons, the adults live underground in burrows which are constructed with strong hind legs and shovel-like processes on each heel. During the rainy season it spends the day at the entrance to this burrow with its eyes and nose exposed. The adults emerge at night to feed on anything that is the correct size. *P. adspersus* hibernates encased in a hard mud shell within its burrow. They are preyed upon by pelicans (Mitchell 1946) and the nile monitor (Loveridge 1950).

(6) Bocage's burrowing frog, *Leptopelis bocagei* (Gunther) (Fig. 14.2F), goes a step further in independence of water, which can be so unreliable on the Chilwa floodplain. This frog is the only burrowing member of an otherwise arboreal genus. It is compact and toad-like in form (53–60 mm), thick skinned and without the usual *Leptopelis* adhesive toe disks. The hind legs are shortened and the back feet have a large flattened tubercle to assist in digging. It forages for terrestrial insects and calls from the ground. Mating is also terrestrial at night, the male holding on to the female by means of a sticky secretion from its ventral glands rather than by using rough nuptial pads on the forearms as aquatic frogs do. The eggs are laid during the rains in deep holes in the ground. The development of *L. bocagei* was said not to be known by Stevens (1974), but Linden (1971) has described that of a fossorial species of *Leptopelis*, which may be identical. It was called *L. viridis cinnamoneus* (Bocage) by Poynton

(1964) almost certainly mistakenly, since it had tapering fingers and no digital disks when adult. The taxomomic status is uncertain (Schiøtz 1975, Stevens 1974). The eggs of this fossorial *Leptopelis* are large (8 mm dia.) and yolky and are laid in deep holes in the ground; and they will die if placed in water. After 30 days development inside the egg membrane, the tadpoles hatch in an advanced state and complete their metamorphosis in the water trapped in ground holes. The froglets are bright green and possess finger disks like the arboreal species of the genus. However, within ten days they recapitulate the phylogeny of their arboreal ancestors by losing their disks and turning dark brown like the soil; and then they disappear underground.

(7) The great grey tree frog, *Chiromantis xeramphelina* Peters (Figs. 14.1B and 14.2E), achieves as an adult, an emancipation from water that no other species has been reported to possess – the ability to excrete uric acid, an insoluble nitrogenous excretory product requiring no water (Loveridge 1970). This is an adaptation similar to that of terrestrial reptiles. *Chiromantis* is, however, tied to water to complete its life history because the larval stages complete their development in temporary pools. The adults, sometimes with several pairs in amplexus, produce a foam nest in trees or stout reeds overhanging temporary freshwater rain pools (Pienaar et al. 1976). The surface of the nest hardens while the interior remains fluid preserving a moist environment for the eggs. The eggs hatch in the nest in approximately five days and the tadpoles drop to the water below.

The adults are highly adapted for an arboreal existence as all digits have large adhesive disks and the forelegs have opposable fingers like that of the chameleon. They are well camouflaged being usually pale mottled grey resembling the bark of a tree. They are capable of rapid colour changes, however, and may change from almost white to dark mottled brown depending on the background and the light conditions (Pienaar et al. 1976) Aestivation takes place in the trees (Stevens 1974). The arboreal snakes, the boomslang (*Dispholidus typus*) and the vine snake (*T. capensis*), are the giant tree frog's principal enemies (Sweeney 1971). The adults have been seen on the woodland fringes of the Chilwa basin and in the ditches along the Chilwa–Zomba road, but are not known to occur on the floodplains where trees are generally absent.

(8) The final step towards adaptation to a terrestrial environment is seen in Power's rain frog, *Breviceps poweri* Parker (Figs. 14.1E and 14.2G). It is a small (39–44 mm) fossorial species living on the edge of the Chilwa floodplain. *B. poweri* cannot swim if placed in water and its toes are without webs, with the inner toes extremely reduced. It can, however, inflate its lungs excessively and blow itself up to the shape of a ball (Stewart 1967, Pienaar et al. 1976), and it can float. With tubercles on the inner metatarsals and stout short legs, the animal is adapted for burrowing, very much like a mole. The rain frog spends most of its life underground only coming out after heavy rains. It feeds on ants, termites and beetles. The large yolky eggs are laid in underground cavities where the tadpoles remain inside the egg membranes until metamorphosis (30 to 40 days) (Wager 1965). An aquatic environment is unnecessary for development.

Summary

When lake recessions occur *Phrynobatrachus* and *Ptychodaena* lose a large part of the area of the environment usually available for breeding and tadpole development. *Xenopus, Pyxicephalus* and *Chiromantis*, which depend on rain pools will complete their life-histories only if the pools persist for the time required from breeding to metamorphosis, which is about 9–10 weeks for *Xenopus*, three weeks for *Pyxicephalus* and 3–4 weeks for *Chiromantis*, since it breeds in trees and the tadpoles are ready to grow legs when they drop into pools. Populations of *Xenopus* will be reduced in dry years, but will be later augmented by those which lived in shallow permanent ponds. *Pyxicephalus* is probably most affected if rainfall is intermittent, or initial heavy rains are followed by a protracted dry period. *Chiromantis* is a border-line inhabitant and depends on trees or bushes and reeds as well as pools at the edge of the floodplain for breeding. Reproduction in *Leptopelis* and *Breviceps* will be little affected by low rainfall, but their sources of food would be depleted.

2. Reptiles: adaptations to environmental fluctuations *by* C. O. Dudley

Seven species of reptile which are particularly exposed to the fluctuations of rainfall and lake levels have been selected for discussion. They include one aquatic species, four amphibious, one fossorial and one species which lives on open ground. Nineteen local species (Stevens 1974) are listed in Appendix A.

(1) The soft-shelled turtle, *Cycloderma frenatum* Peters, inhabits the main body of the lake and probably the mouths of the larger feeder rivers (Fig. 14.3C). This is the largest turtle (50–60 cm) in Central Africa and is adapted for a completely aquatic existence. The forelimbs are paddle-like, the hind feet are webbed and the very long neck (nearly equal to body length) has its nostrils at the tip so that the snout can be protruded into the air while the turtle is still buried in the mud or under shallow water. This species has strong accessory aquatic respiration (Bellairs 1969; Rose 1950; Sweeney 1960). The lining of the pharynx is raised into numerous villi and is highly vascular. The pharynx is ventilated by muscles inserted on the large hyoid bones with their origin on the cervical vertebrae. Water is pumped in and out of the mouth up to 16 times a minute without flooding the lungs. Sweeney (1960) states that the soft-shelled turtle can survive on dissolved oxygen for up to 12 hours without breathing air.

It endures the recessions of the lake, however, because it can aestivate in wet mud, burying itself completely below the surface and breathing through its vascular skin. During this period the turtle is completely retracted within its carapace and plastron, the latter having flaps to seal the animal inside. *C. frenatum* can withstand complete anoxia for several hours and can fast for a whole year (Gans & Dawson 1976) and so it could revive after aestivation when the lake recovers. As the skin is permeable to water it cannot survive complete exposure to the air. It is also permeable to sodium ions and should the lake exceed the turtles' own osmotic pressure it would need to excrete sodium ions. Whether this turtle possesses salt-secreting lacrymal glands like many other reptiles is unknown (Gans & Dawson 1976). Its principal food is the large mollusc, *Lanistes ovum* which was plentiful before the lake dried and has since

254

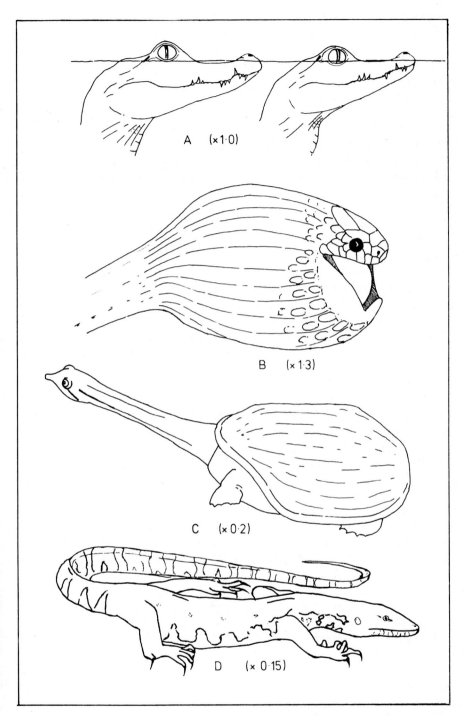

Fig. 14.3 Reptilia of the Lake Chilwa area: A. *Crocodylus niloticus*, immature (adapted from Carr 1968 and Rose 1950); B. *Dasypeltis scabra*, eating an egg (after Carr 1968); C. *Cycloderma frenatum*, immature (adapted from Rose 1950); D. *Varanus niloticus* (after Broadley 1972). (Original diagrams reduced to two thirds.)

re-emerged from aestivation in the mud (Chapter 9). Small fish and aquatic insects are also included in its diet (Sweeney 1960).

(2) The smaller (20–30 cm), amphibious serrated terrapin, *Pelusios sinuatus* (Smith), occupies parts of the swamp and river marshes. The plastron is hinged anteriorly so that when this 'flap' is closed with retracted head and neck twisted sideways, the animal is completely protected. It feeds on fish, frogs, and insects as well as on the floating water lettuce, *Pistia*, which lines the swamp channels and lagoons. This terrapin is able to remain submerged for up to two hours because of the specialized properties of its haemoglobin (Gans & Dawson 1976). During the dry season when the water level recedes, it burrows into the mud.

(3) The nile crocodile, *Crocodylus niloticus* Laurentia (Fig. 14.3A), is now very rare in Lake Chilwa. Yet, only a few years ago one just north of Kachulu harbour killed a scoop-net fisherman (The Times, Blantyre, Aug. 16th 1971). Although it may not have been extremely abundant in the last century due to the lake's extensive *Typha* swamp, the crocodile's rareness now is undoubtedly due to intensive hunting over the last 50 years, partly for human protection and partly for the value of the hide in the skin trade. The remaining crocodiles are confined to the river deltas within the swamp. These areas may be its refuge during dry phases of the lake. Guggisberg (1972) reports that crocodiles may aestivate by digging deep into the mud or may migrate to wetter areas during periods of water shortage and very likely Chilwa crocodiles can cross the sand bar to Lake Chiuta.

The crocodile is not uricotelic (excreting uric acid) like terrestrial lizards, snakes and tortoises and so requires water for urea excretion (Bellairs 1969). Their eggs, which have calcareous shells are porous like those of birds, are laid and incubated on the land in sand or mud. The young feed on amphibians and insects while the adults feed principally on catfish, though occasionally taking small mammals.

(4) The nile monitor, *Varanus n. niloticus* (L.) (Fig. 14.3D), is still fairly common in the more uninhabited areas of the lake shore and stream banks and is never far from water (Howard-Williams 1973). Individuals around Lake Chilwa seldom are longer than one metre. The young monitors feed on insects and frogs while the larger individuals feed on eggs of all kinds, young terrapins and other reptiles. Where close to villages, such as on Nchisi Island, monitors will often kill chickens. Monitors can move considerable distances overland and, during periods of drought in the lake, they undoubtedly move up river courses and to isolated lagoons.

(5) The African python, *Python sebae* (Gmelin), can still be found in small numbers in the vicinity of streams, rivers and lake edges in the Chilwa Basin in which it swims with great speed. This is the largest of the African snakes but it would be unusual indeed to find one greater than three metres in length in this area nowadays. In the early part of the century the snake was common, and in 1930, Chidiampiri Island (north of Nchisi Island) was declared a protected area and re-named Python Island. Since then the plains and islands have become

increasingly inhabited by man and Python Island, for which protection was not implemented, was eventually de-proclaimed in 1950 (Hayes 1972). Today the African python is uncommon although occasionally individuals are trapped in gill nets within the swamps (Howard-Williams 1973). Immature individuals feed on frogs and toads. Adults prefer warm blooded animals such as dassies, hares, cane rats, and small antelopes. During the dry season the snake may aestivate two to four months without eating or drinking (FitzSimmons 1962).

(6) The lowland viper, *Atheris superciliaris* (Peters), is a common snake of the swamp margins and floodplains. It inhabits rodent burrows in the more sandy soils and its main food includes amphibians and small rodents (Stevens 1973). This viper illustrates a peculiar specialization in snakes living in an area of variable rainfall and, thus, variable food supply. The female possesses special receptacles for the retention of sperm after mating, and fertilization may be delayed more than a year until the season is favourable and there is abundant food for her offspring. Additional flexibility in reproduction is available as the female can retain her eggs within the oviduct for a variable amount of time, laying the eggs only after the rains arrive. Eggs hatch immediately after they are laid and the young snakes feed on froglets. Stevens (1973) noted signs of retraherence, i.e. withdrawal from activity, in captive adults and suggested that this snake may aestivate, perhaps in disused rodent burrows.

Inhabiting the same rodent burrows as *A. superciliaris*, is the egg-eating snake, *Dasypeltis s. scabra* (L) (Fig. 14.3B). This species mimics the low-land viper in both colour pattern and terrestrial habits. This pattern is not genetically fixed perhaps since specimens from nearby Mulanje District mimic the night adder and are arboreal (Stevens 1974).

(7) The puff adder, *Bitis arietans arietans* (Merrem), favours the drier edges of the floodplains, the secondary grasslands and woodlands. It is largely independent of water and feeds predominantly on rodents rather than amphibians. The species is viviparous, the young developing in placenta-like conditions within the mother (Bellairs 1969). The eggs are, thus, protected from predators and environmental hazards. Packard, Tracey & Roth (1977) reviewed the relationship between climate and reproduction in reptiles and maintained that viviparity is characteristic in regions of low and/or variable rainfall. Because of its high reproductive potential (20–80 brood) the puff adder remains fairly common in the area. According to FitzSimmons (1962) the puff adder aestivates during unfavourable conditions. This snake is only of moderate length (90–100 cm) but is extremely stout, powerful and poisonous.

Summary

The aquatic reptiles, briefly described, demonstrate adaptation in respiration and aestivation to recessions in the lake. The crocodile and lizard show different degrees of dependence on water, and are capable of seeking other waters when Lake Chilwa dries. The amphibious locomotion of the python requires water for hunting when young, but the lowland viper has withdrawn from water and lives in burrows, while dependent on amphibians for food. Finally the puff adder thrives on the Chilwa plains because it has carried the reproductive adaptation, foreshadowed in the viper cited, to the logical conclusion of vivi-parity.

3. The decline of the mammalian fauna *by* C. O. Dudley

Until the end of the nineteenth century the natural vegetation of the Chilwa basin was a varied one, ranging from densely-forested mountains to woodland scrub on the foothills and open savanna woodland on the plains, to the grass-lands of the floodplain to the marshes, swamp and lake (Jackson 1972, Howard-Williams 1977). Within this area the larger mammals were as diverse and numerous, at times, as anywhere in tropical Africa. Among explorers who commented on the diversity of its mammalian fauna, some approached from the west and south, such as Drummond (1903) and one, at least, walked along the east coast from the Indian Ocean to the well-wooded sand bar (O'Neill 1884).

This section on mammals cannot be restricted to the lake/swamp/floodplain ecosystem alone, since the larger mammals require an extensive heterogeneous area. They were not tied to the waters of Lake Chilwa, although 'they relished its salty taste' (Drummond 1903). They roamed the hills and plains and wandered north to Lake Chiuta and Lake Amaramba. Even hippos are known to move about 12 km away from permanent water and to remain in pools until the onset of the dry season. Since the larger mammals shared the ecosystem under review in this monograph, their decline is considered relevant to these studies on change.

Table 14.1 lists the larger mammals and rodents reported to have occurred in the Chilwa basin in the 19th and 20th centuries (Maugham 1929, Lyell 1923). Those which have been seen more recently on the floodplain and in the swamps have been singled out in the Table: Hayes in 1925 (pers. comm.), Garson (1960), Sweeney (1965), Howard-Williams (1973) and Stead & Dudley (1977). The factors which have contributed to the reduction of the 28 species of mammals to the few species which still survive in the water or in the swamps and floodplain will be discussed.

3.1 *Changes in the environment of the mammalian fauna*

The period under review is the last hundred years. The recessions of Lake Chilwa and the high levels of 1859 and 1978 have been discussed in Chapter 3. The area influenced by the lake has remained within the 4–5 metre terrace on the shore which defines the outline of our ecosystem. Since other permanent waters were available to mammals in the ten permanent streams, as well as the neighbouring Lakes Chiuta and Amaramba to the north, the recessions of the lake cannot have reduced the fauna to a marked extent. Their distribution will, however, have been affected at times, such as in 1883 when O'Neill saw large numbers only on the plains to the northwest of the sandbar near Lake Chiuta, after the marked recession of 1879. On the other hand, changes in the environment, brought about by man, did contribute to the decline of the fauna both directly (through the use of the gun) and indirectly through man-made changes in the vegetation.

The N'yanja people who lived on the Chilwa plains practised 'shifting

cultivation' until the middle of the twentieth century, when increasing alienation of the land and larger populations induced the practice of permanent
settlement. The size of the population on the plains had fluctuated from time to
time, depending on invasions, emigrations, the slave trade and immigrations
(see Chapter 16). Until the arrival of the white settlers and missionaries in the
latter half of the nineteenth century, man and animal had lived together in a
certain measure of 'equilibrium'; they did not possess the means, nor the desire
to exterminate species of mammals.

Table 14.1 Larger mammals and rodents reported from the Lake Chilwa basin below 1000 metres
in the 19th and 20th centuries. Those species marked with an asterisk (*) still exist in small numbers.

*Yellow Baboon	*Papio cynocephalus* (L.)
*Vervet Monkey	*Cercopithecus aethiops* (L.)
Side-striped Jackal 1	*Canis adustus* Sundevall
*Cape clawless otter	*Aonyx capensis* (Schinz)
*Civet-cat 1	*Viverra civetta* Schreber
*Water Mongoose 2	*Atilax paludinosus* (Cuvier)
*Spotted Hyaena 1	*Crocuta crocuta* (Erxleben)
Lion	*Panthera leo* (L.)
Leopard	*P. pardus* (L.)
African Elephant	*Loxodonta africana* (Blumenbach)
Bushpig	*Potamochoerus porcus* (L.)
Warthog	*Phacochoerus aethiopicus* (Pallas)
*Hippo 2	*Hippopotamus amphibius* L.
Grey Duiker 3, 4	*Sylvicapra grimmia* (L.)
Oribi	*Ourebia ourebi* (Zimmermann)
Klipspringer	*Oreotragus oreotragus* (Zimmermann)
*Reedbuck 2, 3, 4	*Redunca arundinum* (Boddaert)
Waterbuck	*Kobus ellipsiprymnus* (Ogilby)
Impala	*Aepyceros melampus* (Lichtenstein)
Sable Antelope 4	*Hippotragus niger* (Harris)
Lichtenstein's Hartebeest 4	*Alcelaphus lichtensteini* (Peters)
Eland	*Taurotragus oryx* (Pallas)
Bushbuck	*Tragelaphus scriptus* (Pallas)
Greater Kudu 4	*T. strepsiceros* (Pallas)
Nyasa Brindled Gnu	*Connochaetes taurinus johnstoni* Sclater
African Buffalo	*Syncerus caffer* (Sparrman)
Black Rhino	*Diceros bicornis* (L.)
Burchell's Zebra	*Equus burchelli* (Gray)
*Gerbil 5	*Tatera leucogaster* (Peters)
*Common rat 5	*Mastomys natalensis* (Smith)

1 Garson 1960	4 Hayes 1925
2 Howard-Williams 1973	5 Hanney 1965
3 Sweeney 1965	

The African culture in Malawi incorporated a type of 'animal conservation'
system (Hayes 1972). It was demonstrable in the seasonal control of fishing by
the Chiefs (Vaughan 1978) and observed in 1967 in the recession by R. G. Kirk
(pers. comm.). There was a long-established custom among the N'yanja, the
permanent inhabitants of the plains, of adopting the name of an animal as a
symbol. Any animal so adopted was protected from being killed, since by
tradition, a man was not allowed to kill animals of his own, or a neighbour's
clan name; and the number of hunters was limited to the line of inheritance of

those who possessed the 'magic' for hunting (Schoffeleers cited in Hayes 1972). Even so, permission to hunt had to be obtained from the Chief who allowed the killing of only one animal at a time. The people of the plain were fish-eaters, so that hunting supplemented their normal diet only on special occasions.

The Chief's control of hunting broke down under the impact of 'foreign' hunters: first Arabs started the ivory trade, followed by Yao invaders from the north (see Chapter 16) and then missionaries continued (Robertson 1970), and finally the British Colonial Administration removed control of 'game' from the Chiefs to its own administration in 1891.

From this date, the *status quo* of man and mammals was drastically changed. In the eighteenth and nineteenth centuries the Chilwa plains had been under-populated (Hetherwick 1887, Drummond 1903) and probably mammals had flourished without very much disturbance. When peace was restored, the N'yanja people returned to the plains and the white settlers opened large farms on the Shire Highlands, on the watershed of the lake. From then the mammalian habitats were slowly changed. The well-wooded hills and plains would have favoured the browsers: kudu, bushbuck and rhino and the mixed-feeding herbivores, such as the elephant and eland, and the smaller buck: impala, duiker and klipspringer. The plains supported grazers such as gnu, zebra, oribi, sable antelope and buffalo. Hippos were numerous around the edges of the lake and in the swamp and streams, and waterbuck grazed the reeds. In the hilly areas warthog and bushpig fed on roots and bulbs; and insects, seeds and fruits supplied the monkeys and baboons. Leopard and lion were the carnivores on whose kill the hyaena depended, and otters fished the streams (Sweeney 1959). Areas of woodland were slowly cleared for 'shifting cultivation', or by seeking timber for the framework of mud-houses and for fuel. Woodland and forests were cleared more rapidly for large-scale farming and later for afforestation with pine, cedar and eucalyptus. Secondary grassland grew and regenerated woodland developed sparsely.

3.2 *The record of the destruction of the mammalian fauna*

This rich mammalian fauna was exterminated during the first thirty years of the British Colonial Administration of Nyasaland. The Chilwa area was only 30–40 km from the Administrative Capital of Zomba and the farms on the Shire Highlands. The custom among officials, settlers and visitors was to spend weekends hunting game on the Chilwa plains, with the best firearms that could be imported (Maugham 1929). The rising impact on the big game is reflected in the official ivory export figures during the last decade of the nineteenth century. By 1893 ivory accounted for 83 per cent of the country's export trade (Baker 1961). Johnston (1897) estimated that almost as much was being exported illegally and the Chilwa region adjacent to Mozambique and the Indian Ocean must have played a part. Between 1891 and 1904 a total of almost 73 tonnes of ivory, representing perhaps up to 3000 elephants was officially exported. By 1899 elephant was already becoming scarce, and trade in ivory fell remarkably quickly. Similar export trends were reported on a smaller scale for teeth of hippo and rhinoceros horn. Johnston (1894) had advocated the 'merciless extermination of hippo' and by 1897, he was able to report that they were no longer a source of danger to navigation on the Shire River (Johnston 1897) and they were only occasionally seen in Lake Chilwa.

The first animal known to have become extinct was the Nyasa Brindled Gnu, an antelope limited to the short grass plains around the perimeter of the lake (Figure 14.4). The decline was noticed by the Administrator Johnston (1897) 'The gnu seems to diminish rapidly before the European sportsmen and in places where it was once abundant (it) is no longer met with...'. In that year the Administration proclaimed two protected areas in Nyasaland and one of them was the Phalombe plain, south of Lake Chilwa. This action was unpopular with the local N'yanja people, because it removed good agricultural land from use. The area was deproclaimed in 1902.

Fig. 14.4 The Nyasa Brindled Gnu, *Connochaetes taurinus johnstoni*, an early 20th century casualty in colonial hunting (from an early sketch by H. H. Johnston 1897).

The annual Handbook of Nyasaland (Murray 1908) reported that with the exception of the elephant, game was still plentiful in the district, and professional hunters such as Lyell (1912) were recommending the Colony as an ideal place for the Big Game Hunter, where game laws were the best (for the hunter) of anywhere in Central Africa (Fig. 14.5).

By 1922, the game populations were deteriorating throughout the Southern Region (Murray 1922). No hippo was recorded from Lake Chilwa and only small numbers of kudu, sable antelope, reedbuck and hartebeest were still to be found in the forested Chikala Hills to the northwest of the lake and in the Sombani woodlands to the southwest. One might have thought that the Chikala Hills were fairly inaccessible in those days, but according to G. D. Hayes (pers.

Fig. 14.5 The White Hunter with his trophies. Sketch from a photograph by Sutherland (1912).

comm.) 'the old Zomba/Malakotera military road passed through' so that with guns handy and easy targets available, the mammals suffered.

Game was scarce even in the Chikala Hills by 1930 and only warthog, bushpig, baboon and bushbuck and the occasional lion or leopard were seen on the western slopes of the Chilwa basin (Murray 1932). Sweeney (1965) after extensive flights over the floodplains of the north and south of Lake Chilwa could report seeing only 15 reedbuck, 1 bushbuck and several duikers. During the last ten years of intensive field work on the floodplain, swamp and lake, the only mammals recorded have been hippo, otter, reedbuck and water mongoose (Howard-Williams 1973) and hyaena (Hobbs 1976).

Hanney (1965) in a survey of rodents of Malawi, found large colonies of the gerbil, *Tatera leucogaster* (Peters) on the floodplain of Chilwa and the common field rat *Mastomys natalensis* (Smith) around houses near the lakeshore. These days, rodents are not unimportant, since in times of lake recession they are zealously hunted by digging by the people who suffer from shortage of fish (see Chapter 18). It is the custom to hunt rodents in the dry season every year, by burning the grass and digging.

The eastern plains of the Chilwa basin, Mozambique suffered a similar decimation of the mammalian fauna, perhaps less intensively, since when lion and leopard are occasionally seen around Zomba mountain, it is usually reported that they have come over from Mozambique across the sand bar. This has happened several times in the last ten years.

Another possible cause of the decline of the larger mammals has to be considered, namely, the possibility of disease. Tsetse fly, which carries a protozoan causing trypanosomiasis (sleeping sickness in man and n'gana in animals) had been mentioned in reports, but in relation to sickness in man (Duff 1903) but *Glossina mortisans* is restricted to dry bush areas. An outbreak of rinderpest, which affects ruminants, occurred in Central Africa in the later nineteenth century, but it was claimed to have by-passed Nyasaland (Johnston 1897); and Hobbs (1976) after examining historical records agreed that the 'epidemic had little direct effect on Nyasaland's faunal communities'.

In spite of this record of destruction and lack of conservation, the wildlife of the Chilwa plain has not been entirely lost to Southern Malawi. Since 1947 the National Faunal Preservation Society urged the various governments to pursue an active conservation policy. Game Reserves and National Parks have been established, the latest of which is Liwonde National Park, declared in 1972. It is only 40 km northwest of Lake Chilwa, beyond the Chikala Hills. Today, thanks to protection, this Park is populated by most of those species which nineteenth-century explorers reported in the Chilwa basin (Dudley & Stead 1976, Stead & Dudley 1977).

3.3 Pastureland for cattle

Cattle were first introduced into this area in 1880 (MacDonald 1882) and instead of the mixed herds of game, only cattle in increasing numbers now find pasture on the Chilwa plains (see Chapter 17). The richness of the plains (and lake) will have suffered from loss of nutrients when the droppings of the game were no longer available to the Chilwa system. McLachlan (1971) attributed an important role to the release of nutrients from dry grass and hippo and elephant dung from the Lake Kariba shoreline. The present increase in cattle on the Chilwa plains may alleviate this situation.

Grazing by a mixed community of herbivores results in high levels of nutrients in the soil and thus high primary productivity of the grasslands (Bell 1969, Vesey FitzGerald 1969). By means of a successional grazing pattern, wild herbivores develop a mosaic of mixed species of grass, some palatable and others less so. A variety of palatable grasses still exists on the Chilwa plains today (see Chapter 17). The composition of the grasslands may change as a result of this ecological simplification of grazing by one species only, but how far degradation might proceed is not certain.

In a recent study in the Serengeti of Tanzania (McNaughton 1978), the effect on the pasture grasses of unselective feeders like buffalo and gnu were compared with that of selective feeders such as zebra and gazelle. It was found that, when isolated, the non-selective grazers preserved the mosaic of grass species rather better than did those which selected mainly palatable grasses. Cattle are non-selective grazers and perhaps when numbers are well managed, they may not degrade the pasture irrevocably.

The natural rotation pattern, enforced on cattle management by the inundation of the floodplain in the rainy season, allows both floodplain and drier grassland to rest seasonally, as described in Chapter 17. However, Vesey FitzGerald (1970) observes that 'At present the perimeter grasslands (of the Chilwa area) show signs of being overgrazed by cattle which, however, hardly use some of the natural pastures on the floodplain.' This is probably the result of unplanned herding patterns and the location of dip tanks and kraals where the cattle are sheltered overnight on the way to market. Perhaps intensive, well-managed husbandry on the Chilwa plains will succeed in maintaining a good standard of productivity on those parts of the plain and floodplain which are set aside for stock. The trend has been, and still is, however, for more and more land to be put under cultivation of staple crops for human consumption. This is not unexpected in an area where fishing is highly productive and meat-eating is a habit which has not been cultivated, nor can yet be afforded. Cattle will be maintained to increase the number of oxen used as draught animals and for exports of cattle products to urban centres.

4. The birds of the Chilwa area *by* D. E. Stead & G. G. M. Schulten

The East African lakes, swamps and marshes with their rich flora and fauna provide many habitats for large and varied populations of birds. Lake Chilwa is no exception to this and has over 150 species of resident birds. Several species of birds of the northern hemisphere occur in the eastern parts of Central Africa and some species remain at Lake Chilwa for a short period on their way to or from wintering grounds further south; others stay in the area during the palearctic winter (Moreau 1972). At Lake Chilwa thirty-seven species of Palearctic birds have been observed, of which twenty-two are regular visitors between September and April. In addition, the area is visited by some species which can be considered as inter-African migrants. A check list of species after Schulten & Harrison (1975), including early records (Belcher 1930, Benson 1953), is given in Appendix A. Throughout this section English common names (after Roberts 1940) are used and the scientific names are given in the Appendix.

The Chilwa area may be divided into the following bird habitats: (a) the open water of the lake, (b) the *Typha* swamp, (c) the marshes, (d) the floodplain grasslands, (e) disturbed areas, including rice paddies and man-made 'lagoons'.

A description of the vegetation and environmental features of these areas is given in Chapter 7; further details of the floodplain will be found in the discussion of the locust habitat in Chapter 15.3 and in relation to the pastures for cattle in Chapter 17.6.

Three factors which cause changes in bird populations will be discussed: (1) the seasonal rise and fall of the lake level, which has its greatest impact on the floodplains that are inundated in the wet season and dry out completely in the dry season; (2) the very variable lake levels – those of the major recession in 1966–68, when successive seasons with low rainfall culminated in the lake bed drying completely; and also, the minor recession of 1973, which was followed by several years with maximum lake levels higher than had been recorded before the major recession; (3) the increasing intensity of rice cultivation on the floodplain during the last ten years which has provided a rich source of food for ducks and seed-eaters, so as to increase their populations beyond those usually found in natural areas of swamp and lake.

4.1 *Open water*

The open water is largely used by fish predators (Table 14.2). The Pink-backed Pelican is quite common, usually fishing close to the swamp. The White Pelican was considered common before the lake dried out in 1966–68, but is now very rare (Schulten & Harrison 1975). The Grey-headed Gull is numerous, especially in Kachulu Bay, the major fishing centre. Two terns occur at present, during the years of high lake level: the migrant White-winged Black Tern is very common from September to April, and in non-breeding dress it is very difficult to distinguish from the Whiskered Tern, which was known to breed at Lake Chilwa in 1924. There were no further records until 1960. Several birds in breeding plumage were seen in November in 1966 and 1967 and from April to November in 1969 and 1974. Benson & Benson (1977) regard the species as nomadic rather than migratory and it is interesting to note that its presence at Lake Chilwa coincided with phases of lake recovery after low lake levels in 1922, 1960, 1969 and 1973.

The single Malawian record for the Gull-billed Tern is from Lake Chilwa (Schulten 1972). The African Skimmer is an occasional visitor, but the lake is very probably too turbulent for this species to fish effectively.

The Reed Cormorant is the only common cormorant. The White-fronted Cormorant is uncommon, since it prefers clearer water as in Lake Malawi. Probably the turbidity of Lake Chilwa is responsible for the exclusion of other species such as the African Darter, which has rarely been seen.

Various duck species frequent the open water but are more often seen in the lagoons and river mouth areas. The Pygmy Goose and White-backed Duck are more restricted to the larger lagoons and swamp fringes than the other common species. The White-backed Duck appears to have declined in recent years and has not been recorded recently. Schulten & Harrison (1975) cite a local estate owner who stated that this species was the most numerous in 1937 and in the late 1940s was still numerous. A number of flocks were seen just after the lake

refilled in 1969 at the north end of the lake (C. Howard-Williams pers. comm.)
Adults and immatures were observed in December 1970 (Schulten & Harrison 1975).

4.3 *The Typha swamp*

The difficulty of access to the *Typha* swamp causes a problem to investigators of the avifauna. Although special attention was paid to swamp, no heronries

Table 14.2 Common birds of the major habitats of the Lake Chilwa Area.

1. Open water	2. Typha swamp
Pink-backed Pelican	Sedge Warbler
Pelecanus rufescens	*Acrocephalus schoenobaenus*
Grey-headed Gull	African Reed Warbler
Larus cirrocephalus	*A. baeticatus*
White-winged Black Tern	Lesser Swamp Warbler
Chlidonias leucoptera	*A. gracilirostris*
Reed Cormorant	King Reedhen
Phalcrocorax africanus	*Porphyrio porphyrio*
White-backed Duck	Little Rush Warbler
Thelassornis leuconotus	*Bradypterus baboeculus*
	Red-shouldered Widow
	Euplectes axillaris
	Black Egret
	Egretta arderiaca
	Black Crake
	Limnocorax flavirostris
	Moorhen
	Gallinula chloropus
3. Marshes	
Black Egret	Little Egret
Egretta arderiaca	*E. garzetta*
Fulvous Tree-Duck	Purple Heron
Dendrocygna bicolor	*Ardea purpurea*
White-faced Tree-Duck	Grey Heron
D. viduata	*A. cineraea*
Hottentot Teal	Black-headed Heron
Anas hottentota	*A. melanocephala*
Redbill	Sqacco Heron
A. erythrorhynca	*Ardeola ralloides*
Southern Pochard	Glossy Ibis
Netta erythrophthalma	*Plegadis falcinellus*
Moorhen	Openbill Stork
Gallinula chloropus	*Anastomus lamelligerus*
African Jacana	Red-billed Quelea
Actophilornis africanus	*Quelea quelea*
Black Crake	Orange-breasted Waxbill
Limnocorax flavirostris	*Amandava subflava*
African Water Rail	Red Bishop
Rallus caerulescens	*Euplectes orix*
The Great White Egret	Black-backed Cisticola
Egretta alba	*Cisticola galactotes*
Yellow-billed Egret	
E. intermedia	

4. **Floodplain grassland**	5. **Birds of Prey**
Wood Sandpiper	Fish Eagle
Tringa glareola	*Haliaëtus vocifer*
Marsh Sandpiper	African Marsh Harrier
T. stagnatilis	*Circus aeruginosa ranivorus*
Greenshank	Yellow-billed Kite
T. nebularia	*Milvus aegyptius*
Ruff	Lesser Kestrel
Philomachus pugnax	*Falco naumanni*
Little Stint	
Calidris minuta	
Curlew Sandpiper	
C. ferruginea	
Painted Snipe	
Rostratula benghalensis	
Ringed Plover	
Charadrius hiaticula	
Kittlits Plover	
C. pecuarius	
Blacksmith Plover	
Vanellus armatus	
Long-toed Plover	
V. crassirostris	
Red-winged Pratincole	
Gareola pratincola	

have ever been found. The Black Crake and King Reedhen are quite common, especially along the fringe adjacent to the marshes. Several species of reed warblers occur within the swamp (Table 14.2). These include the Sedge Warbler, African Reed Warbler and Lesser Swamp Warbler. The migrant Great Reed Warbler and the Little Rush Warbler are probably more common than the few records suggest. The Red-shouldered Widow is very common in dense *Typha* swamp. Many marshland species probably utilize the swamp as a resting area, such as the Black Egret which, from 1969–73, occurred along the swamp/lake fringe.

4.3 *The marshes*

This area has the most diverse vegetation and consequently bird diversity is high (Table 14.2). Marshland around Lake Chilwa is greatly affected by fluctuating lake levels, which must seriously affect populations of certain bird species. The role of neighbouring lakes as a refuge during the dry phases has not been described. Lakes Malawi and Malombe lie to the northwest and Lakes Chiuta and Amaramba are very close to Lake Chilwa to the north and do not dry out, and they may form a refuge for waterfowl. Visual observations by C. Howard-Williams and A. J. McLachlan in January 1969, just after the lake started refilling, gave the impression that there were huge migrations of duck from the south into the rising lake. The extensive swamps and marshes of the Lower Shire Valley may have been one refuge in the south when Lake Chilwa dried.

The most common duck species at present is the Fulvous Tree Duck which has probably been numerous for many years (Schulten 1974). The White-faced Tree Duck and the Hottentot Teal are also common. On most visits, Redbill and Southern Pochard are seen. Populations of the latter have apparently declined in the last seven years. In 1969 and 1970, Schulten & Harrison regarded it as the second most numerous species.

The Moorhen and African Jacana are common in the lagoons and on water lily pads. In the marsh grassland the secretive Black Crake and African Water Rail are regularly observed. Other species which are probably more common than records suggest are Baillon's Crake, the Striped Crake, and one or more species of Flufftail. Lesser Moorhen, King Reedhen and Lesser Reedhen also occur.

Three egret species use the marsh grassland: the Great White Egret, Yellow-billed Egret and the Little Egret. Of the larger herons the Purple Heron is the most common, and the Grey Heron and Black-headed Heron are often seen. Recent breeding records exist for these two species in tall blue-gum trees near Kachulu Harbour. The Squacco Heron and Black Egret are present, the latter especially in inundated grassland. Perhaps the most obvious marsh-land bird at the present time, when the lake level is very high, is the Glossy Ibis. This was also common from 1971–73, but after the low lake level in 1973, their numbers appeared to decline. Now flocks of about 300 individuals are seen regularly around Kachulu Bay. Glossy Ibis is probably seriously affected by decreases in lake level due to loss of a suitable feeding ground.

The Openbill is the only stork. They were observed eating the large aquatic snails *Lanistes ovum* by Schulten & Harrison (1975). It is quite possible that they also eat the schistosomiasis vector, *Bulinus globosus*. This snail occurs mainly in the swamp lagoons, particularly on *Nymphaea* spp. and in the small harbours at the ends of boat channels, where the Openbill is often seen feeding (see Chapter 10).

The large tracts of grassland marshes are frequented by several species of seed eaters. Flocks of Red-billed Quelea pose a potential threat to adjacent rice-fields. The Red-shouldered Widow, Orange Waxbill, Red Bishop and Black-winged Bishop are numerous. The Black-backed Cisticola is the most usual small insectivore.

Burning of the swamp and marsh is an annual practice throughout the dry season, perhaps more so just before the onset of the rains, usually in November. A recent aerial survey (November 1977) indicated that at least 10 per cent of the swamp and marsh had been burnt. Howard-Williams & Walker (1974) suggested that a third of the area might be burnt in some years. To some degree this, undoubtedly opens up the swamp to birds, facilitating roosting and possibly breeding for a number of species such as Egrets and Glossy Ibis. However, the food supply of many species may be adversely affected.

4.4 *The floodplain grassland*

The inundated grasslands provide an important feeding ground for the African and Palearctic migrant waders (Table 14.2). The effects of periodic low lake levels on the food of the waders has not, however, been studied in detail. The arrival of the Palearctic species in September–November coincides with the seasonal decreasing low levels of the lake, which leave large muddy flats pock-marked with cattle and pig spoor. These flats are rich in cattle dung and their salinity is high. The onset of the rains in November and December and the consequent filling of the lake floods this area and the waders appear to move away. Lake levels are high at the end of the rains from April to June during the northward migration, and waders are not common at this time. Dry phases probably create larger areas of suitable feeding grounds. More migrant waders were certainly present in February and March 1974, after the minor recession of 1973, than in the succeeding years when the lake continued to rise. The Greater Flamingo appears to be rare in years of high lake level, although large flocks occurred during the drying phase and filling phases of the lake (1968–70). Their decline may be associated with the change in the distribution of chironomid fauna (described in Chapter 9) when the lake rose higher.

Although the size of the wader populations may be affected by the changes in lake levels, diversity is little affected. The most common species is the Wood Sandpiper. The Marsh Sandpiper, Greenshank and Ruff appear to have decreased disproportionately since 1974, and are now seen much less frequently, more usually in August. Other regular visitors include the Little Stint, Curlew Sandpiper, Green Sandpiper and Painted Snipe. The Painted Snipe was common on the floodplain in 1969–70. One of the most conspicuous birds at that time during the dry seasons of 1969–73 was the Red-winged Pratincole sometimes found breeding. It feeds on grasshoppers and locusts.

In 1974 the Ringed-Plover was regarded as being fairly common, but since then has been seen much less frequently. The Great Snipe, which regularly occurred in rice-fields in 1972 (Schulten & Harrison 1975) has not been recorded recently in the inundated grasslands. However, several species always considered uncommon or rare still appear in most years. These include the Terek Sandpiper, Sanderling, Ruddy Turnstone and Avocet. The Black-winged Stilt is considered to be migratory, but is known to breed occasionally.

Another bird which appeared at the time of the rising lake level after the minor recession of 1973, was the Blacksmith Plover, especially around Kachulu Bay. It has not been recorded since 1975 but the Long-toed Plover appears much more in the high-level years of the lake. Schulten & Harrison (1975) regarded both species to be fairly common before and after the lake level declined.

A variety of birds of prey hunt in the Lake Chilwa basin. Resident species include the Fish-eagle, which is not as abundant as around Lake Malawi, and the African Marsh Harrier. The inter-African migrant Yellow-billed Kite is frequently seen in August and the Lesser Kestrel in some years.

Schulten & Harrison (1975) have listed only one species of vulture – the White-backed Vulture. Large numbers of this species were seen with White-headed Vultures and a few Hooded Vultures around cattle killed by hyaenas on the sandbar in 1975. The number of birds exceeded 50 individuals, but such occasions are now rare since there are few hyaenas (see section 14.3). Vultures are not resident at Lake Chilwa. The Cape Grass Owl may be a fairly common species in tall grassland. Sweeney (1965) records 20–25 individuals being flushed by a low-flying helicopter.

4.5 Disturbed areas

In recent years large parts of the floodplain have been irrigated from the perennial rivers, Likangala, Naisi and Domasi for intensive cultivation of paddy rice, and many species of seed-eaters frequent these fields. It had been reported that rice-fields were sometimes heavily damaged by ducks and geese, and so a study of the foods of several duck and goose species was undertaken from January to July 1972 (Schulten 1974). The stomach contents of 76 birds, belonging to seven species were examined: Sixteen food types were recognized of which ten were seeds from certain common plant species. Rice seeds were found in only eleven birds, with the Fulvous Tree Duck accounting for seven of them.

The Southern Pochard fed mainly on water-lily seeds (*Nymphaea* spp.) and on other seeds when they became available, such as *Typha domingensis* the dominant swamp plant, *Scirpus littoralis* in November, and *Paspalidium germinatum* in August and seeds of a leguminous tree, *Aeschonomyne pfundi* in the marshes. In November when the marshland was dry, the underground corms of *Nymphaea* were eaten. Only two pochards had eaten rice seeds in September, and these were probably gleanings after the harvest, since the seeds appeared burnt.

The Fulvous Tree Duck fed mainly on the seeds of *Ipomoea aquatica* (a wild 'water-spinach' of the same genus as the sweet potato) which grows in the marshes, and on *Typha*, *Paspalidium* and *Nymphaea* seeds. The White-faced Tree Duck fed mainly on *Paspalidium*, *Typha* and *Scirpus* seeds and some unidentified 'grass' leaves, probably rice, were found in the stomach.

The Hottentot Teal and Red-billed Teal showed a clear preference for seeds of the grass *Sacciolepis africana*, although the former also took water beetles (*Hydrophilidae* and *Gyrinidae*). Three Hottentot Teal individuals had fed almost exclusively on the *Ostracod*, *Heterocypris giesbrechti* which are very common in the flooded mud at the shoreline. They had also eaten some insect larvae and pupae (*Ceratopogonidae*) and adult beetles (*Hydrophilidae*). The White-backed Duck fed on *Paspalidium* and *Nymphaea* seeds, while the Pygmy Goose had eaten mainly *Nymphaea* seeds.

About 25 birds were shot in the rice-fields but on the whole they preferred the associated plants in the irrigation channels and ponds, which were in flower and seed at the same time as the rice was ripening. Schulten (1974) suggests that ducks may do some mechanical damage to the rice plants while landing or taking off. He considered that more data were needed for a proper evaluation of duck damage to the rice-fields. The African Spoonbill and the Black-headed Heron use the rice-fields for feeding. Reports of the food of ducks and geese in Zambia (Douthwaite 1974, 1977) and in Kenya (Watson, Singh & Parker 1970) show that the seeds of many marsh plants are eaten, including *Vossia cuspidata* which is very common in the Lake Chilwa area and that *Algae* are also important food components. The White-faced Tree Duck and Hottentot Teal of Chilwa included *Spirogyra* and similar algae found on *Typha* stems in their diet (Schulten 1974).

There is an increasing bird problem in the rice paddies caused by the Red-billed Quelea and Red-headed Quelea. They do not breed at Lake Chilwa, their nesting areas being in the southern part of Mozambique and the Northern Transvaal of South Africa (Ward 1971). Besides Queleas, the rice-fields are used for feeding by flocks of Bishop birds and Weaver birds. When the rice is ripe and awaiting harvesting, it is a familiar sight to see rice-fields dotted with children, scaring away from sunrise to sundown, any bird which threatens to land on the rice-fields.

4.6 Summary

The rich diversity of bird species in the Chilwa area is clearly related to the variety of habitats, (open lake, swamps, marshland, floodplain, ponds and lagoons) which provide a range of food, shelter and breeding sites. Fewer species than expected fish mainly in open water, because some cannot tolerate the turbidity and turbulence of the lake.

Marked differences in the numbers and species of birds have been noticed in relation to the availability of food and shelter with the varying extent of the floodplain habitat during the major and minor recessions of the lake. The seasonal change in the inundation of the floodplain is accompanied by the visits of migrants some of which are waders which stay over the dry season. It appears that the recent years of high lake maxima have increased the area of inundation, but kept the floodplain under water for longer. The drier years offer a greater area for residents.

Nomadic birds appeared for a few years just after a specially low lake level. But when the lake was very low from 1967–68, the fish-eaters such as Pelicans and mollusc eaters such as the Open-billed Stork were very reduced in number. Such a drop in numbers appeared not to occur in a minor recession such as that in 1973.

After a dry period, the ducks and geese soon re-occupy the area because their retreats are not far away, either to the north in the Lake Chiuta swamps or in the southern swamps around the Lower Shire River.

The irrigation schemes for rice cultivation have attracted many seed-eaters of the floodplain, some of which may cause economic damage.

References

Amphibia

Balinsky, B. I. & Balinsky, J. B. 1954. On the breeding habits of the South African bullfrog, *Pyxicephalus adspersus*. S. Afr. J. Science 51:55–58.

Blake, D. K. 1965. The 4th Umtali Museum expedition to Mocambique: November–December, 1964. J. Herp. Assoc. Rhodesia (1965), 23/24:31–46.

Grobler J. H. 1972. Observations on the amphibian *Pyxicephalus adspersus* Tschudi in Rhodesia. Arnoldia 6(3):1–4.

Linden, I. 1971. Development of *Leptopelis viridis cinnamoneus* (Bocage) with notes on its systematic position. Zool. Africana 6:237–242.

Loveridge, A. 1950. History and habits of the East African bullfrog. J. E. Afr. Nat. Hist. Soc. 19: 253:255.

Loveridge, J. P. 1970. Observations on nitrogenous excretion and water relations of *Chiromantis xeramphelina* Peters. Arnoldia 5(1):1–6.

Mitchell, B. L. 1946. A naturalist in Nyasaland. Nyasaland Agric. Quart. J. 6:1–47.

Moore, J. A. 1964. Physiology of Amphibia. Academic Press, N.Y. 654 pp.

Pienaar, U. de V., Passmore, N. I. & Carruthers, V. C. 1976. The Frogs of the Kruger National Park. National Parks Board of South Africa, Pretoria. 90 pp.

Poynton, J. C. 1964. Amphibia of the Nyasa–Luangwa region of Africa. Senckenberg. biol. 45 (3/5):193–225.

Russell, I. J. 1971. Function of the lateral line system in *Xenopus*. J. exp. Zool. 54:621–658.

Schiøtz, A. 1975. The Tree frogs of Eastern Africa. Steenstrupia, Copenhagen. 232 pp.

Stevens, R. A. 1974. An annotated check list of the amphibians and reptiles known to occur in south-eastern Malawi. Arnoldia 6(30):1–22.

Stewart, M. M. 1967. Amphibians of Malawi. State University of New York Press. 163 pp.
Wager, V. A. 1965. The Frogs of South Africa. Purnell and Sons, South Africa. 242 pp.

Reptiles

Bellairs, A. 1969. The Life of Reptiles. Vols. 1 and 2. Weidenfeld and Nicolson, London. 590 pp.
Broadley, D. G. 1972. The Reptiles. In: The Bundu Book, Mammals and Reptiles. Longmans, Rhodesia, pp. 93–117.
Carr, A. 1968. The Reptiles. Life-Time International. N.Y. 192 pp.
FitzSimmons, V. E. M. 1962. Snakes of Southern Africa. Purnell and Sons, Cape Town. 423 pp.
Gans, C. & Dawson, V. R. 1976. Biology of the Reptiles. Physiology V. 566 pp.
Garson, M. S. 1960. The geology of the Lake Chilwa area. Geol. Survey Dept. Bull. 12. Government Printer, Zomba, Malawi. 67 pp.
Guggisberg, C. A. W. 1972. Crocodiles. David and Charles: Newton Abbot. 203 pp.
Hayes, T. D. 1972. Wild Life Conservation in Malawi. Soc. Malawi J. 25:22–31.
Howard-Williams, C. 1973. Vegetation and Environment in the Marginal Areas of a Tropical African Lake (Lake Chilwa: Malawi). Ph.D. thesis, University of London. 312 pp.
Packard, P. C., Tracey, C. R. & Roth, J. J. 1977. The physiology and the evolution of viviparity within the class Reptilia. Biol. Rev. 52:71–105.
Rose, W. 1950. The Reptiles and Amphibians of Southern Africa. Maskew Miller, Cape Town. 240 pp.
Stevens, R. A. 1973. A report of the lowland viper, *Atheris superciliaris* (Peters), from the Lake Chilwa floodplains of Malawi. Arnoldia 6(22):1–22.
Stevens, R. A. 1974. An annotated check list of the amphibians and reptiles known to occur in south-eastern Malawi. Arnoldia 6(30):1–22.
Sweeney, R. C. H. 1960. The Chelonia of Nyasaland Protectorate. Nyasaland Soc. J. 13:35–50.
Sweeney, R. C. H. 1971. Snakes of Nyasaland: with new added corrigenda and addenda. A. Asher and Co., Amsterdam. 200 pp.

Mammals

Baker, C. A. 1961. Nyasaland, the history of its export trade. Nyasaland Soc. J. 15:7–35.
Bell, R. H. V. 1969. The use of the herb layer by grazing ungulates in the Serengeti. Symp. Brit. Ecol. Soc. 10:111–124.
Drummond, H. 1903. Tropical Africa. 11th edn. Hodder and Stoughton, London. 228 pp.
Dudley, C. O. & Stead, D. 1976. Liwonde National Park. I. An Introduction. *Nyala* 2(1):17–28.
Duff, H. L. 1903. Nyasaland Under the Foreign Office. George Bell and Sons, London. 422 pp.
Garson, M. S. 1960. The Geology of the Lake Chilwa area. Geol. Surv. Nyasaland, Bull. 12. Government Printer, Zomba, 67 pp.
Hanney, P. 1965. The Muridae of Malawi. J. Zool. 146:577–633.
Hayes, G. D. 1972. Wildlife Conservation in Malawi. Soc. Malawi J. 25(2):22–31.
Hetherwick, A. 1888. Notes on a Journey from Domasi Missions Station Mt. Zomba to Lake Namaramba. Aug. 1887. Proc. Roy. Geog. Soc. 10:25–32.
Hobbs, J. 1976. An Ecological Appraisal of Liwonde National Park (Malawi) B.Sc. thesis, North London Polytechnic (unpublished). 120 pp.
Howard-Williams, C. 1973. Vegetation and Environment in the Marginal Areas of a Tropical African Lake (L. Chilwa). Ph.D. thesis, University of London (Unpublished). 312 pp.
Howard-Williams C. 1977. A checklist of the vascular plants of Lake Chilwa, Malawi, with special reference to the influence of environmental factors on the distribution of taxa. Kirkia 2: 563–579.
Jackson, G. 1972. Vegetation of Malawi. In: S. Agnew & M. Stubbs (eds.) Malawi in Maps. Univ. London Press. 38–39.
Johnston, H. H. 1894. Report by Commissioner Johnston of the first three years of administration of the eastern portion of British Central Africa (Presented to both Houses of Parliament by command of Her Majesty) H.M.S.O. London. 43 pp.
Johnston, H. H. 1897. British Central Africa. Methuen and Co., London. 544 pp.
Lyell, D. D. 1912. Nyasaland for the Hunter and Settler. Horace Cox, London. 116 pp.

Lyell, D. D. 1923. Memories of an African Hunter. Fisher Unwin, London. 268 pp.

MacDonald, D. 1882. Africana or the Heart of Heathen Africa, II. Mission Life. Simpkins Marshall, London. 371 pp.

Maugham, R. C. F. 1929. Africa as I Have Known It. J. Murray, London. 372 pp.

McLachlan, S. M. 1971. The rate of nutrient release from grass and dung following immersion in lake water. Hydrobiologia 37:521–530.

McNaughton, S. J. 1978. Ungulates and feeding selectivity influences on the effectiveness of 'plant defence fields'. Science 199:806.

Murray, S. S. 1908. A Handbook of Nyasaland. Government Printer, Zomba. 300 pp.

Murray, S. S. 1922. A Handbook of Nyasaland. Government Printer, Zomba. 314 pp.

Murray, S. S. 1932. A Handbook of Nyasaland. Government Printer, Zomba. 310 pp.

O'Neill, H. E. 1884. Journey from Mocambique to Lakes Shirwa and Amaramba. Proc. Royal Geog. Soc. 6(11):632–647; 6(12):713–725.

Robertson, O. H. 1960. Trade and suppression of slavery in British Central Africa. Soc. Malawi J. 13(2):16–21.

Schoffeleers, M. 1972. Animals and clan names cited in G. D. Hayes, 1972. Wildlife Conservation in Malawi. Soc. Malawi J. 25:22–31.

Stead, D. & Dudley, C. O. 1977. Liwonde National Park. II. The Mammals. Nyala 3(2):29–38.

Sutherland, J. 1912. The Adventures of an Elephant Hunter. MacMillan, London. 324 pp.

Sweeney, R. C. H. 1959. A Preliminary Annotated Check List of the Mammals of Nyasaland. Nyasaland Society (Society of Malawi), Blantyre, Malawi. 72 pp.

Sweeney, R. C. H. 1965. Notes on some birds seen in the Lake Chilwa Region. Soc. Malawi J. 18(2):55–58.

Vaughan, M. 1978. Uncontrolled Animals and Aliens: Colonial Conservation in Malawi. History Department Seminar Paper, University of Malawi. 16 pp. (unpublished).

Vesey-FitzGerald, D. F. 1969. Utilisation of the habitat of the buffalo in Lake Manyara National Park. E. Afr. Wildl. J. 7:131–145.

Vesey-FitzGerald, D. F. 1970. The origin and distribution of valley grasslands in East Africa. J. Ecol. 58:51–75.

Birds

Belcher, C. F. 1930. The Birds of Nyasaland. Technical Press, London.

Benson, C. W. 1953. A checklist of Birds of Nyasaland. Heatherwick Press, Blantyre, Malawi.

Benson, C. S. & Benson F. M. 1977. The Birds of Malawi. Montfort Press, Limbe, Malawi.

Douthwaite, R. J. 1974. The ecology of ducks on the Kafue Flats in Zambia. Bull. Internat. Waterfowl Res. Bull. 38:80–84.

Douthwaite, R. J. 1977. Filter-feeding ducks of the Kafue flats, Zambia, 1971–1973. Ibis. (In press.)

Howard-Williams, C. & Walker, B. 1974. The vegetation of a tropical African Lake: Classification and Ordination of the Vegetation at Lake Chilwa. J. Ecol. 62:831–854.

Moreau, R. E. 1972. The palaearctic-african bird migration systems. Academic Press. London and New York. 384 pp.

Roberts, A. 1940. Birds of South Africa. Fourth Edition. Revised by McLachlan, G. R. & Liversidge, R. 1978. Trustees of the John Voelcker Bird Book Fund, Cape Town. 660 pp.

Schulten, G. G. M. 1972. A sight record of the Gull-billed Tern Gelochelidon nilotica at Lake Chilwa (Malawi). Ostrich: New Distribution Data 3.42:138.

Schulten, G. G. M. 1974. The food of some duck species occurring at Lake Chilwa, Malawi. Ostrich 45:224–226.

Schulten, G. G. M. & Harrison, G. 1975. An annotated list of birds recorded at Lake Chilwa (Malawi, Central Africa). Soc. Malawi Journal 28(2):6–30.

Sweeney, R. C. H. 1965. Notes on some birds seen in the Lake Chilwa region. Soc. Malawi Journal 18(1):55–58.

Ward, P. 1971. The migration patterns of Quelea quelea in Africa. Ibis 113(3): 275–297.

Watson, R. M., Singh, T. & Parker, I. S. C. 1970. The diet of duck and coot on Lake Naivasha, E. Afr. Wildl. J. 8:131–141.

Insects of the floodplain

H. R. Feijen, H. D. Brown & C. O. Dudley

1. Pests of rice on the Chilwa plain *by* **H. R. Feijen**

1.1 *Development of new rice schemes*

At the end of the 1960's and the beginning of the 1970's several irrigated rice schemes were started in the Chilwa plain. The Mpheta Rice Scheme in Domasi, fed by the water of the Domasi River, started with 100 hectares and it is intended that its size grows to 500 hectares. The Khanda Rice Scheme, at the north side of Mt. Mpyupyu is irrigated by the water of the Naisi River. It covers approximately 80 hectares. Several schemes are fed by water of the Likangala River: the Likangala Rice Scheme with 180 hectares (intended size 400 hectares), the Njala Rice Scheme (100 hectares) and some smaller, peripheral schemes like Mwambo Rice Scheme (57 hectares). Most of the schemes are quite close to the lakeshore, and the lower part of the Likangala Rice Scheme is only 3 km away from the shore (Fig. 15.1).

Fig. 15.1 View of Rice-field in Khanda Rice Development Scheme (photo: C. M. Schulten-Senden).

A fairly rigid programme for working hours each day and simultaneous timing of cultivation practices have been introduced. The schemes use the smallholder system, each farmer cultivating about 0.8 hectare, and most of the work is done by hand. The main variety used is Blue Bonnet, although in the Khanda scheme the local 'Faya' variety of *indica* has also been used. Variety trials are still being carried out. Yields vary from 3–5 tonnes per hectare with one crop per year and the average yield in schemes with double cropping is close to 6 tonnes per hectare (see Chapter 17).

Nurseries are sown at the start of the rainy season in the beginning of December. Transplanting starts in the second half of December, but if rains have been late and the rivers are low, it sometimes does not start before the 15th of January. Transplanting continues into the second half of January and sometimes into the first half of March. Harvesting takes place from the end of April until the second half of June. Due to the limited supply of water only some parts of the schemes, especially Mpheta, can be used for a second crop. Maize and wheat have also been tried as a second crop, but without much success.

In the experimental fields fertilizer is applied at the rate of 82 kg ha^{-1} nitrogen and 34 kg ha^{-1} phosphate, but in farmers' fields the rate of application is usually less. Before sowing, seeds are disinfected with an organic mercury compound (methoxy ethyl mercuril chloride). There are, however, doubts as to the effectiveness of this treatment and it might be discontinued. In January 1971, an outbreak of *Diopsis macrophthalma*, a stalk-eyed fly, occurred in the Likangala Rice Scheme. It was at that time considered necessary to control the outbreak by aerial spraying of a 75 per cent DDT solution at a rate of 1.12 kg ha^{-1}. This was unsuccessful and due to this lack of success, the spraying was then repeated with phosphamidon. In the 1971/72 season some fields were sprayed with diazinon against the Rice Beetle *Trichispa sericea*. In the 1972/73 season nurseries attacked by the Rice Beetle were hand-sprayed with phosphamidon or dimethoate.

The presence of such intensive culture near the lake shore will certainly have an effect on the general ecology of Lake Chilwa in the future, since the schemes are still expanding. The main causes of these effects will be diversion of water by draining the river water away from the lake, more especially at the beginning of the rainy season. There will be a slow accumulation of poisons from the use of fungicides (especially those containing organic mercury compounds) and from persistent pesticides which will be washed into the lake. The large-scale use of DDT close to this shallow, endorheic lake met with opposition, especially from members of the Lake Chilwa Co-ordinated Research Project (Schulten 1971). Stimulated by this controversy on the use of DDT, a research study was started on the pests of rice in April 1971 and it ended in August 1975. The investigation received financial support from WOTRO (Netherlands Foundation for the Advancement of Tropical Research) and the Research Committee of the University of Malawi.

1.2   Pests of rice

The ecology of the rice pests was studied by regular sampling of flies, eggs and rice stems (for larvae and pupae). The economic importance was investigated by extensive sampling in the field and by insecticide and simulation trials. Data about other pests of rice were collected to establish the relative importance of *Diopsis macrophthalma*. A list of the more common pests of rice in the Chilwa plain is given below, as well as some information about their relative importance.

278

1.2.1 Diptera

Diopsis macrophthalma Dalman 1817 (syn. *D. longicornis* Macquart 1835 and *D. thoracica* Westwood 1837) is the most numerous insect pest of rice and it is the only stalk-eyed fly of any economic importance in the Chilwa plain (Fig. 15.2). This fly has been recorded as a pest of rice from most parts of tropical Africa (Descamps 1957, Crossland 1964, Bakker 1974) and in West Africa it is even regarded as the major pest of rice (Morgan & Abu 1973).

Fig. 15.2 Diopsis macrophthalma Dalm. the stalk-eyed fly, main pest of rice in the Chilwa floodplain (photo: H. R. Feijen).

The flies spend the dry season in shady places along the banks of streams and pools where they can be found in large swarms. At the beginning of the rainy season, they move towards the rice-fields. When a second crop is grown in the year, flies are attracted there, but only in low numbers.

Eggs are laid singly on stems or leaves of the rice plant, and usually no eggs are laid on rice higher than 75 cm. The larva enters the stem via the leaf sheath and develops there. The feeding of the larva causes the yellowing of the terminal leaf ('dead heart'). The next two or three terminal leaves of the stem may also be turned into 'dead hearts'. The occasional attack of older rice can cause 'white heads'. Except when the larva attacks seedlings, it remains in the same stem until pupation. Just before pupating the larva can move towards another stem or even another hill. This may be a defence mechanism against predators and parasites which would be attracted by the 'dead hearts'.

In the Chilwa plain it was found that small rice seedlings can be killed by larval attack. In one season, approximately 10 per cent of the seedlings were destroyed in this way.

The influence of *D. macrophthalma* on transplanted rice is more difficult to assess. In the field, it was found that heavily attacked fields had more stems per hill than sprayed or less attacked fields. Simulation experiments in some detail, however, showed that attack by *Diopsis* larvae can have both positive and negative effects on the growth of rice plants, depending on the time of attack, the level of attack and on the general growing conditions, such as the amount of fertilizer and the sizes of rice 'hills'. The effects also vary with the variety of rice used.

In general it can be said that only under conditions of poor soil and heavy and prolonged attack can *D. macrophthalma* be considered a pest of rice in Malawi. A paper on the importance of *D. macrophthalma* in Malawi is in press (Feijen 1979). In the Chilwa Rice Schemes, this pest is to a large extent controlled by three egg parasitoids, *Trichogramma kalkae*, *T. pinneyi* (Schulten & Feijen 1978) and *Trichogrammatoidea simmondsi* (Nagaraja, in press), and by a pupal parasitoid *Tetrastichus diopsisi* Risbec. It should be emphasized that parasitism of the eggs of the pest which can reach up to 90 per cent, dropped to below 2 per cent in nurseries sprayed with phosphamidon. Obviously the parasites of the rice pest were much more susceptible than the pest, *Diopsis*, itself.

D. apicalis Dalm., which sometimes formed up to 20 per cent of the diopsids present in the rice schemes, only lays its eggs on 'dead hearts' caused by other insects (mainly *D. macrophthalma*), a phenomenon which has also been recorded by Scheibelreiter (1974). Larvae of *D. macrophthalma* may be preyed upon by larvae of *D. apicalis*.

Pachydiplosis oryzae (Wood-Mason), the rice gall midge became especially numerous at the end of the prolonged rainy seasons (1973/74 and 1974/75 in the fields which were planted later. The larvae induce 'onion-shoots'. Three larval parasitoids kept this pest under control.

Another Dipteran, a rice leaf-miner, *Hydrellia* sp. and about 10 other species of small leaf-mining or secondary stem-boring Diptera were found. None of these, however, was of much importance (see Appendix A).

1.2.2 Lepidoptera

Seven lepidopterous stem-borers were found which are listed in the check list, but they were of little importance as pests, since they are well controlled by many egg, larval and pupal parasitoids. Lepidopterous leaf eaters were of even lesser importance, although seven species were found (see Appendix A).

1.2.3 Coleoptera

In the period 1971–75, several outbreaks of the rice beetle *Trichispa sericea* occurred. In the 1971/72 season, it was a local outbreak in some fields in the 'heading' stage of rice and about one hectare was almost destroyed. In the 1972/73 season heavy attacks were seen in a number of nurseries in January and several hectares became useless. Some fields were planted with seedlings from infected nurseries with very bad results: a drop in yield of about 50 per cent. In April of the same year older fields were attacked, resulting in about 20 per cent loss of yield. Small outbreaks also occurred in the 1973/74 and 1974/75 seasons. Various parasites of the beetle were found.

The only other beetle, which could be sometimes found in high numbers was the meloid, *Epicauta velata* Gerst. These beetles were usually feeding on the weeds among the rice growing in the fields, but they were observed sometimes feeding on the flowering rice. Several species of *Chrysomelidae* were also commonly found, but never in large numbers.

1.2.4 Orthoptera

Zonocerus elegans Thumb., the elegant grasshopper, was often numerous in and around the rice schemes, as well as *Chrotogonus hemipterus* Schamm. Both preferred various weeds. Other grasshoppers were of no importance. The mole cricket, *Gryllotalpa africana* (P. de Beauv.) was often observed and a parasitoid of this species was collected. No red locusts were seen.

1.2.5 Hemiptera

The green rice bug *Nezara viridula* (L.) was regularly found, but it never became numerous.

1.3 *Pest control*

In general it has been concluded from this four-year study that improved methods of rice culture are the best ways of preventing insect attack in the rice schemes of the Chilwa plain. These involve some of the measures in the Taiwanese programme which require to be strictly enforced over the whole area.

(i) The long duration varieties such as 'Faya' should not be used and early maturing varieties should be selected to avoid the spread and build up of pests.

(ii) No overlap of crops should be allowed in order to avoid prolonging the period when rice is available to pests and fungus.

(iii) Early transplanting should be practised and transplanting done during a short period and simultaneously over the whole area.

(iv) There should not be too much water in the fields. Water should be better controlled to avoid fungus disease and it should be noted that rice beetle populations are also associated with too much water.

Chemical control measures are in general not advisable as these interfere with the well-established natural control of the parasitoid systems. The parasitoids prove to be much more sensitive to insecticides than the insect pests themselves. The most abundant pest, the stalk-eyed fly *Diopsis macrophthalma* is anyway not likely to cause economic damage as rice plants can compensate for the damage done by this larval borer by tillering. The larvae are not affected by superficial insecticides. At the moment, it is considered necessary to apply insecticides only against local outbreaks of the rice beetle, *Trichispa sericea*.

2. Lake Chilwa and the red locust *by* H. D. Brown

The red locust *Nomadacris septemfasciata* (Serville) (Fig. 15.3) is a tropical African savanna species restricted to floodplain grasslands. Its more important habitats are characterized by natural mosaics of tall and short grasses, with bare patches of soil, which apparently provide favourable combinations of oviposition and food-shelter habitats.

Fig. 15.3   Nomadacris septemfasciata (Serville) the red locust (photo: H. D. Brown).

Although small non-swarming populations of red locust breed and persist over wide areas of Africa without giving cause for concern, in a few well-defined source areas, known as outbreak areas, swarms can develop from scattered populations and give rise to widespread locust plagues, which last for long periods. Only about 8 such areas of varying importance are known in tropical Africa and one of these is Lake Chilwa in Malawi. Lake Chilwa can however be regarded as a low grade outbreak area because in forty years only one small upsurge of *Nomadacris*, with a small swarm, has been recorded (Sweeney 1962).

Ever since the first exploratory attempts to trace the origins of the last 1930–45 red locust plague, Lake Chilwa, because of its obvious similarities with the key outbreak areas of Lake Rukwa and the Mweru wa Ntipa, has been regarded with suspicion and was one of the areas singled out for intensive study as long ago as 1934 (Michelmore 1934). This resulted in a reconnaissance survey in 1935 by Lea (1938), who concluded that it was not an outbreak area. However, heavy concentrations of locusts, no doubt part of the declining plague, were encountered in Chilwa and Lea used the term 'incipient swarms' to describe them. From about 1892–1910 an earlier red locust plague also covered southern Nyasaland but its origins were not known (Sweeney 1962).

After the last plague had collapsed in 1945 there were definitely no reports of swarms until 1962 when a small swarm was reported from Namanga, on the northwest plain, near Chikala Hill (IRLCS 1962). In subsequent control operations a dense concentration about 1 km² in area was sprayed from the air with 85 per cent carbaryl, but about 10 km² of the plain was lightly infested and left untreated (IRLCS 1963). Vesey-FitzGerald reported that the Chilwa floodplains were similar to the Lake Rukwa floodplains which he described in 1955 (Vesey-FitzGerald 1955) and that the area had all the characteristics of a red locust outbreak area (IRLCS 1962). Despite regular aerial surveillance of the area at this time no further upsurges were observed.

The important locust areas are located to the north of the lake (near Nayuchi village) below the sandbar adjoining the Mozambique border and along which the railway line runs, and to the northwest (near Mposa village) south of the Liwonde National Park near Chikala Hill, and where the 1962 swarm is presumed to have originated (Fig. 15.4) In these areas the *Typha* swamp (*T. domingensis*), which is the dominant floristic element of the Lake Chilwa system, gives way to extensive floodplain and alkaline grasslands, which occupy about 390 km². The former is dominated by *Sporobolus pyramidalis*, *Eragrostis* spp. and *Hyparrhenia rufa*, and the latter by *Diplachne fusca* and *Panicum repens* (Chapter 7). East and south of the Chikala Hills a zone of marsh, dominated by the sedges *Cyperus procerus* and *Cyperus articulatus* forms a band between the *Typha* swamp and the floodplain grassland. The *Typha*

Fig. 15.4 The red locust breeding area east of the Chikala Hills on the Chilwa plain (photo: N. Lancaster).

provides a satisfactory roosting habitat for the adult locusts during the dry season, especially when the grasslands are burnt off: the unburnt islands of *Typha*, which are more resistant to burning, then serve as a foci for adult concentration.

Berreen (1970) described the oviposition sites as sandy patches raised just above the water level on the floodplain and between the floodplain grassland and sedge community in burnt areas. Hatching took place in January and the early instar hoppers fed on grasses on drier ground just above the boundary of the flooded area. They were also to be found in patches of village maize bordering the grassland. Newly fledged adults appeared in March and dispersed into both the drier and wetter parts of the plain. During the period April to May, most of the adults moved into the *Typha* community, probably as a result of the rapid drying out of the grassland community and its subsequent destruction by fire. Locusts moved to the edge of the burnt areas for mating and oviposition during October and November when the rains commenced.

During times of locust abundance dead locusts collected by local villagers make their appearance on the local markets and provide useful evidence of impending population increase in the plains. The locusts are plucked at dawn from the tops of the *Typha* plants where they roost at night. Sacks thus filled are carried on foot 30 km to Zomba market, where they are sold at prices that reflect their relative abundance and ease of collection.

Although Stortenbeker (1967) showed insect predators like dragonflies (*Odonata*) and robberflies to be important mortality factors in the early hopper instars in the Lake Rukwa area, no corroboration of this was obtained from Lake Chilwa. *Odonata* investigated here preyed largely on mosquitoes and damselflies (Berreen unpublished data) but this will probably depend on the relative abundance of the prey as these are facultative predators.

Samples of red locusts collected to see whether they were in a solitary or swarming condition, from Chilwa in September 1973, a year of minor lake fluctuation, yielded E/F ratios (elytra/femur) of 1.87 (range 1.74 to 2.10) for males (n = 30) and 1.84 (1.71 to 1.97) for females (n = 27) which are consistent with solitary phase measurements for this locust species (Rainey et al. 1957). This sample was noted for its dark pronotal markings, probably associated with the burnt background of the *Typha* habitat.

During the sixties and the seventies Chilwa was periodically visited by locust specialists from the International Red Locust Control Service in Zambia and latterly by local entomologists from the Agricultural Research Stations of Bvumbwe and Makoka in Malawi, who conducted ground and aerial surveys of the floodplains at irregular intervals. Locust numbers have tended to remain low apart from a small concentration in July 1969 in the area between Chikala Hill and the swamp. In March 1977 the northern grasslands were heavily flooded and locust numbers were low. Lake levels have remained high during the seventies because of good rains (Chapter 3). However, in neighbouring Mozambique, in the Buzi-Pungwe River delta area conditions favoured locust breeding and swarms were produced over four successive seasons from 1972 on.

Non-swarming populations of red locust have long been known from the summit of the Zomba plateau and also occur sporadically in the Ndindi Marsh of the Lower Shire Valley.

284

Upsurges of red locust in the plains bordering the main outbreak area of Lake Rukwa have been found to be correlated with dry periods when the lake waters receded (Gunn 1973). However, a similar sequence did not occur at Chilwa and during a major recession of the lake in 1967/68 the locust population remained relatively low. The reasons for this are not known, but it does appear that the relatively limited extent of the floodplain grasslands, owing to more prolonged and deeper flooding conditions which are largely responsible for the great development of the *Typha domingensis* community, is a key factor in determining the relative unimportance of Chilwa as a red locust habitat. Hopefully the considerable wealth of information now available on the Chilwa swamps will in time be extended to the red locust where more intensive investigations on the locust itself may help to explain its inability to swarm more frequently there. Situated within 30 km of a University research centre, Chilwa presents a unique opportunity for carrying out field research on this notorious locust pest.

3. The dung beetles of the western grasslands of the Chilwa area *by* C. O. Dudley

Two factors which have affected dung beetles (*Coleoptera*: *Scarabaeidae*) on the Chilwa plains are discussed in this section: (1) the relative decrease in the availability of mammalian dung since domestic cattle have replaced the mixed herds of larger mammals, which were exterminated from 1890–1930 (see section 14.3); (2) the annual inundation of the floodplain, where the cattle feed only in winter.

Halffer & Mathews (1966), who have described the taxonomy of African dung beetles, hold the view that they were first forest species feeding on decaying vegetable material, lying on the forest floor, in the same way as beetles in other families of *Scarabaeoidea*. Early in their evolutionary history, coprophagy (eating dung) became the dominant feeding adaptation. As a companion development, the manipulation of dung, including burying, tunnelling or rolling became a feature of some species. When the coprophagous scarabs invaded the savanna during the Tertiary Period, in the presence of very varied types of large mammal excrement, adaptive radiation occurred rapidly.

Africa, the continent with the largest and most varied populations of savanna mammals, has probably the most diverse scarab fauna. Malawi, in spite of the loss of much of its larger wildlife (except in Game Reserves), still has over 150 species, which have been described by Ferreira (1967) and a general ecological account is given by Dudley (1977). An analysis of the role of dung beetles in facilitating the retention in the soil of the nitrogen in cattle faeces was made by Gillard (1963) and quoted at some length by Roux (1969). A number of different types of 'manipulators' of dung are involved and large patches of newly-voided dung can be buried in four days or so. Subsequently a great deal of investigation of dung beetles in Africa has resulted in their export to Australia, since Australian scarabs are adapted to burying the hard pellets of marsupials and are unable to remove the large soft dung of introduced cattle (Ferrar 1973). The scarabs described on the highveld of South Africa by Gillard (1963) include many more genera than are found on the Chilwa floodplain.

The most important ecological activity of dung beetles is their role in the recycling of organic nutrients in savanna ecosystems. Fifty to ninety per cent of food consumed by herbivores is not assimilated and is passed out as faeces. Thus, dung removal and its subsequent incorporation into the soil maintains the vegetative growth necessary for herbivore populations – both domestic and wild. As an auxiliary benefit, dung beetles improve soil texture and water retention and constantly raise subsoil to the surface. In Serengeti National Park over 500,000 kg of dung are produced and buried each day (Birch 1971). On local managed pastures, the amount of dung buried may be over 1235 kg ha^{-1} per year during the wet season.

Dung beetles also act as important biological control agents by removing breeding sites for flies and, through their activities, destroying parasitic worms which infect wild mammals and cattle. The latter activity is important in cattle-raising areas.

The new adult, after emerging from the soil, spends a variable period feeding on fresh dung either on the surface or, after burying it, in the ground. This feeding builds up fat reserves for later oviposition activities. The majority of dung beetles do not roll a ball of dung but bury the dung directly beneath or immediately adjacent to the original dung mass. The beetle then has access to the dung mass by an underground tunnel. The ball rolling species roll the dung ball up to several metres from the original dung mass before burial.

Eventually, depending on the species and length of the season, the adults stop constructing feeding tunnels and begin excavating oviposition tunnels. These tunnels are often more elaborate, though of the same basic plan. Brood balls of dung are constructed within a brood chamber and eggs are deposited inside the balls. The adults usually leave the tunnel, after closure with soil, to begin new excavations. All immature stages are passed inside the brood ball, and in a number of genera, brood care is common, e.g. *Heliocopris*.

3.1 *The Lake Chilwa plains*

This study has been almost entirely restricted to the plains near Kachulu Harbour. The 'natural' grasslands near Kachulu are quite narrow while the 'secondary' grasslands are extensive, being several kilometres wide. The secondary grasslands were created as a result of cutting down trees for domestic purposes, building and firewood, and through the traditional shifting cultivation in which exhausted farm land was abandoned until it recovered, and new areas were cleared for crops. The secondary grassland is dotted here and there with species of coppiced woody plants common to the remnant woodlands found further from the lake shore. During the year much of the natural floodplain grasslands are seasonally inundated while the secondary grasslands being above the 4–5 m terrace (Chapter 2) drain fairly rapidly. The soils vary in a mosaic pattern from alkaline clays to more acidic sands (Howard-Williams & Walker 1974). Wild ungulates no longer occur, being replaced in the area by several thousand cattle of the small Zebu breed of *Bos indicus* (Schmidt 1969).

Writers of the late 19th century (Drummond 1903; Maugham 1929) described the Lake Chilwa area as one containing a dense and diverse population of wild mammals, including elephant. Except for the seasonally inundated grasslands, the area was light to heavily wooded. Since these early accounts human settlement around the lake has grown immensely and the environment has been radically altered.

There are plans by the Malawi Government to develop the cattle resources of the plains on a more rational and intensive basis. One of the purposes of this investigation was to examine the dung beetle fauna in light of this proposed increase in cattle.

3.2 The scarab fauna

There are several features of the scarab fauna of the Kachulu grassland that are of interest. The features are particularly striking if this fauna is compared with that of a dairy ranch (Rathdrum Farm), a short distance from Lake Chilwa. Rathdrum is an area of partially cleared woodland with a small herd (100) of large well-fed cattle and deep, sandy clay soils. The farm is less than 16 km from Kachulu and 135 m higher in altitude.

Table 15.1 compares the two scarab faunas. The Rathdrum fauna was richer in species, less dominated by a single species and showed a greater average number and a much wider range in size. The same species was not always dominant at both sites but varied with the emergence patterns during the year. In general the picture of the Kachulu fauna is that of an impoverished one. What may be the reasons for this?

Table 15.1 A comparison of the number, size and dominance of the dung bettle species found at Kachulu on the floodplain and Rathdrum Farm 16 km inland.

	Kachulu	Rathdrum Farm
Habitat type	natural grassland	secondary grassland/ woodland mosaic
Altitude	630 metres	765 metres
Number of species	16	57
Species size ($\bar{X}$) range	(0.8) 0.2 −1.5 cm	(1.1) 0.2 − 5.0 cm
% frequency first 3 dominant species	71 : 20 : 3	36 : 26 : 17
Numerically dominant species – descending order	*Onthophagous depressus* Har. *O. gazella* (F) *Euoniticellus intermedius* (Reiche) *Liatongus militaris* (Cast.)	*L. militaris* *E. intermedius* *O. gazella* *O. vinctus* Er.

Initially the fauna was probably richer throughout the region, including both Rathdrum and Kachulu. With the removal of the woody cover and the disappearance of most of the wild mammals this richness declined at Lake Chilwa, stabilizing at a low level and based entirely on the small droppings of Zebu cattle. Rathdrum maintained its richness at a much higher level since the managed herds produce larger droppings and the remnant woodlands still harbour a few of the smaller antelopes.

Historically, the number of dung beetle species on the seasonally inundated grasslands near Kachulu was, perhaps, never as high as that of the immediately surrounding areas. The seasonally high water table and alkaline soil conditions probably restricted the fauna during the wet part of the year. However, during the dry season, small populations of different species from the secondary grasslands may temporarily colonize this area. Successful colonization by dung beetles of the middle to larger sizes (1.0–5.0 cm) does not occur. The larger genera, such as *Catharsius* and *Heliocopris*, which occur at Rathdrum Farm and bury the dung the deepest (Fig. 15.5), cannot live on the floodplain because the high water table prohibits deep tunnelling. These genera also breed once a year and require one year to develop, so that even if they confined their breeding to the dry season, their larvae would not survive inundation of the plain.

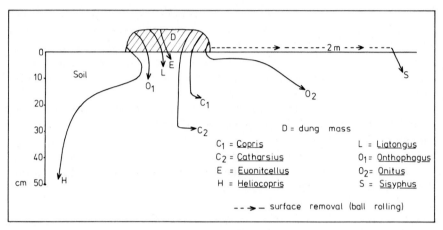

Fig. 15.5   A diagrammatic general pattern of dung resource exploitation by dung beetles.

Zebu cattle on the Kachulu floodplain seem to have smaller droppings than the supplementary-fed cattle on Rathdrum Farm. This too might limit the colonization by larger dung beetles with a short life history, since there might not be enough food for larvae as well as adults.

There is evidence from studies at Rathdrum that changes in the vegetation of secondary grasslands on the Chilwa plains, after ground had been cleared of trees, farmed and allowed to lie fallow, may also have affected the species of dung beetles through the changes in the microhabitat. Many species which were associated with the larger mammals of the original woodlands disappeared along with the demise of the game in the early twentieth century. The relative importance of the specific type of mammalian dung and the qualities of the micro-environment for the scarab beetle fauna are not known.

Small droppings have an impact on the smaller species as well by increasing competition for a limited food resource. This does not seem to occur at Rathdrum where some dung pats are only partially used. Small pats also dry out quickly further intensifying this competition. Under such conditions only a small number of species are likely to be successful. A few of the more successful aggressive species (*Onthophagous depressus*, *O. gazella*, *Euoniticellus intermedius*, *Liatongus militaris*) (Fig. 15.6) show strong numerical dominance over the rest, as they use resources which in other areas (e.g. Rathdrum) are normally shared by many more species. It is interesting to note that for the most part the same species dominated the fauna at both sites; the order of dominance and the density differed.

In the Kachulu fauna only one species of dung roller (*Sisyphus* sp.) is known.

288

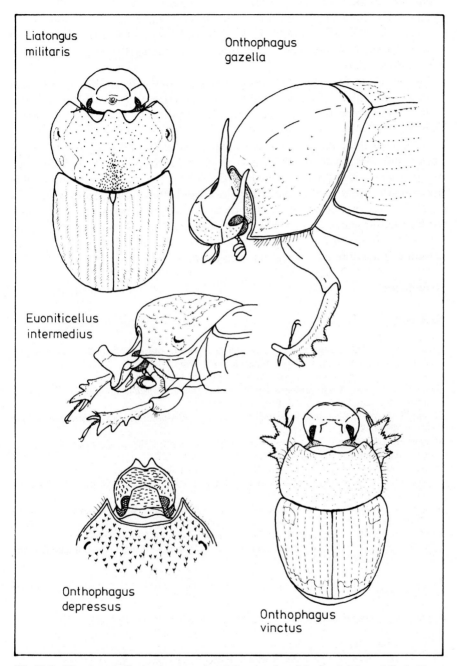

Liatongus
militaris

Onthophagus
gazella

Euoniticellus
intermedius

Onthophagus
depressus

Onthophagus
vinctus

Fig. 15.6 The numerically dominant species of dung beetles of the floodplain, and *Onthophagus vinctus* Er. from Rathdrum Farm.

Yet the species of this group would be expected to do better in the more open areas such as grasslands. One possible explanation is that Lake Chilwa is the location of a large and diverse bird population (Schulten and Harrison 1975). Ibises are known to feed readily on scarabs and dung rollers are more exposed to this type of predation. This *Sisyphus* is a small species (0.5 cm) and probably escapes detection.

On the whole, the Kachulu scarab fauna is strongly affected by both physical (water) and biological factors (food resources, competition, vegetation). The fauna is small and undiversified and dominated by two or three aggressive species.

In spite of the paucity of the Kachulu fauna, unburied dung is unlikely to accumulate as it has in other regions of the world with low scarab diversity, such as Australia (Bornemissza 1960) or California (Ferrar 1973). Where this accumulation does occur, pasture is reduced and intensive measures have to be taken to correct this situation (i.e. dung beetle introduction). At Kachulu, during the wet season any unburied dung is broken up and transported into the lake when the water rises and then recedes over the grassland. During the dry season termites remove any remaining dung.

References

Rice pests

Bakker, W. 1974. Characterization and ecological aspects of rice yellow mottle virus in Kenya. Agric. Res. Rep. 829. Agric. Univ. Wageningen. 152 pp.
Crossland, N. O. 1964. New rice pests in Swaziland. Wld. Crops 16:51.
Descamps, M. 1957. Contribution a l'étude des Diptères *Diopsidae* nuisibles au riz dans le Nord-Cameroun. J. Agric. Trop. Bot. appl. 4:83–93.
Feijen, H. R. 1979. Economic importance of the rice stem-borer (*Diopsis macrophthalma*) in Malawi. Expl. Agric. 15 (in press).
Morgan, H. C. & Abu, J. F. 1973. Seasonal abundance of *Diopsis* (Diptera, Diopsidae) on irrigated rice in the Accra plains. Ghana Jnl. agric. Sci. 6:185–191.
Scheibelreiter, G. 1974. The importance of *Diopsis tenuipes* Westwood as a pest of rice, based on a comparison of the egg-laying behaviour of *D. tenuipes* and *D. thoracica* Westwood. Ghana Jnl. agric. Sci. 7:143–145.
Schulten, G. G. M. 1971. The use of DDT at Lake Chilwa. University of Malawi (cyclostyled). 5 pp.
Schulten, G. G. M. & Feijen H. R. 1978. Two new species of *Trichogramma* (Hymenoptera; Trichogrammatidae) from Malawi; egg parasitoids of *Diopsis macrophthalma* Dalman (Diptera; Diopsidae). Ent. Bericht. 38:25–29.

Red Locusts

Annual Reports of the I.R.L.C.S. 1962, 1963. Unpublished. Mbala, Zambia.
Berreen, J. M. 1970. The floodplain. In: M. Kalk (ed.) Decline and Recovery of a Lake. Govt. Printer, Zomba, Malawi, pp. 40–41.
Gunn, D. L. 1973. Consequences of cycles in East African Climate. Nature 242:457.
Lea, A. 1938. Investigations on the red locust in P.E.A. and Nyasaland in 1935. Sci. Bull. No. 176. 27 pp.
Michelmore, A. P. G. 1934. Summary of results and programme for future investigation on the red locust. 3rd Int. Locust Conf., London. Appendix 11:96.
Rainey, R. C., Waloff, Z. & Burnett, G. F. 1957. The behaviour of the red locust in relation to the topography, meteorology and vegetation of the Rukwa Rift Valley, Tanganyika. Anti-Locust Bull. 26:1–96.

Stortenbeker, C. W. 1967. Observations on the population dynamics of the red locust in its outbreak areas. Pudoc. agric. Res. Rep. No. 694. 118 pp.

Sweeney, R. C. H. 1962. Red locust in Nyasaland. Min. Nat. Resources Bull, Zomba. 2:1–13.

Vesey-FitzGerald, D. F. 1955. The vegetation of the outbreak areas of the red locust in Tanganyika and Northern Rhodesia. Anti-Locust Bull. 20:1–31.

Dung beetles

Birch, M. C. 1971. Status of dung beetles (Coleoptera: Scarabaeidae) in the ecology of the Serengeti National Park, Tanzania. Research proposal, Serengeti Research Institute, Tanzania (unpublished).

Bornemissza, G. F. 1960. Could dung-eating insects improve our pastures? J. Austral. Inst. Agric. Sci. 26:54–56.

Drummond, H. 1903. Tropical Africa. 11th edition, Hodder & Stoughton, London. 228 pp.

Dudley, C. O. 1977. The natural history of dung beetles (Coleoptera: Scarabaeidae) with special reference to Malawi. Nyala 3(1):38–47.

Ferrar, P. 1973. The Council for Scientific and Industrial Research Organisation dung beetle project. Wool. Tech. & Sheep Breeding 20:73–75.

Ferreira, M. C. 1967. Os Escarabídeos de Moçambique, I. (Subfamílias Scarabaeinae e Coprinae). Revta Ent. Moçamb. 10:5–778.

Gillard, P. 1963. An ecological study of grazing intensity on *Trachypogon*-other species grassland. Ph.D. Thesis, University of the Witwatersrand (unpublished).

Halffer, G. & Mathews, E. G. 1966. The natural history of dung beetles of the sub-family Scarabaeinae (Coleoptera: Scarabaeidae). Folia Entomol. Mexicana, 12–14:1–312.

Howard-Williams, C. & Walker, B. H. 1974. The vegetation of a tropical African lake: Classification and ordination of the vegetation of Lake Chilwa (Malawi). J. Ecol. 62:831–854.

Maugham, R. C. F. 1929. Africa as I have known it. Murray, London.

Roux, E. 1969. Grass: A Story of Frankenwald. Oxford University Press, London. 212 pp.

Schmidt, R. 1969. Cattle in Southern Malawi. Soc. Malawi J. 22(2):57–72.

Schulten, G. G. M. & Harrison, G. 1975. An annotated list of birds recorded at Lake Chilwa. Soc. Malawi J. 28:6–30.

Part 3. The people of the Chilwa area

5 The influence of history on Lake Chilwa and its people

B. Pachai

The influence of history on Lake Chilwa and its people

1. The early phase of human settlement and activity

Lake Chilwa and its plains were at one of the crossroads of Central Africa until the end of the nineteenth century. Its history has been influenced by events which went beyond the rise and fall of its water levels. Its human settlement was made up of a number of migrations each bringing with it features which were either constructive or destructive or which carried elements of both. Besides the impact of migrations, the policies and preferences of the colonial administration and its successor, the Republic of Malawi, have also left their imprint on the area and its peoples.

The Shire Highlands is the most ethnically conglomerate part of Malawi. The last ten thousand years before the present saw the Shire Highlands and the rest of the country pass through two cultural periods: the tailend of the Late Stone Age which accounted for some eight thousand years before the birth of Christ and the Iron Age which was roughly coterminous with the Christian era (Pachai 1972). During this period the red and brown soils of the Shire Highlands and the black soils of the lowlands produced crops while the abundance of wild life provided, in the words of Desmond Clark (1972), 'one of the richest environments anywhere in the world for hunting and gathering people'. To these must be added the ready availability of fish in Lakes Malawi, Malombe, Chilwa and Chiuta, the tsetse-fly-free belt between the Luangwa Valley in the west and these lakes in the east. The easy access routes into the country from north, south, east and west completed the attractions for hunters, fisherfolk, herders and cultivators who entered it in that approximate order. Agnew (1970) sums up these environmental advantages which, by their combined effect, led to growth and provided a cushion against disaster: 'In its ecology of a land and water symbiosis the country (Malawi) offers an environment of increment whether for hunter-gathering societies, fisherfolk, cattle-keepers or swidden cultivators. Implicit in this assertion is that survival rates would be higher than in less favourable habitats.'

The first settlers to inhabit the country in the latter part of the Late Stone Age were small-statured hunting people not to be confused with the Bushmen or San, who eventually made their homes in South Africa, but persons who belonged to the Nachikufan culture (bored stones and polished stone axes with traces of pigments) first found in caves in northern Zambia indicating continuous habitation from the earlier half of the first millenium B.C. onwards (Rangeley Papers 1950). These small-statured people 'appear either to have merged with the local Bantu or to have been killed off, as recently as one hundred years or less. Tradition has it variously that they painted, were connected with the 'Batwa' ceremonies, had big heads, wore beads, used bows and arrows, carried their grindstones about with them, had no stock, lived by rivers and in caves.'

After extensive field investigations, Rangeley argued in 1956 that the Nachikufan small-statured, stone-age hunter-gatherers were in turn supplanted by pygmoid persons who were skilled smelters of iron (Rangeley Papers 1956).

These iron smelters were largely absorbed by the third stream of immigrants, the Bantu speakers, to whom they passed on their skills of iron smelting. These later pygmoid persons he described as the Kafula, distinguishing them as an Iron-Age people. Though, like the earlier Nachikufans, they were hunters, nomads and gatherers they lived in more sophisticated houses: in 1948 a Kafula village was identified on the Dowa–Lilongwe border where the dwellings were made of burnt bricks; the huts of poles and mud were constructed first and then fired (Pachai 1978). Some of the Kafula kept cattle and possessed pottery and became distinguished 'by their habit of making holes in the lobes of their ears.' These cattle keepers were also referred to as the Pule, Lenda or Katanga.

Thus, in the period of about eight thousand years, spanning the latter part of the Late Stone Age and the Early Iron Age (8000 B.C. to 100 A.D.) the following cultural groups settled in the country:

(1) Pre-Bantu: Nachikufan (or Batwa) stone age, hunter-gatherer-fisherfolk.
(2) Pre-Bantu: Kafula – iron age, hunter-gatherer-fisherfolk.
(3) Proto-Bantu: Kafula (or Pule, Lenda, Katanga) – hunter-pastoralists-fisherfolk.

It was during the end of the Later Iron Age or about 1200 A.D. and the following centuries that the Bantu-speaking migrants arrived on the scene. Central and southern Malawi became settled by Maravi peoples (a collective designation for a related people also known separately as Chewa, Mang'anja Nyanja, Mbo, Chipeta, Nsenga, Chikunda, Ntumba and Zimba) while in northern Malawi in the area between the Dwangwa River and the Songwe River, the country was settled at different times in this period by the Tumbuka, the Tonga, the Kamanga, the Henga, the Phoka, the Ngonde and the Lambya, as well as a number of smaller groups (Fig. 16.1).

In the southern part of the country in which Lake Chilwa and its environs fall, the Bantu-speaking Maravi migrants consolidated their settlements from their permanent headquarters established at the south-western lakeshore of Lake Malawi, in the area of Mankhamba in present Dedza district and spread in all directions including the Dwangwa River in the north, the Luangwa in the west, the Zambezi and Shire areas in the south-east. By 1750 Maravi chieftainships were widespread and local and territorial trade was active in places of favourable geographical location. A Portuguese traveller journeyed through this territory in 1616 on his way from Tete to Kilwa on the East African coast (Fig. 16.3) and recorded the names of some chieftainships, while Portuguese reports as early as 1572 described the crops produced in the lower Shire region and the trade between the Portuguese and the Mang'anja. There was, however, no contact in this early period between the Portuguese and the interior of Malawi. That was to come much later. A letter by Father Aloysius Mariana dated 1624, written after he had spent some 10 years in Mozambique, gives some information on Lake Malawi (which to the Portuguese appeared to be an attractive route to try to reach the elusive kingdom of Prester John): 'Its width is between 4 and 5 leagues (24 km) or more, and so in some parts one cannot see the land on the other shore. Of this part there is no information, but the Africans,

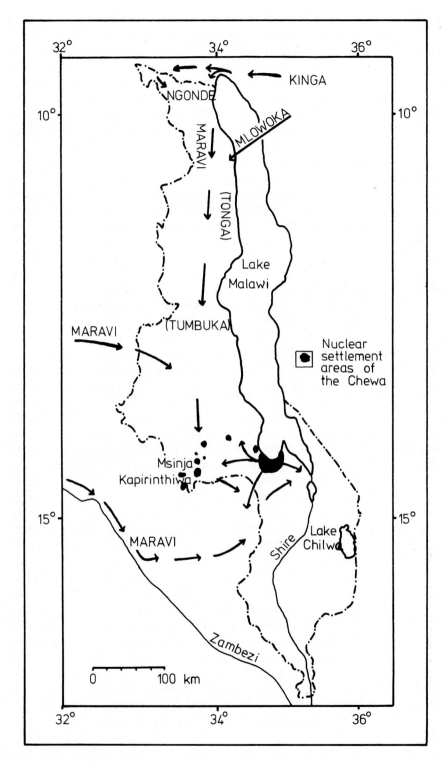

Fig. 16.1 Malawi *c.* 1200 A.D., showing early migration routes of the Maravi and other peoples (redrawn from Pachai 1973).

according to what they tell me, neither navigate nor enter into trade. The whole of this lake is strewn with desert islands in the lee of which those who sail on it shelter. It has a lot of fish; it is 10 "brassas" (about 22 m) in depth, and for this reason it has large waves resulting from the force of the winds which normally accompany the two monsoons, which occur also along the coast of Mozambique and which pass from there and are found over the lake in April and May. They say that the people living on the shores do not lack maize or wheat, and lack even less ivory which exists in abundance and is very cheap, (a good thing for those who wish to take advantage of the situation) . . .' (Beccari 1912).

The advantages in the country waiting to be exploited were underlined by Father Mariana and the shores of Lake Chilwa were similarly blessed. Though 'the oldest clear evidence of man's presence in Malawi' is traceable to a unique butchery site dated between 50 000 and 100 000 years old in northern Malawi, no such sites have been unearthed elsewhere in the country but 'there is no reason to suppose,' says Desmond Clark (1972), 'that other sites in undisturbed context with fossils will not be found when a systematic search of the Later Pleistocene alluvial deposits in the larger river valleys on the plateau and of the lake basins of Chilwa and Chiuta . . . is carried out.'

It is known that the advantages of the country were utilized by the Late Stone Age dwellers whose occupational sites in caves, rock shelters and open sites have been excavated and traced to the period 8000 B.C. to ± 200 A.D. and by Iron Age dwellers, whose sites have been excavated and dated from Karonga in the north to Kapeni in the south, giving an Early Iron Age date of third century A.D. at Phopo Hill in the north and a fourth century A.D. date at Nkope in the south. Later Iron Age sites in both the northern and southern parts of the country produce dates ranging from the eleventh century A.D. to the fifteenth century A.D., and Robinson (1972) concludes that: '. . . the available evidence may indicate that some of the first Early Iron Age immigrants to enter Malawi did so from the north, perhaps by way of the Songwe Valley. Groups may have travelled down both sides of Lake Malawi. The possibility that the lake itself was used by at least some of the immigrants seems worth considering. Although there is as yet no direct evidence that this did happen, it was an obvious choice for people who may already have been used to lakes and rivers. Evidence from Nkope shows that the inhabitants were well able to exploit the lake for food which would have been difficult without canoes.'

One serious lacuna in the state of present knowledge of the Iron Age in Malawi concerns the country between Lake Malawi and the Indian Ocean, the territory of Mozambique bordering on Lake Chilwa. About 1200 A.D., or when Maravi chieftainships were first being established in various parts of central and southern Malawi, an advance group of Lomwe, called the Lolo or Kokhola, were believed to have migrated westwards in the vicinity of Lake Chilwa and made contact with the dominant Maravi. By the sixteenth century A.D. these members of the Lomwe peoples lived with the Mang'anja group in the lower Shire area under Chief Lundu, a descendant of the royal family of the Karonga, founding father of the Maravi dynasty. In the seventeenth century the kingdom of Lundu expanded eastwards to include the country of Bororo which was in fact an extension embracing present Mozambique as far as the eastern seaboard. This included Lake Chilwa and a large portion of southern Lomwe country, leaving the main Lomwe cluster and the other eastern peoples, the

300

Makua and the Makonde, north of this line (Nurse 1972). If this was so, Lake Chilwa and its environs had already become a crossroads for the infusion and diffusion of external influences from as early as 1200 A.D. to about 1800 A.D. by which time another group of peoples, reputed to be traders, also became involved in the area, introducing yet another feature of early external contact between the peoples of Lake Chilwa and the eastern seaboard in the early phase of human settlement. These were the Yao who are generally associated with waves of destructive nineteenth-century migrations and settlement in the central and southern parts of Malawi but who, before that, were already engaged in a number of activities including agriculture, smelting and forging iron, hunting and fishing, making bark cloth and such items as hoes, axes, knives, razors and mat needles. One of their clan, the Chisi, built forges and furnaces and earned a reputation in the making of iron goods which, together with tobacco and skins, were exchanged with coastal traders for calico, guns and gunpowder, beads, trinkets and hardware. They also traded with the Maravi and obtained salt, ivory and cattle from them. It is possible that in the pre-nineteenth-century invasion era, the Yao traded with the Maravi at various points along the four lakes and in the intervening territory between the lakes and the sea (Abdallah 1919). This was a period of peaceful trade and the interaction in which the peoples in the Lake Chilwa area, among others, participated. The organization and extent of this interaction have yet to be investigated in Malawi and in Mozambique. If, as Abdallah points out, the market place for this early period of interaction was at Ng'ombo on the extreme southeast shore of Lake Malawi, the distance between it and the environs of Lake Chilwa (100 km) would not appear to be prohibitive (Alpers 1972). In the mid-eighteenth century, ivory was the mainstay of the trade with Malawi while the Arabs and the Portuguese were the main customers (Pachai 1973).

In this section of the early phase of human settlement and activity it is possible to conclude that the area of Lake Chilwa and its environs was settled by different cultural and ethnic communities for centuries before 1891; that the land abounded in game and parts had good soil for crop cultivation. It was also good cattle, elephant and fishing country. It had, therefore, the potential for expansion and survival, aspects which became more crucial with the advent of the later phase to which we must now turn.

2. The later phase of human settlement and activity

Of the fourteen main ethnic groups in Malawi today (Chewa, Lomwe, Nyanja, Mang'anja, Ngoni, Tumbuka, Ngonde, Yao, Sena, Tonga, Lukwa, Lambya, Wemba and Nyakyusa), three resulted from migrations in the nineteenth century, the Ngoni, Yao, and Lomwe (Fig. 16.2). The fact that there are no identifiable ethnic boundaries separating the groups would suggest that a thorough mixing of peoples has taken place in settlement areas under different chieftainships. Chiefs' areas were settled by any number of representatives from different ethnic groups even if certain villages were predominantly settled by members of a single ethnic group. A blurring of identities has also taken place, especially of minority groups who sought security and advancement through association with dominant groups.

Of the three migratory groups of the nineteenth century the Ngoni were the only ones without any form of previous contact with their hosts. Two Ngoni divisions, the Jere and the Maseko (Fig. 16.2) entered Malawi around 1836–37 and, after a period of meandering in search of permanent settlements, established themselves in different parts of the country – the Jere settling in the north about 1855 and the Maseko in the central region around 1867. For the purposes of our present study, it is worth noting that the Jere and Maseko Ngoni did not remain intact for long in their chosen homes in Mzimba or Ntcheu but that fragmentation soon led to their settlement and influence over a larger part of the country extending from north to south. Jere offshoots spread

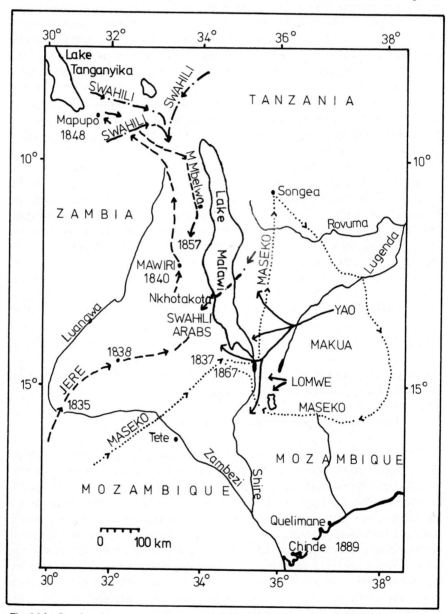

Fig. 16.2 Immigration routes of three ethnic groups in the nineteenth century: Ngoni (Jere and Maseku), Yao and Lomwe (redrawn from Pachai 1973).

302

to Dowa and Nkhotakota while Maseko offshoots spread to Dedza, Lilongwe, Mwanza, Mchinji, Thyolo and Chiradzulu (Pachai 1973). While it has been argued that the Ngoni factor was destructive in the wake of its military domination and that villagers were terrorized into submission or forced to flee, the negative picture is balanced by the stabilizing influence of the Ngoni presence in the mid-nineteenth century when refugee or invading Yao communities also entered the country occupied by the Maseko Ngoni. The Ngoni offered shelter and security from slaving parties and practices.

The Yao refugees and invaders of the mid-nineteenth century were in circumstances different from those of their peaceful ancestors centuries before. Victims of pressures behind them, the Yao found outlets in forced migrations into Malawi. The period of peaceful trade of the early eighteenth century received a serious setback when Lake Malawi, and the hinterland it served, was opened to foreign traders towards the close of the eighteenth century and became involved in the trans-shipment of slaves and ivory across Lake Malawi. Further south, the position was extremely difficult for the people of the Lake Chilwa area: 'To the south and east of the Lake (Malawi) and between it and Lake Chilwa, the Yao, infiltrating as colonisers since 1830, brought their chiefs into the web of coastal commerce. They had been long-distance traders since the previous century. Powered by guns and adopted into Islam in 1890, the Yao became the long arm of the Indian Ocean slave trade preying upon and occupying the land of the undefended "Nyanja' of the lakeshore and Shire Highlands' (Agnew 1972).

Between the 1830s and the 1860s, or about the very time when the invading Ngoni were in search of their new homes, four of the ten original Yao clusters in Yaoland entered Malawi as invaders and as refugees, in war as well as in peace. The general picture which gained official credibility in 1891 was that all the Yao were inveterate slave traders and raiders, whose presence and activities in Malawi were prompted solely by their interest in the slave trade. While some Yao leaders did indeed maintain an active interest in the trade it became necessary for the British administration in 1891 to label all the Yao as incorrigibles, who had to be brought to their knees by military conquest. Such a view was strenuously resisted by the Anglican missionaries who worked among the Yao in Mozambique: '. . . we do not admit that they are bad people at bottom. We believe them to be distinctly contaminated by coast influence, which has acted only too surely, on their quick receptive natures . . .' (Chauncey Maples 1893). The first British Consul and Commissioner-General, Harry Johnston, received official approval for his wars of conquest, largely because he attributed all the sins of the time to 'foreigners' like the Ngoni and the Yao. For convenience, he decided to wipe out the Yao resistance first. He reached this conclusion despite the experience of the Livingstonia missionaries who had received land and permission in 1875 from Mponda, a Yao chief, to open up a mission at Cape Maclear and the Blantyre missionaries who had received similar support in 1876 from Kapeni, also a Yao chief. In the first few months of setting up his new administration Johnston was at war with a number of Yao chiefs, mainly in the region south of Lake Malawi and certainly in the environs of Lake Chilwa: Chikhumbu, in Mulanje and Mponda, Makandanji, Zarafi, Kawinga and Makanjira in present Zomba, Kasupe and Mangochi districts. These wars lasted five years (1891–96) and probably represented one of the most disturbed periods in the history of Lake Chilwa and its peoples, from the point of view of peace, law, order and stability.

The country to the south of Lake Chilwa leading to the Fort Lister gap between the Machemba and Michesi peaks of Mulanje mountain was a popular land route for slave caravans heading for the coast. Mangochi, to the northwest of the Chilwa plains, situated between Lakes Malawi and Malombe, represented another popular slave route which headed eastwards for the coastal disposal points across the Chilwa/Chiuta sand bar. The other popular Lake Malawi disposal points, Nkhota Kota, Chilumba and Karonga were too far away to affect the Chilwa area. With two of the five disposal points so perilously close to Lake Chilwa and the Chilwa plains being familiar ground for slaving caravans, the effects on the area must have been devastating.

The effect of slave raids was mitigated by two contemporary events: the first of these was the setting up of British rule with the ultimate intention of creating law and order; the second was the opening of a mission and school on the Chilwa plains among the Lomwe peoples in 1894. This takes us to the third of the migrant communities of the nineteenth century to enter Malawi, the Lomwe. Various written accounts refer to Lomwe peoples living in Malawi in the nineteenth century: Livingstone found a Lomwe village at the confluence of the Shire and Lirangwe rivers in 1859, situated to the west of Zomba mountain; in 1861, missionaries found some Lomwe residents at Magomera, south of Zomba town; a government official, Duff MacDonald, observed Lomwe settlers along the Malawi–Mozambique border area between 1882 and 1894. By 1891 when British rule was begun there were already some 100 000 Lomwe in Malawi in the Shire Highlands. By 1945 the figure had risen to over half a million (Chilivumbo unpublished data, Baker 1961).

One of the early larger concentrations of Lomwe workers on European private estates was that on Songani Estate on the outskirts of Zomba town. This estate was opened in 1893 by R. S. Hynde and R. R. Starke for the purpose of growing tobacco. Though Africans had been growing tobacco before 1893, Hynde concluded that Europeans would not be attracted to the local variety grown. He imported seeds from Virginia in the United States and distributed these among his African tenants. Though the tobacco was smoke-cured from fire burning in the middle of the huts, the cultivation was very successful. When Songani Estate was taken up by the Blantyre and East Africa Company in 1901 and fire-curing was introduced, tobacco cultivation in Zomba increased rapidly (Rangeley Papers 1956). With the expansion of the plantation economy in the wake of British Protectorate rule, migration of Lomwe workers from across the border also increased. There was another reason, too, why thousands of Lomwe persons crossed over from the 1890s onwards: the whole of Lomwe country was at last opened up to the Portuguese administration between 1897 and 1907. The advent of this administration introduced forced labour, heavy taxation, physical brutality and plunder of life and property. Some Lomwe chiefs resisted while others collaborated. In the end both suffered the same fate. The final result was that people fled to the British Protectorate of Nyasaland.

'A few years later there was famine in the country. Whatever little food there was such as cassava and sorghum, was plundered by the Boma sepoys. By this time there was no alternative for the poor Alomwe people but to leave the

304

country and to settle in neighbouring British territory. Some of them entered the country through the Chilwa plain and settled in Zomba and Chiradzulu Districts; others came through the North and South of Mlanje and settled round Mlanje and Cholo Districts. Thus these areas became occupied by the Alomwe people.' (Bandawe 1971.)

In the context of these developments, British missionary endeavours in Portuguese Catholic domains stood no chance of lasting success. But the attempt was made and the missionary impact on Lomwe life must be recognized. It began by chance: in May, 1894 Blantyre was devastated by a locust plague in the thick of the harvesting season. Anticipating serious food shortage in 1895, three missionaries of the Church of Scotland, Blantyre Mission, set out for Lomwe country to the east of Lake Chilwa to purchase food crops (Rankin 1896). In August, 1894 they reached a point east of the southern extremity of Lake Chilwa where they observed that the 'surface bore evident marks of having at one time formed part of the bed of the lake. They found the earth impregnated with salt: the lake overflows much of the country in the wet season, and from the soil the natives extract, by a simple process, a coarse salt of a dirty grey color. The number of salt-earth heaps at the side of the houses gives the village a rather uncomfortable appearance.'

In two days the missionaries had no difficulty in purchasing 5000 pounds in weight of rice (Rankin 1896) and had they stayed longer more would have been easily obtained. This draws attention to the indigenous cultivation of rice. The expedition also described the ravages of the slave trade and the reception accorded them by friendly Lomwe. One Mlomwe boy, then aged seven, was one of those who met the mission party. His name was Lewis Mataka Bandawe, whose memoirs tell of this trip. Two years later, in 1896, after a year of negotiations, the Church of Scotland Blantyre Mission opened three mission stations in Lomweland, east of Lake Chilwa.

The three mission stations were allowed to function for six years only. In 1902 the Portuguese authorities directed that they be closed. Soon after, the buildings were set on fire by the Portuguese authorities and razed to the ground. By then the attachment of the Lomwe communities to these missions was such that many more ventured into Malawi.

In the overall story of life on Lake Chilwa and on its plains extending in all directions, it is possible to say that the area and its peoples were never in a state of complete isolation from the almost continuous traffic of influences. Its geographical situation was such that it, too, was a form of mini-corridor in relation to the huge lake to its north, to the Zomba and Mulanje mountains flanking the area in the east and west, to the Shire Highlands with whose economic and political history it was tied and to Zomba town and district adjacent to the Chilwa area. When European elephant hunters shot game in the Shire valley in the 1870s they also visited the Lake Chilwa plains; when the consulate was built at Zomba in 1887, the influences reached out for miles around as did the influences of Zomba and Domasi Missions of the Church of Scotland in the 1890s (Watson 1955).

Because of its low-lying situation with a comparatively dry and enervating climate and less fertile soil the Lake Chilwa plains did not experience the land pressures which Malawians have suffered from in colonial times in the Shire Highlands, in places such as Mulanje, Thyolo, Blantyre, Chiradzulu and

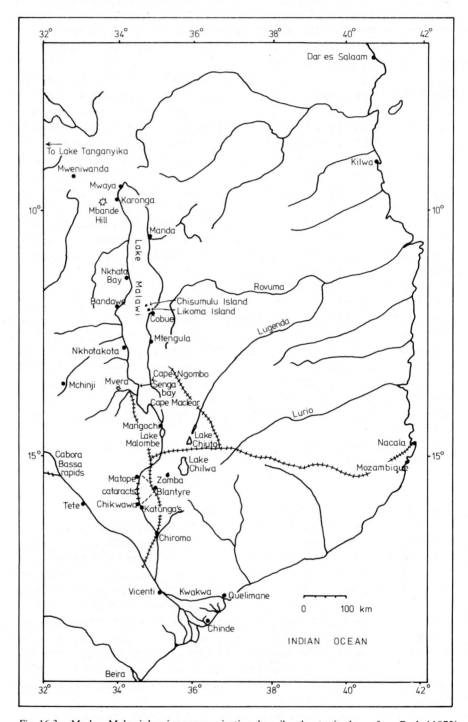

Fig. 16.3. Modern Malawi showing communications by rail and water (redrawn from Pachai 1972).

Zomba where white settlers grew tea and tobacco. Perhaps it was advantageous in the long run that the northern extension of the railway line was built on the central, south to north, route from Blantyre to the lake, having the effect of fending off alienation of land on the Chilwa plains. The closeness of Lake Chilwa to Zomba and the links between Zomba and Blantyre in colonial times rendered it unlikely that the economic development of the Lake Chilwa area could have been arrested through inadequate transport and marketing facilities, although these towns remained inaccessible to the inhabitants of the Chilwa plains except on foot. In more recent times, an interesting development was the opening of the rail link in July, 1970 from Nkaya junction to Liwonde and from Liwonde to Nova Freixo and Nacala on the Indian Ocean. This line which passes between Lakes Chilwa and Chiuta has brought the railway very close to Lake Chilwa and has already influenced the lives of the people on the Kawinga plain (Figs. 16.3 and 2.8a).

The mixed ethnic origin of the people of the area is a boon to performance since ethnicities need not come between organization and individuals. At the present time, for example, in Chief Mwambo's area in Zomba district there are five ethnic groups juxtaposed under a single traditional authority: Lomwe, Chewa (Nyanja), Yao, Ngoni and Sena. A student conducting field investigations a few years ago noted: 'It is generally difficult to say for sure where each of these tribes are mainly settled' (Takula 1974). There is much in common in their cultures (Chapter 17) and this ease of association is important, since economic development, social cohesion and political stability need the factors and forces that unite rather than those which divide in the process of modernization.

References

Abdallah, Y. B. 1919. Chikala Cha Wayao ed. & transl. M. Sanderson. Zomba.

Agnew, S. 1970. Factors affecting the demographic situation in precolonial and colonial times. Seminar Paper 2977, University of Malawi, Zomba (unpublished).

Agnew, S. 1972. Environment and History: The Malawian Setting. In B. Pachai (ed.) The Early History of Malawi. Longman, London, pp. 28–48.

Alpers, E. A. 1972. The Yao in Malawi: the importance of local research. In B. Pachai (ed.) The Early History of Malawi. Longman, London, pp. 168–178.

Baker, C. 1961. A note on Nguru immigration to Nyasaland. Soc. Malawi J. 14:41–42.

Bandawe, L. M. 1971. Memoirs of a Malawian (ed.) B. Pachai. CLAIM, Blantyre, Malawi.

Beccari, Camillo. 1912. Rerum Aethiopicarum Scriptores Occidentales. 12, Rome.

Chauncey Maples 1893. The Yao People. Nyasa News 2. Blantyre, Malawi.

Chilivumbo, A. B. People on the Move: A Study in the processes of immigration, integration and adaptation of Alomwe Peoples into the Malawi society (unpublished).

Desmond Clarke, J. 1972. Prehistoric Origins. In B. Pachai (ed.) The Early History of Malawi. Longman, London, pp. 17–27.

Nurse, G. T. 1972. The People of Bororo: a lexicostatistical enquiry. In B. Pachai (ed.) The Early History of Malawi. Longman, London, pp. 123–135.

Pachai, B. 1972. (ed.) The Early History of Malawi. Longman, London. 454 pp.

Pachai, B. 1973. The History of a Nation. Longman, London. 324 pp.

Pachai, B. 1978. Land and Politics in Malawi 1875–1975. Limestone Press, Queens University, Kingston, Ontario, pp. 11–29.

Rangeley Papers. 1950. In J. Desmond Clarke to W. H. J. Rangeley, File 1/1/1. Malawi Society. Blantyre, Malawi.

Rangeley Papers. 1956. The History of the Tobacco Industry of Nyasaland. File 1/3/6. Malawi Society, Blantyre, Malawi.

Rankin, W. 1896. A Hero in the Dark Continent. Memoir of the Rev. W. Affleck Scott, Edinburgh and London.

Robinson, K. 1972. The Iron Age in Malawi: a brief account of recent work. In B. Pachai (ed.) The Early History of Malawi. Longman, London, pp. 49–69.

Takula, W. H. 1974. The History of Chief Mwambo with special reference to the Lomwe Tribe. History Research Paper, University of Malawi, Zomba (unpublished).

Watson, J. H. E. 1955. Some historical notes on Zomba. Soc. Malawi J. 8:58–71.

7 The people and the land

Swanzie Agnew

The people and the land

Malawi is one of the smaller states in East Africa, but its population density is about tenfold that of the neighbouring states of Mozambique, Tanzania and Zambia, which share common boundaries with it. The total population in 1977 was just over five and a half million (Population Census 1978). The area of Malawi, including land and water, is 118,485 km². The Lakes Malawi, Malombe, Chilwa and Chiuta cover 21 per cent of the total area and in addition there is another 5.5 per cent of wetlands, including marshes, swamps and deltaic areas of rivers and the seasonally flooded grassland of low-lying areas, leaving 92,989 km² of inhabitable land, 58 per cent of which is arable (Agnew & Stubbs 1972).

So high a proportion of water and wetland gives considerable importance to fishing in the general economy of this lakeland country, and provides the opportunity for the cultivation of paddy rice. Alternatively the floodplains may be used for natural grazing. When development of these areas is undertaken a careful appraisal has to be made in allocating areas for staple food crops, cash crops such as rice and cotton and for cattle. This problem is highlighted in the Chilwa basin, where periodic fluctuations in lake level give a measure of uncertainty to the fishing industry, so that support for the people must be derived from the alternative development of the agricultural potential of the lacustrine plain, which, as will be seen below, is quite considerable.

1. Inhabitants of the Chilwa basin

The Chilwa basin, an area of 7500 km², occupies parts of the three Administrative Districts in the Southern Region of Malawi: Kasupe in the north, Zomba in the west and Mulanje in the south (Fig. 17.1). These parts are under the immediate control of eight Chiefs, whose names are used to denote areas. Table 17.1 indicates that the total population of these Chief's areas in 1977 was 382 895 persons, and of these 25 000 or 6.5 per cent was concentrated in the fishing villages of the floodplains, in a belt about 8 km wide around the shores on north, west and south and at Chinguma in the northeastern peninsula. The number of people in the fishing villages is almost three times as high as it was in 1966 and is comprised, to a large extent, of temporary male soujourners connected with fishing (Population Census 1972, 1977). These men belong largely to the farming villages of the plains.

The location of Lake Chilwa, bounded by the Shire Highlands on the west watershed, is potentially advantageous to the economy of the peoples of the plains, since the highest densities of population in Malawi, with over 500 persons in settled areas, rising to 1000 km⁻² in a few dormitory areas, occur there (Population Census 1972). These densely settled parts contain the majority of the wage earners in Malawi. There is also a dearth of agricultural land in the Shire Highlands for subsistence farming and for cultivation. The Lake Chilwa Fisheries already plays a large part in supplying food to the urban people, and the prospect of markets for farm produce is good, but as yet little developed (Coleman 1976).

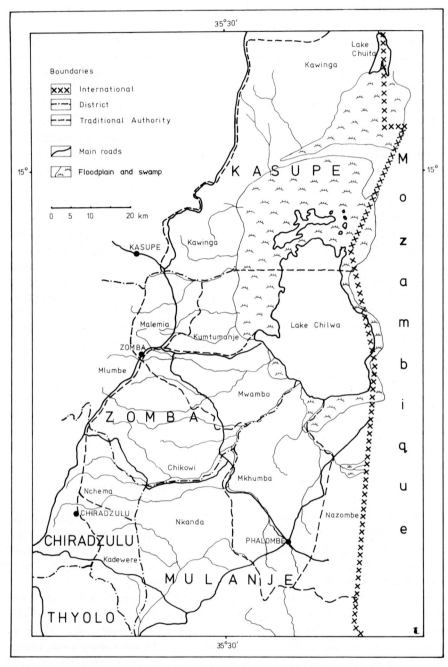

Fig. 17.1 Administrative Districts and Traditional Authorities around Lake Chilwa.

Despite the fact that the Shire Highlands, during the colonial period, was economically the most advanced of the regions of Nyasaland, the Chilwa basin remained little affected by modernization. It became, however, the reservoir for population overspill. Immigrants who were unable to find land elsewhere sought free unoccupied land on the Chilwa plains, and the more enterprising attempted to enter the money economy through fishing and fish trading (Agnew 1972). Large changes in the intensity of fishing and fish trading have taken place in the last ten years or so, as will be seen in Chapter 18.

Table 17.1 Population : 1966 and 1977 Censuses : Chilwa areas, Malawi.

District	Traditional Authority – T.A.	Population T.A's 1966	Population T.A's 1977	Population, E.A.* touching on L Chilwa – 1977
Zomba	Mwambo	60 935	72 793	7 286
	Mkumbira	2 015	5 195	5 195
	Kumtumanje	30 750	40 829	1 666
Kasupe	Kawinga	42 801	54 052	6 033
	Mlomba	11 581	20 087	– (marsh land)
	Mposa	7 104	9 895	289
Mulanje	Mkhumba	89 971	122 048	2 760
	Nazombe	51 796	57 996	2 118

*E.A. Census enumeration areas.
Source: National Statistical Office, Zomba.

In an ethnographic context, Malawi forms part of the Central African complex of matrilineal societies with little tradition, prior to the nineteenth century, for cattle keeping. Fish was an accepted form of relish in a diet low in animal protein. Historically, dried fish and salt, had long been commodities for long distance travel (Fagan 1969). The Chilwa people played a part in revictualling the slaving caravans in the eighteenth century and later in supplying fish to the settlements on the highlands in the savanna woodlands (Bandawe 1971).

The historical origin of the ethnic groups on the Chilwa plains has been described in Chapter 16. Chilivumbo (1969) has explained the similarities in culture of these people as follows: 'Four main ethnic groups live in the area and these are: Yao, Lomwe, Mang'anja and Nyanja (Chewa). Although linguistically somewhat different, their religious practices, educational system, their land values, their kinship system and familial social structure are virtually the same. They are all matrilineal societies (i.e. inheritance is through the mother's line). This cultural similarity has facilitated inter-ethnic marriages, so that the tribal divisions are breaking down.'

The organization of villages on the plains depends on matrilineage. Large families based on the mother's relations, i.e., her sisters and their husbands, her daughters and their husbands and the small children of all, tend to stay together in clusters of houses. This forms the basis of the 'extended family', the foundation of the society. All members of it have a serious responsibility for the welfare of all its members. Even when a son or daughter leaves village life to make his or her way in the professions in the urban area, the first duty is the welfare of the extended family. The eldest son of the oldest woman is the leader

313

of this group. Several extended families related through an older female ancestor constitute one matrilineage and a group of matrilineages together inhabit a village with a Village Headman, responsible to a sub-chief or Chief.

Males who may belong to a different ethnic group, through the uxorilocal custom of living on his wife's land, can thus join a matrilineage by marriage. The children acquire the ethnic affiliation of their mother. In practice a woman's eldest brother is the caretaker of her land and the custodian of her children.

Among these people the concept of private ownership of land is non-existent; the land belongs to the people as a whole. By custom, the Chief (or his deputy) gives land to a woman when she marries and her husband leaves his mother's family to live on his wife's land. The 'right of land use' is recognized and husbands obtain this right through their wives and may build houses and grow crops. The 'right of use' of the land can be passed on to the next generation through the daughters, when they marry. There is no buying or selling of land, although it may be transferred (by the Chief) or inherited.

Since 1967, this type of 'right of use' of customary land has been superseded by land tenure of a different kind in those small areas set aside at present for special development by Government. The new type of tenure is dependent on the satisfactory use of the land. Within a Development Scheme, land may be allotted by a committee to men or to women who accept responsibility for growing crops according to set agronomic practices. The land may be forfeited if proper use is not made of it, but it may also be inherited when satisfactorily worked.

2. Settlement and housing

The distribution of the population of the Chilwa plains is determined by the extent of the water during the wet season. Dry-siting of housing is provided by the micro-relief, such as: sand spits and beach bars of the former and present lake shores, the levees of river channels and deltaic streams, the high slopes of the plains, or along roadsides.

Settlements therefore may occur in alignments, such as along the Phalombe River and Chiuta sandbar, or they may be spread evenly in a well-drained area such as that of the Phalombe plain in the south or the middle Domasi basin in the north. The edges of roads and serviceable tracks to markets and trading centres for surplus crops are becoming increasingly favoured for housing.

There is an element of impermanence in settlement distribution in those areas that are particularly affected by the fluctuations of lake level. Settlements that are dependent on fishing are particularly ephemeral. During periods of high fish production, temporary hutments along the shores proliferate. A scale of permanence ranges from the reed shelters erected on *Typha* platforms, *zimbowera*, in the swamps to the mud-plastered houses of the permanent villages, to those with sun-dried clay bricks and iron roof, to the brick and concrete housing of the personnel of the Fisheries, Agriculture and Veterinary Departments and for some retired men of means (Figs. 17.2a, b, and c).

Traditional houses are built by a man and his neighbours from local materials. First a rectangular framework made of fairly straight poles is erected with stout posts at the corners and for the midrib of the roof. Branches are

314

Fig. 17.2a *Zimbowera*, a temporary hut built in the swamps (photo: A. J. McLachlan).

Fig. 17.2b A traditional house on the Chilwa plain with *nkokhwe* (photo: N. Lancaster).

315

Fig. 17.2c   A. 'modern' type house made with sun-dried clay bricks on the Chilwa plain (photo: J. Lancaster).

interwoven horizontally and vertically to form an effective framework of laths for mud plaster. The supporting structure is allowed to dry for many weeks or months. Then a clay plaster is applied to both sides to make walls about 15 cm thick. The roof is thatched, usually with dried *Hyparrhenia* grass to give an overhanging eave, acting as a drip-course, to protect against splash erosion at the foot of the wall. The eaves also provide shade for children at play or women at domestic work. While being constructed, one or two small windows may be left, although light is obtained mainly from the single door, facing the main yard or roadway. Within the traditional house, boxes may be used for storage, and clothing and utensils may be hung from beams and posts. Mats are laid out at night for sleeping on the stamped earth floor. Separate accommodation is built for boys and for girls past infancy. A cook/eating house may also be erected, so that the family comes to occupy a group of huts. A set of woven bamboo storage bins, *nkokhwe*, are constructed, in which the harvest will be stored. These are raised above ground for dryness, and against infestation by ants and rats. Small stock may be kept under the raised platform. In good weather, meals are taken outdoors, the family seated around the cooking pots in the yard.

The traditional method of house construction is slowly being replaced by more modern housing, with walls of large sun-dried clay bricks, which are not plastered (Fig. 17.2c), and may be fitted with window frames and doors. The roof may be thatched or made of corrugated iron. Such houses are often larger and contain tables, chairs and beds.

The degree of mobility of the people in response to fishing may be gauged from the count of the population of Nchisi Island, who are both farmers and fishermen. In 1966 at the height of successful fishing, the census enumeration established a population of 2000 people; but in a recount in 1968, when the lake was dry, only 779 persons were recorded in the four 'permanent' villages (Agnew 1970). The number of people on the island had increased to almost 3000 in 1977 in the years of the high lake level (Census Report 1977). At Kachulu harbour village there were 800 residents in 1966, which dropped to 186 in 1968, when fishing had ceased. Many buildings were in a state of collapse. Such was the direct effect of the drying and recovery of the lake on the settlement.

The migration of the fishermen to the lake in years of high level draws to the fishing centres a host of other people. These include men hired by those fishermen who own their own canoes, fish traders, who own bicycles, fish processors, women sellers of beer and flour, pedlars of medicines and charms and ferry service men who work canoes or motorized craft to carry passengers and goods across the lake from Kachulu to Chinguma or to Nchisi or Tongwe Islands (Phipps 1973).

3. Agricultural potential

Four main categories of land for primary production may be broadly identified. The first two are characterized by excess water during the rains. This land is suitable for both rice-growing at the height of the seasonal flooding and for stock grazing when the water level falls at the approach of the dry season and a succession of grasses and marsh plants offer pasture. The other two categories represent land for hoe cultivation. The first constitutes the well-drained soils of the drift-plain, where permanent agriculture based on fallow rotation is practised. In these areas a sandy soil occurs which cakes after rain and may develop a subsoil of lateritic gravel or hardpan. The second type of dry land cropping occurs within the lacustrine plain, where rapid changes in soil composition and groundwater level demands selective planting and variations in height of ridges for drainage, when maize, millet, cassava, beans and groundnuts are grown.

The response to these variable soil and water conditions is demonstrated by the distribution of the main cash crops shown in the map (Fig. 17.3). It will be seen that although indigenous rice-growing is restricted to the periphery of the marshes in seasonally inundated land, there are four irrigated Rice Schemes based on perennial rivers. Groundnuts and tobacco (the latter farmed by whites as well) are on higher ground and governed by soil types and ground water levels. Cotton has become more widespread, because it can withstand drought and porous soils, better than the other two crops. A survey of cotton-growing on the southern Chilwa plain was made by Chikwapulo (1971) (Fig 17.4).

On the Chilwa plains may be seen a transition from subsistence farming to the rising importance of cash cropping and a money economy in which cattle are very slowly taking a more important place. The most backward economy is seen in the Kawinga area in the northern Kasupe District (population 54,000) and the northern marshland fishing communities of Mlomba and Mposa

(20,000 and 9895 inhabitants). The people living in these areas have little contact with modern life since the only road is impassable for half the year, and markets are inaccessible at the time when crops are ripe. In the dry months trade in locusts is an interest, since the area had been a minor potential 'outbreak' area of the red locusts for many years. Locusts are picked off the *Typha* plants, where they roost at night, and carted in large sacks on foot along winding paths to Zomba market (Chapter 15). Until the railway, from Blantyre through to Nacala in Mozambique across the sandbar, was opened in 1970 there was little contact with the world. Since then a trading centre has been

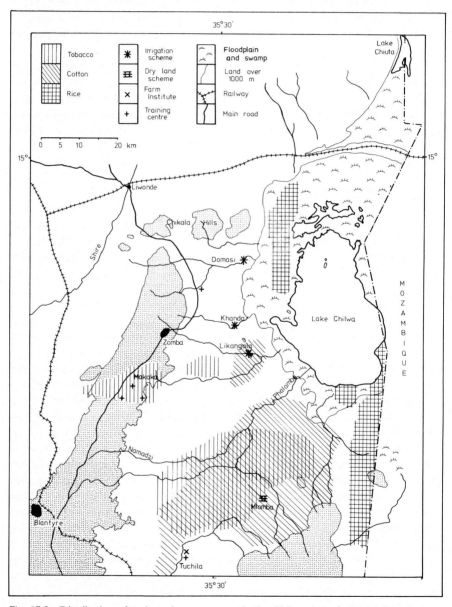

Fig. 17.3 Distribution of main cash crops grown in the Chilwa Area (redrawn from Surveys Department Maps).

318

opened where surplus produce may be sold for cash to the Government organization of the Agricultural Development and Marketing Corporation (ADMARC), thus stimulating the production of surplus farm produce for the first time. The results of an agro-economic survey in 1972 in this area are presented in section 17.4.

In contrast to this isolated situation, there are the Areas of Mwambo and Kuntumanje in the central, Zomba District (population 117 817), where the

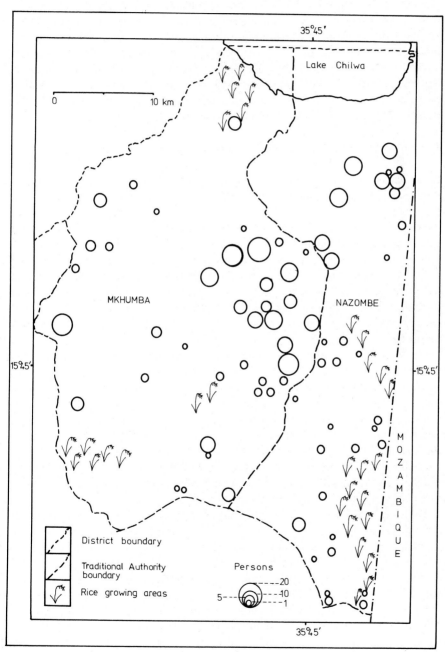

Fig. 17.4 Distribution of cotton growers on the Phalombe plain (after Agnew 1973).

people have had access to Zomba markets and further afield, since an all-weather road to Zomba from Kachulu harbour was built in 1967. In addition International Aid has encouraged the intensive cultivation of rice, with the greater part of the crop sold for cash to ADMARC, bringing about an increase in income for those engaged in irrigated farming.

In the southern, Mulanje District, the more densely populated Mkhumba Area (population 122 048) has the advantage of the two perennial rivers and marshland. A cotton Development Scheme has been commenced and a piped water supply to some villages completed. It too has access to a large market amongst the workers on the tea plantation around Mulanje. The incentive to produce for sale has grown since coffee, tea and tobacco plantations have attracted seasonal labour on a large scale to the Shire Highlands, from the beginning of the 20th century. To illustrate the effect of 'development', a study of a rice project will be described in section 17.5, in contrast to that of the traditional economy detailed in the next section.

4. Semi-subsistence farming on the northern plain, Kawinga area

In preparation for a future development plan covering the lacustrine plain, an investigation was conducted over a period of two years, 1970–71, on the Kawinga plain (Agro-economic Survey 1972). Thirty-nine households were studied daily at four villages just south of the new railway line on the Chilwa/Chiuta sandbar in Chief Kawinga's area. The findings of this survey give an account of subsistence and semi-subsistence peasant agriculture with an incipient cash economy. These general characteristics are applicable to all land-based villages on the Chilwa plains, even though emphasis may shift between fishing, stock-keeping and agricultural interests according to the local ecological conditions, the proximity of the village to the lake and the accessibility of markets.

This case study is broadly representative of differences in standards of living, the variation in size of land holdings and the kind of farming done by the rural people within a small area, when there is no injection of capital from an outside source, such as from fishing or from Development Aid. Of the 39 households investigated, comprising 160 people, 29 were headed by men and ten by women, whose husbands were absent as migrant labourers. The size of the household or 'garden' families, who shared the same 'pot', ranged from one to eight persons, but the majority of households catered for three or four persons. In the various age groups, 74 were children under the age of 15 years (46 per cent) and 18 were relatives above that age.

Table 17.2 shows that holdings vary widely in size, and this is common throughout the lake plain. The mean size of holding was 1.92 hectares, which is slightly larger than the national mean, and considerably higher than in the more heavily populated areas of the Kawinga District further from the swamp to the north, where 0.4 hectare is the average holding. But over half the households had less than 2 hectares. The area under investigation was deliberately chosen to include villages where proximity to the swamp had encouraged rice farming. The large size of holding in these villages depends on the free availability of the floodland for cultivation, which was hitherto not much used. Occupation is at present mainly restricted by the nature of the intractable soil, which makes

320

heavy work for hoes and which requires intensive labour for planting and weeding.

Table 17.2 also indicates the mean number of 'gardens' for each holding, which is 3.3, a little higher than the average for the country as a whole. Some gardens are on drier ground where maize, cassava and millet are grown and those further away from the villages on the wet soil are used for rice. Some may be as far as 5 km distant. The man/land ratio, i.e. the available land per adult (with children under 15 years calculated as half a unit) varies with the size of the holding. The basic requirement in African agriculture is considered to be 0.4 hectares per head or 2 hectares per family of 4–5 people, where maize is the staple food. In this case no surplus maize can be produced for sale, and, in this study, 50 per cent of the families fell into this category. When the holdings are between 3 and 4 hectares, there is greater latitude for successful small-holder farming, but in the four Kawinga villages only five out of thirty-nine families achieved this. When the holding is 6 hectares, the farmer is of quite a different kind. He can produce a surplus for sale and has entered the cash economy. This depends on his use of more modern practices, such as hiring a tractor or an ox and the employment of paid labour. In the sample villages only one man fell into this category. Table 17.2 illustrates, by the sizes of the holdings, a gradient from complete subsistence economy to a cash economy.

Table 17.2 The relationships of size of holding, number of gardens and man/land ratio in 39 households of 4 villages in Kasupe District (1970–71).

Area hectares	No. of families	Mean size of holding hectares	Mean no. of gardens	No. of adults	Mean no. adults per holding	Man/land ratio
0–1	8	0.8	2.1	19	2.4	0.8
1–2	13	1.5	3.4	40	3.1	1.2
2–3	13	2.1	3.5	49	3.8	1.5
3–4	4	3.5	4.3	16	4.0	2.2
5–6	1	5.8	5	5	5	2.9

Source: Agro-Economic Survey No. 9. Lake Chilwa 1972.
Government Printer, Zomba, Malawi.

It may be questioned how it is that in a country where land is given and not sold that there should be a four to six fold difference in the sizes of holdings. The answer appears to be that a woman may obtain as much land as she and her husband can work. If her husband has obtained capital enough by savings from wage labour as a migrant or domestic worker, or as a fisherman or civil servant, he can afford to work his land with the assistance of hired labour and to use an ox for ploughing or even to hire a tractor for a day. The employment of labour raises the output, all of which is sold to the Government agency ADMARC. Thus profits will grow, more help can be employed, and more land can be worked. Cash will be earned and children will go to school and acquire more knowledge of agriculture practice. Extension pamphlets will be read and more innovations will be possible.

The main crops grown were maize, sometimes interplanted with cassava, and rice. Cassava is the 'insurance' crop to be consumed if others fail. About one half of each holding was planted to the food staple, maize, on plots ranging from 0.4 to 1.2 hectares. The average yield of 2172 kg ha^{-1} for these villages is considerably higher than that of Kawinga area further north and west, which is more highly populated (with smaller holdings and poorer soil). On the average, in Kawinga area 'gardens' yielded only 904 kg ha^{-1}. No fertilizer was used in this area. The yield from the very few holdings where land was fertilized was much higher than in the four sample villages, namely 3478 kg ha^{-1}.

Yield of maize varied with the age of the grower: older householders obtained far better crops than younger men. A similar tendency towards a relation between yield level and growers' ages was also found with rice farming. The implication here is that when a man considers himself 'retired' from migrant labour, his efforts are then concentrated on farming. The yield of maize also varied with the size of the garden and that obtained from under one hectare was doubled in the largest 'gardens'.

About 25–40 per cent of each farmer's land in the survey was under rice. The yield of rice per hectare about equalled that of maize, less than 2000 kg ha^{-1} in the small gardens and over 3300 kg ha^{-1} in the larger gardens. Even the higher yield is less than that urged in the Guide to Agricultural Production in Malawi, which states that using the indigenous Faya variety (as people do in the Kawinga villages) yields of up to 4320 kg ha^{-1} of rice, worth K150 (approximately £100 Sterling) can be grown, provided that seeding is done in lines, which reduces the weeding time by half (Ministry of Agriculture 1977). The yields of Kawinga are similar to those in irrigation schemes in the Northern Region of Malawi.

The organization of an average working day of seven to eight hours may be relevant to understanding why the yield of farm produce is low. Only 34 per cent of the time is, on the average, spent on work connected with farming by both men and women, although, at times, many more hours are devoted to planting or weeding. About 39 per cent of the time is spent on 'domestic work' by both men and women, although the man's work is more concerned with house maintenance. Women walk long distances to fetch water every day and pounding maize may take one and a half days per week. About 12 per cent of the time is spent helping neighbours and on social obligations; 3.5 per cent of the time only is spent on livestock care and that is mainly the responsibility of the young men and boys; 1.5 per cent of the time is spent on fishing, mainly by men and boys, 7 per cent of the year may be devoted to paid work. School is attended mainly by children of the larger farms, but also by boys over 15 and some men as well. Significant in 'domestic work' is the 9.5 per cent of the time spent on the care of the sick.

The farm production depended almost entirely on the effort of those responsible for the holdings, in other words, the man and his wife, rather than on extra labour from hired men or relatives, or from children whose contribution is insignificant. Girls from the age of about ten years participate in looking after younger children, cooking and fetching water and boys herd cattle and help the uncles and fathers.

The amount of time spent on growing rice is about five times that required for maize, but the months of intensive work for each crop alternate. Maize demands time from November to January and again in April while rice is more demanding in January to March and May to August. Cassava requires the least attention and occupies time in April, July to September and in November.

4.1 *Monetary income and expenditure*

A measure of the subsistence level of the economy in thirty-one of the thirty-nine households is given by the average cash income, which was below K20 per annum. In the other eight households, five earned up to K100, and 3 over K100.

These figures do not include the value of food grown for home consumption which appeared just adequate, since hardly any maize was bought or sold by those with low incomes. Data showed that even the 'bigger' farmers sold maize on local markets (possibly to fishermen) but for less than K20 in the year, and only 15 kilo was sold to ADMARC. Excess cassava, beans, groundnuts, peas, vegetables and fruit was sold or bartered at local village markets. The most important source of income, such as it is, is from the sale of paddy rice to ADMARC. Of 30 households who grew rice, only 19 households sold it. The price was 8 tambala (about 6 pence in Sterling currency) per kilo for grade A and 6 tambala per kilo for grade B. Rice was sold mainly by farmers who had grown less than one hectare of this crop. Twenty householders did not grow rice because of the difficult and hard labour involved and 85 per cent of the sales were from patches under 1 hectare in size. Sometimes chickens, fish and beer and, surprisingly in one case, four bulls were sold. Where exchange of produce occurs it is usually by barter. One surprising sale of 48 bags of cassava for K96 illustrates the potential for this crop in the money economy in this area.

The survey observed that comparatively few cattle (68) were kept by 8 households only, although grazing was available throughout the year and cattle could be used to supplement labour to obtain an income by sale and for improved diet. This may be because the keeping of cattle is a 'foreign' practice to the ethnic groups of the area. Over 400 poultry were counted and these are consumed, bartered or sold. The ownership of 2 trained oxen for ploughing, and one improved bull to increase the supply of milk by the herd, shows a small beginning in the integration of cattle with other farming operations.

Fish were an important resource only at subsistence level for the people in this case study; its sales amounted to less than those from local beer. Fish, however, are available freely in the marshes and the lake and are caught by using home-made traps, while beer entails the use of home-grown grain and equipment.

Payment for hired labour may be mostly in kind and only ten households hired men and that for only a few weeks. The mean hourly rate of pay is, usually, 5 tambala (3 pence) per hour for agricultural work, such as ridging, clearing, weeding, planting. But a skilled canoe builder might get 27 tambala an hour and the one tractor driver earned K3.50 an hour. This may be compared with the wages of a domestic worker in town, who works for 12 months instead of one or two and may earn K240 a year or more, which amounts to 10 tambala or 6 pence an hour.

The main items of expenditure are first, taxes which all men must pay. Salt, soap, clothes, paraffin, matches and sugar are the main items bought by most families. One man only bought a pen and books, and a few purchased spare parts for bicycles and medicines. There was also a certain amount of purchase of maize, rice and cassava in the local markets as well as fruit and vegetables. It is possible that the same families were sellers on other occasions when their crops were ready. Only seven larger households bought fertilizer and 13

323

purchased farm implements, which are normally made at home. Only five households found money to pay school fees.

In conclusion, the Agro-economy Survey (1972) summed up the situation as follows: 'This area is a remote part of the country, although the distance to Zomba (town) is only 112 km away. Road connections between the area and the main road are difficult and sometimes impossible during the rains. Roads and tracks are flooded frequently. Bad communications will have a negative effect on the spread of new ideas, innovations, modern techniques etc. The survey area is largely dependent upon subsistence agriculture, with annual cash earnings below K25.00 per annum. The most progressive farmers, applying fertiliser and using oxen . . . have a store . . . or other business. This group obtains the highest yield per acre . . . Land is not a limiting factor, but farmers fail to plant large acreages because the soil is very hard.'

'Rice farmers should be encouraged to use ox-drawn implements, which should be supplied by "tractor and ox-units" and (possibilities to bring this about) should be investigated. Cassava, which requires little labour input, is not widely grown. It should be encouraged for sale to National Oil Industries Ltd. or other traders. Farmers should also be encouraged to keep cattle, for there is extensive unused grazing land, . . . and to use fertiliser on their maize.' In this way, it is implied, more cash would become available for schooling, attendance at clinics and the buying of amenities to improve the quality of life.

5. Rice production and the response to planned change in the Zomba district: a case study

This section, in contrast to the former one deals with the progress in rice farming for cash in the central part of the Chilwa basin near permanent rivers. Besides enabling the farmer to improve his standard of living by means of a cash income, the production of rice, based on controlled irrigation, is one of the means by which foreign currency is earned and under-utilized wetlands are profitably used. In the Statement on Development Policies for the decade (1970–1980) the Economic Planning Division envisages the expansion of paddy grown rice from 10 000 tonnes in 1970 to 110 000 tonnes in 1980 in the whole country of which 30 000 tonnes would still be derived from the traditional rain-fed cultivation in seasonally flooded land. The projected expansion is dependent for the decade on trade agreements, controlled production and a comprehensive investment for the whole country of K16 million in rice and mixed vegetable produce as a second crop (Statement of Development Policies 1972).

Rice production is thus seen as the most direct method of increasing the income of farmers in lakeside areas, where the present cash crops of tobacco and cotton, grown in many other areas, are environmentally unsuitable to a fluctuating water table. Areas for major irrigation projects have been evaluated in the three Regions of Malawi. Two types of scheme have been suggested: single crop cultivation on controlled drainage lines, best described as run-of-the-river schemes, suitable for peasant participation at a cost of K99 per hectare; and engineer-constructed polders for double cropping at a cost of K740 per hectare. Local people and full-time settlers such as Young Pioneers are offered plots in both kinds of scheme. One of the incentives in attracting local farmers into irrigated rice production is the lessening of labour through the use of tractors to work the very hard soils, and threshers to speed up the preparation of the rice for market. Water control will also space out farming operations and prevent losses.

The introduction of high-yielding *indica* varieties, however, instead of the local Faya, long-grain type, requires new techniques in cultivation. In this, Malawi has been fortunate in obtaining technical aid from Taiwan, in which Chinese agronomists teach the required husbandry for successful rice and vegetable cultivation grown at different seasons.

In the Chilwa basin, between the Likangala and Domasi Rivers, four Development Rice Schemes have been established: the Khanda, Likangala, Mpheta and Njala Projects (Fig. 17.3). These schemes depend upon the streams which drain the high rainfall on the plateaux immediately west of the basin. These rivers provide suitable gradients and a high seasonal flow, well suited to run-of-the-river irrigation for a single rice crop a year.

The northern and southern parts of the Chilwa basin do not lend themselves to controlled irrigation, and in these parts, traditional rain-fed rice cultivation persists, as described above in the four more northern villages in Kawinga, northeast of the Chikala Hills.

5.1 *Organized rice production in a cash economy*

Irrigated rice farming under strictly controlled conditions is an innovation in Malawi (Chilivumbo 1969). The first scheme was started in 1967 on the Likangala River, where approximately 3 hectares were prepared by Taiwanese, using tractors. Twenty-two plots of 0.15 ha were linked to a canal system, served by four water pumps. The farmers were then invited to take up plots. At first there was a marked reluctance by villagers to participate. After a great deal of persuasion, 16 persons joined the scheme with Chief Mwambo and his son and several kinsmen, leading the way. Since the response was so small, six Young Pioneers were directed to join the scheme, having already undergone training to be initiators of modernization in agriculture.

After the success of the first harvest, two other schemes were started. In 1968, 91 males and 16 females registered in the three schemes which then covered 60 hectares; and 45 of the men had been fishermen before the lake recession of 1967–68. Three of the men were single and traditionally would not have held land. These features already depart from 'custom'. The labour on the rice plots included 42 wives but only 17 holders indicated the use of children or grandchildren on the plots.

Unit plots of 0.15 hectare were enclosed and about six of them constituted an average holding, worked by two people. The Government accepted the recommendation of the Taiwanese advisers that the principle of 'effective use' of land should be applied to the farmer (or to his kinsmen if the plot has been transferred) instead of 'right of use'. This meant that forfeiture would be exacted when laxity of land husbandry practice or failure to develop the plot was noted. A wife became the sole beneficiary of the sale of the crop if the husband did not participate in the work. Lapsed holdings would be offered to others, there being no limit to the size of the holding 'owned' as long as 'effective use' was guaranteed. These rules too were contrary to tradition in the peasant subsistence economy.

325

By these rules, as the schemes expand a cadre of rice farmers will emerge, who will be able to engage hired labourers. These would be either peasants, or lapsed rice plot-holders, who had been excluded from the new form of land tenure by lack of proficiency or unwillingness to submit to the exacting requirements of full-time farming. The hours and intensity of work expected in the scheme is about double that in the present subsistence economy.

The sociological study carried out by Chilivumbo (1969) was undertaken at an early stage in the development of irrigation farming. He predicted that the rice cash economy would not only increase incomes, but would erode traditional practice. Instead of land belonging to the whole of people, a concept of 'permitted' ownership is established . . . although it is not quite the same as 'private ownership'. The status of a husband is also changed, since he can hold land in his own name and can acquire the right to control the labour of his children, who would otherwise dwell in his wife's village. Simultaneously a woman may achieve independence of her brother who is the titular head of the sorority in the village. If she retains the labour of her children she would obtain independence of both her husband and the matrilineage.

Table 17.3 shows the data for irrigated rice projects on the Chilwa plain in 1977, ten years later than Chilivumbo's study. The original forecast of 800 hectares under irrigated rice by 1975 has been realized. The rice production data of the 1977 summer crop only, indicated 958 hectares 'owned' by 2569 people, 617 of whom were women. The total production was 3550 tonnes, an average yield of 3.7 tonnes per hectare, to this must be added the figures for production from the second (winter) crop which are not yet available (National Statistical Office 1978). About a quarter of the total production was consumed at home, which indicates a change in pattern of the diet of the people, as well as a larger surplus for sale.

Thus it may be seen that one change in agriculture introduced during the last ten years, was actually stimulated by the unreliability of the fluctuating lake level and dearth of fishing in some years. It has brought about sociological changes in traditional village life. Compared with the production of rice in the subsistence economy described above in the four villages in the northern part of the Chilwa basin, the yield under irrigation was on the average 2.6 times that of the poor farmer and just a little more than the 'rich' farmer in the rain-fed Kawinga area.

5.2 *Problems in irrigated rice cultivation*

The growing of irrigated rice on the Chilwa plain has already encountered some difficulties. A fairly rigid programme for working hours and simultaneous cultivation practices have been introduced, which were alien to Malawian farmers. This regimentation is necessary partly because a second winter crop is planned, and partly to control pests. The planting and weeding and cropping is done by hand. The main variety used is Blue Bonnet, a high yielding variety of *indica*. Variety trials are still being carried out, since a shorter maturing variety would make double cropping more feasible. The target for floodplain rice production of 4 883 kg ha^{-1} set by the Agricultural Department (Agricultural Department 1977) is lower than in experimental fields, where fertilizer is applied at a higher rate than on farmers' plots.

Table 17.3 Rice summer crop in Chilwa development schemes, 1977.

| Scheme | Farmers | | Total area hectares | Total tonnes produced | ADMARC tonnes purchased | Home use tonnes | Tonnes ha^{-1} yield |
	No. of Male	Female					
Njala	257	19	92	258	155	103	2.8
Mpheta	976	407	440	1482	1195	287	3.4
Likangala	620	163	357	1598	1344	255	4.5
Khanda	99	28	69	212	175	37	3.0
Total	1952	617	958	3550	2869	682	3.7

A second crop was planted in June, but figures are not yet available.
Source: National Statistical Office, Zomba, Malawi.

The most important diseases were: *Pyricularia oryzae* Cav. (blast), *Helminthosporium oryzae* Breda de Haan (brown spot) and *Corticium* spp. (sheath blight). Considerable damage by neck blast, *Pyricularia oryzae*, occurred in the Khanda Rice Scheme in the 1973/74 season, which had prolonged rains. The fields with the long-duration variety of Faya were especially attacked. The Faya had been specially grown for seeds, but after the attack it could not be used for this purpose. The variety Blue Bonnet was much less affected.

Before sowing, seeds are disinfected with an organic mercury compound (methoxyl ethyl mercuril chloride). There are, however, doubts about its efficacy and certainly about its deleterious effect on fish in the lake and it may be discontinued. The main pests are *Diopsis macrophthalma*, the stalk-eyed fly, and the beetle, *Trichispa sericea*. A thorough investigation of these has been undertaken and is reported in Chapter 15. A number of recommendations have been made which depend on improved methods of rice culture rather than chemical control, suggesting that biological control can be effective (Feijen 1977).

The presence of large-scale intensive cultivation of rice (and cotton) near the lake shores will certainly have an effect in the future on the general ecology of the lake ecosystem, the more so since the schemes are still expanding. The main dangers are the diversion of river water from the lake, more especially at the beginning of the rainy season and the accumulation of derivatives of fertilizers, fungicides (especially those containing organic mercury compound) and persistent insecticides. The schemes are very close to the lake, the nearest one being only 3 km away.

6. Stock rearing

6.1 *The development of stock keeping*

In spite of the extensive natural grazing on the customary land of the Chilwa plains, cattle are not as numerous as one might expect. It is possible that the area could be much more heavily grazed, if managed well. There are, however, historical reasons for the present situation not connected with disease carried by tsetse flies (*Glossina mortisans* (Westwood)) as in so many other parts of Africa.

The country is the domain of matrilineal peoples who originally used 'ash and hoe' cultivation, associated with pig and fowl as small household stock (Schmidt 1969). The payment of 'lobola' or bride-price (as cattle) which cements the marriages among patrilineal people was alien to the uxorilocal custom, where the husband has the use of the wife's land. In addition, the spread of Islam amongst the Yao people, who had been in contact with Arabs for over a century, made the pig taboo and had encouraged goat-keeping for ritual purposes. Amongst such people, cattle remained unimportant and cattle products were seldom found in the diet, nor oxen on the farms.

After 1890, under the British Colonial Administration and land settlement by whites, cattle became increasingly available in the nearby Shire Highlands. Since Mission schooling had been more successful in the northern region, a large number of northerners belonging to the ethnic groups, Ngoni and

Ngondi, who are cattle-keepers by tradition, migrated to the capital at Zomba and became civil servants. Cattle were their usual means of capital investment before banking and post office savings became more general. The salaries paid in Zomba, albeit small by western standards, led to an accumulation of wealth and social prestige which was most often expressed by the purchase of cattle sent for grazing on the Lake Chilwa grasslands in the care of local people. It was well-known that the floodplains were free of tsetse fly and east-coast fever-bearing brown tick, and hunting was steadily clearing the area of predators. These factors assured a greater survival rate of the stock deposited in the care of herdsmen, than would have been the case in their remote northern homes, and the area was near enough to urban residences in Zomba to be supervised from time to time.

From 1960 onwards, cash earnings from prosperous fishermen also took the form of investment in cattle, particularly since capital on the hoof might be an insurance against lake recession and the collapse of cash earnings from fishing. From 1970, the merit of draft animals for ploughing and carting has further encouraged stock ownership. In the towns, tea-drinking and artificial feeding for babies offered a market for the sale of milk and the possibility of cash earnings increases through organized marketing of cattle slaughtered.

The Chilwa cattle were Malawi Zebu, a variety of *Bos indicus* with thoracic humps. They are characterized by wide tolerance of heat, humidity and drought, of flies and ticks and their utilization of very rough herbage. They are small (maximum 300 kg), slow maturing and have a low yield of milk.

In the last ten years, abattoirs have come under the supervision of the Government-supported Cold Storage Co. in the towns and the Animal Health and Industry Department supervises field abattoirs and dip tanks. The Malawi Milk Marketing Depots with pasteurizing facilities are to be increased after the trial period of the Pilot Dairy at Blantyre. Holding grounds and supervised drive routes to main markets are being organized (Fig. 17.5). The increasing purchases of cattle at four cattle markets on the Chilwa plain demonstrates the entry of Chilwa cattle into the cash market (Table 17.4). Cattle route camps are provided during the movement on the hoof to the market at Zomba. Exotic breeds with a higher milk yield are slowly being introduced.

The population of cattle on the customary land of the eight Chief's areas of the Chilwa plains shown in Table 17.5 was 24 093, with an increase of 28 per cent over 1964 in Zomba and Kasupe Districts (National Statistical Office 1978). In spite of this, beef does not figure much in the local people's diet. Perhaps it is eaten on the average, once or twice a year, on special occasions. A very small proportion, perhaps 5 per cent is sold annually to the Cold Storage Co. or to local abattoirs for sale in the towns. Nor is cow's milk used much in the diet. Zebu cattle have not been bred for milk yield.

The number of pigs kept must be about 10 000 (Veterinary Report 1975) and a greater proportion of pork is eaten. There are about 40 000 goats which serve as a local source of meat and milk. Although poultry are often kept, eggs are not consumed by those who need them most – pregnant women and children, although the taboo is very slowly breaking down in the towns, at least. Chickens are not restricted to pens but feed themselves as they roam around, and congregate near the pounding of maize to eat the bran.

Table 17.4 Numbers of slaughtered cattle purchased by Cold Storage Co. on the Chilwa plains 1975–77.

District	Traditional authority	Cattle market	1975	Years 1976	1977
Zomba	Mwambo	Njala	–	8	87
Kasupe	Mlomba	Mikoka	86	172	317
Mulanje	Mkhumba	Mpasa	2	35	76
		Putheya	10	163	257

Source: National Statistical Office, Zomba.

Table 17.5 Cattle numbers on the Chilwa plains, 1964 and 1975.

District	Traditional authority	1964	Years 1975
Zomba	Mwambo Mkumbira	5 258	5 888
	Kumtumanji	4 129	6 362
Kasupe	Kawinga		
	Mlomba Mposa	5 822	7 187
		15 209	19 437
Mulanje	Mkumba	n.a.	1 342
	Nazombe	n.a.	3 314

n.a. not available.
Source: National Statistical Office, Zomba, 1978.

Surprisingly few cattle are converted into oxen as draught animals. About 5000 were listed for the whole Southern Region in the Annual Report of the Department of Veterinary Services (1975), but the number in the Chilwa villages seems relatively low, estimated at about 1000 for the 200 000 farmers of whom the majority do not use oxen. The distribution of cattle in Southern Malawi is shown in Fig. 17.6, which draws attention to the concentration of cattle in the Chilwa basin.

6.2 Pasture grasses

The grasses of the Lake Chilwa basin offer a variety of palatable grasses for cattle (Howard-Williams 1977). At present, a natural grazing rotation operates, governed by the annual rise and fall of the water level. During the dry season, cattle are moved on to the floodplains and swamp fringes to graze, and during the wet season, when the area is flooded, the grasses are able to recover. Flowering normally occurs before the next dry season. The annual flooding of the marginal areas of the lake thus plays an important role in maintaining the productivity of the marginal grasslands.

On the alkaline soils in the south of the lake basin, *Chloris gayana* (Rhodes grass) associated with *Bothriochloa glabra*, *Dicanthium papillosum* and *Cynodon dactylon* form a very palatable pasture. In other parts of the lake basin the

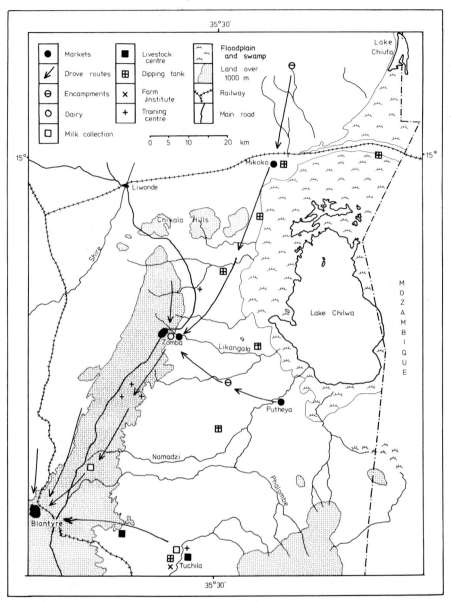

Fig. 17.5 Organization of dip tanks and stock industries in Southern Malawi (redrawn from Surveys Department Maps).

drier floodplain soils support communities of *Hyparrhenia rufa*, *Eragrostis gangetica*, *Cynodon dactylon* and *Sporobolus pyramidalis*, all good palatable grasses except for the last. *Hyparrhenia rufa* is particularly extensive on the northern plains, where, if moderately grazed, it maintains a reasonable pasture.

In depressions and in wetter areas towards the swamp fringes, a number of particularly favourable grazing species occur. *Vossia cuspidata* (Hippo grass)

331

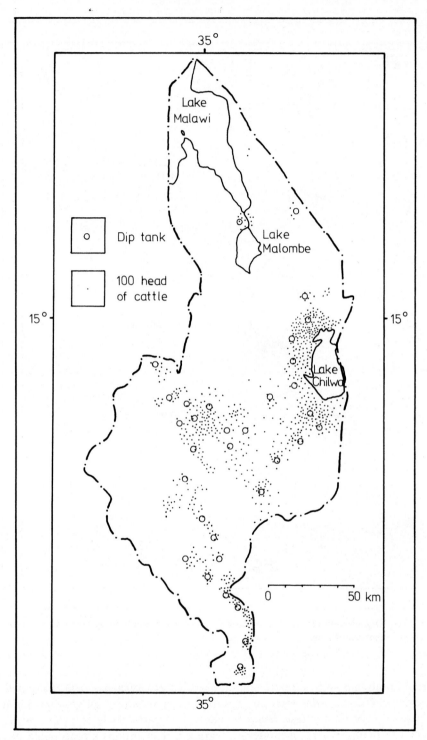

Fig. 17.6 Distribution of cattle in Southern Malawi, illustrating the concentration of cattle in the Chilwa basin (after Schmidt 1969).

332

and *Echinochloa pyramidalis* are heavily utilized, and with their thick stolons are liable to withstand considerable grazing pressure. *Panicum repens* is of particular importance as a fodder grass, and cattle will even wade neck deep into the water to get at this species. The crude protein and phosphorus values of several Chilwa grasses during the dry (grazing) season are given in Table 17.6 below. The values for crude protein are generally lower than for temperate grassland species, and reflect the nitrogen-poor sandy soils of much of the Chilwa floodplain. Crude protein values will, of course, be higher during the wet season, but grazing is not carried out on the floodplain then. Phosphorus values appear adequate for African fodder grasses. On the swamp and marsh fringes, cattle will also graze *Phragmites mauritianus* and *Paspalidium geminatum*. *Typha* is not however utilized.

Table 17.6 Crude protein and phosphorus values of several grasses from the Chilwa plains, collected during the dry season. (C. Howard-Williams unpublished data). Values as % dry matter.

Species	Crude protein %	Phosphorus %
Vossia cuspidata	5.16	0.17
Paspalidium geminatum	12.80	0.36
Hyparrhenia rufa	4.09	0.10
Cynodon dactylon	4.64	0.29
Echinochloa pyramidalis	1.56	0.28
Panicum repens	4.31	0.19
Leersia hexandra	3.28	0.12
Chloris virgata	–	0.30
Eriochloa borumensis	4.00	0.15
Diplachne fusca	4.61	0.13

While the existence of extensive grasslands with many palatable species, coupled with a natural grazing rotation, gives the basis for the development of a cattle economy in the area, grazing land in certain places is already showing stress by overstocking. This is particularly so in areas where herds are confined to common land, and along routes to holding grounds and dipping tanks. Man-made factors such as the proliferation of bicycle tracks and footpaths across heavily grazed areas and the invasion of settlements near roads have decreased the grazing land available. The extension of areas for cultivation is also restricting the extent of the natural grasslands. It is a question for decision of regional land use planning how much pasture land should be preserved for cattle (which most local people still cannot afford to eat) and how much should be devoted to dry-land and irrigated crops, bearing in mind that a 'ley' rotation may be introduced to improve soil structure, as farming improves.

6.3 Investments in cattle

The general importance of cattle, in the area between the Phalombe River and the Sombani embayment, where Mambo village is situated, is illustrated in Fig. 17.7, where in ecological area I in 1967, 4000 head of cattle were being grazed

(Agnew 1973). In 1971, a survey was made of Mambo village in the Phalombe
River delta area 12 km from the fishing villages at Chipeta in Mkhumba's area.
The village is linked by a track on the south bank of the river to a Trading
Centre 5 km upstream and to the landing beaches of the lake. In a population of
469, there were 163 adult males including Lomwe, Yao and Nyanja. Half the
males had lived in the settlement for more than 10 years and others had
married into the village, and its permanent status is indicated by a maize mill,
churches and canteens. A wide variety of crops were grown, especially maize,
sorghum, pigeon peas, beans and vegetables. The rice lands were, however,
yielding poorly because of a change in a channel distributary in the delta after
the dry period of 1967–68. In such alluvial soil, given proper drainage, maize,
pulses and vegetables could provide surpluses for sale. No dry season crops
were grown because of depredation by stock.

There were 531 cattle and 125 small stock owned by 30 households. Four
men herded cattle they did not own. Most of the cattle had been acquired from
profits derived from fishing, and some from the earlier sale of rice. The small
stock had been purchased from sales of fish or crops.

The distribution of cattle in the Chilwa area is not uniform. Further inland 35
km from the water's edge on the Sombani plain, cattle do not rank as highly
(Table 17.7), where a minority of fishermen are cattle owners. In a wider
survey of the lacustrine plain near Zomba, in Chief Mwambo's area,
Chilivumbo (1972) found 71 per cent of the fishermen owned livestock.

The disparity between different areas in sample studies may be accounted for
by a variety of factors. These are: availability of capital, proximity to grazing
grounds, supervision by owners, distance from dip tanks, interest in cotton or
rice farming, and of course, the residual element of unfamiliarity with cattle-
keeping. The last factor is illustrated in Fig. 17.7, in ecological area VI where
the people of this part were mainly refugees from Mozambique and were
particularly remote from modernizing influences affecting the relatively more
sophisticated lacustrine people of the Zomba District.

6.4 Diseases affecting stock

The Department of Animal Health and Husbandry organizes a number of dip

Table 17.7 Cattle owners and Fishermen, Sombani area 1967.

Village (Population)	No. of families interviewed	No. of fisherman	No. of cattle owners	No. of cattle owners as fisherman	No. of cattle
Mkoko (194)	46	36	10	3	97
Likachale (98)	29	13	6	4	75
Mphepo (360)	79	44	12	3	91
Total	156	93	28	10	263

Source: Agnew S. 1973.

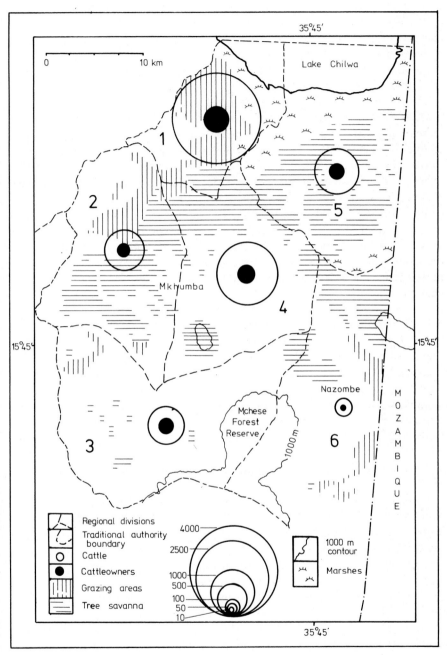

Fig. 17.7 Distribution of cattle on the Phalombe plain, showing densities (after Agnew 1973).

tanks treated with arsenical compounds or toxaphene to control tick-borne diseases. The standard of usage and the possible toxic effects on cattle are monitored. Intensive campaigns are run to vaccinate against tuberculosis in cattle, against Newcastle's Disease and Fowl Pox in poultry, and other bacterial diseases (e.g. *Clostridium* strains) which might be brought into the country when animals are imported. An efficient inspection service is carried out at the main abattoirs, which operate at centres throughout the country. Control of field abattoirs is also practised locally.

The cattle on the Chilwa plains are mostly the original Zebu cattle, which are more resistant to diseases and climatic variation than the exotics which are increasingly being imported to improve the quality of the national herd. Bacterial disease such as T.B. is low. Rabies sometimes occurs among cattle (34 cases in the whole country in 1973) and there is a continuous vigorous campaign every year to eliminate it. 20 000 stray dogs were eliminated in the Southern Region alone in 1973 (Annual Report for Veterinary Services 1975). Fascioliasis (liver fluke) carried by a snail *Lymnaea* sp. is a major disease, and 30–50 per cent of beef livers from the whole country are condemned annually at the abattoirs because of liver fluke. However, the low rainfall area of the Chilwa plain and the periodic recessions of the lake level, which dry out parts of the floodplain more quickly than usual, limit the number of snail vectors.

6.5 Development policy and problems

The number of cattle on the north and central parts of the Chilwa plains has increased by 28 per cent in eleven years (Table 17.5). This is in accord with the national policy of increasing and improving the national herd. The Government is aware that improvement in the livestock industry must keep pace with the meat requirements resulting from the rising standard of living of the increasing urban population. It is intended that stock rearing should be integrated with other types of farming so as to use residues as fodder, utilize manure as fertilizer and oxen as draft animals (Statement of Development Policies 1972).

The marketing system had already developed in 1976 so as to make the country self-sufficient in meat except for mutton and lamb. The objective now is to raise the quality of beef production as well as the amount. This will depend on social attitudes to animal husbandry and better means of care of the animals during transport, as well as on introducing new breeds to markets.

A concerted attack can be and has been made on disease by the Department of Animal Health. The use of artificial insemination will gradually improve the quality of beef and milk and castration of poor sires will provide oxen for farming. The over-riding problem of management of open range customary land still remains to be tackled. Since grazing is free and communally exercised, some means must be sought to impose an acceptable system of rotation, without introducing fencing. There may be a tendency towards over-stocking the land and there is a real possibility of encroaching on land best used for rice production or other crops. These problems are later likely to be particularly acute in the Chilwa basin.

7. Dangers of the use of insecticides in the Chilwa basin *by* Margaret Kalk

In the last decade, the Chilwa basin has seen the initiation of small-scale development in rice schemes where insecticides and fertilizers are used. Some proportions of all chemicals used in such farming will eventually accumulate in Lake Chilwa, because it is a closed drainage basin. The possibility of this danger was alerted by Schulten (1971) who drew attention to the spraying of DDT, by aeroplane, upon the rice fields around Lake Chilwa. Not only was the insecticide inappropriate for the purpose to which it was put (to destroy fly larvae within the stems) but more significantly for the lake, it was applied in the rainy season. Some was, no doubt, carried into the lake by the large number of streams and irrigation channels flowing at this time. DDT is well known to be a most persistent toxin. It could do immeasurable damage in a closed basin.

At present, insecticides other than DDT, such as carbaryl, are being used on rice pests and their accumulation in the lake is not expected, since they are non-persistent. But mercury compounds are used in the rice nurseries. In Chapter 15, Feijen put forward suggestions for alternative control of the main insect pest, the stalk-eyed fly, *Diopsis macrophthalma*, by improving cultivation methods, so as to allow free play for its destruction by the natural parasitoids, which have been fully described.

Another potential source of pollutants, which might be injurious to fish in the lake are the insecticides used on a large scale at the Cotton Research Institute farm at Makoka, where run-off of rain water from the fields drains towards the Likangala River to Lake Chilwa. At the Institute it has been demonstrated that yields of cotton in Malawi may be reduced by half if insect pests, such as the Sudan boll worm, *Diparopsis castanea*, the American boll worm, *Heliothis* sp. the pink boll worm, *Pectinophora* sp. and the spiny boll worm, *Earias* are not controlled. DDT has been found to be the most effective insecticide against *Heliothis* and carbaryl against *Diparopsis*. It has therefore been recommended to all cotton farmers that spraying with a mixture of carbaryl and DDT would be most effective. The method of ultra-small-volume spraying under pressure is being introduced (Mowlam 1974). An attempt to reduce the amount of insecticide used depends on scouting for eggs, to determine the degree of likely infestation. Female sex-attractant pheromones, used to bait traps, which attract males, have been employed by the Institute and their common use, in order to be able to concentrate spraying on a small area, would enable farmers to estimate the probable incidence of pests earlier and with less labour than looking for eggs and to use less insecticide. But large-scale control with pheromones appears too expensive (Annual Report Makoka 1975).

A small Cotton Development Scheme of approximately 2000 to 3000 hectares is at present projected for the Phalombe plain. It has been estimated that 12–19 tonnes of DDT would be used each year, but that levels reaching the lake would be unlikely to be more than 19 kg per year (G. K. C. Nyirenda pers. comm.). This may be only a beginning, and so it is worthwhile examining the data from investigations of DDT accumulation in fish carried out through the Cotton Research Institute in Malawi.

The level of insecticide residues in the run-off from Makoka farm towards the lake are indicated by a small-scale 'model' of a closed basin, the Makoka Farm dam which is being monitored. DDT residues are high and those absorbed by different species of fishes have been analysed (Annual Report, Makoka 1977). Table 17.8 shows that more DDT residue is retained in the gut than in the edible muscle of the fish species studied. The highest retention in the gut was, however, in *Barbus* (the *matemba* of Lake Chilwa) in which the gut contained an average of 0.88 mg kg^{-1} DDT residue. This fish is dried and eaten whole including the gut (Chapter 12), so that all its contamination would be passed on to the consumer. In addition, this species is the most important commercial fish in Lake Chilwa (see Table 18.3).

Table 17.8 Total 'DDT' Residues in Fish Samples in mg kg^{-1} (F=Flesh; G=Gut).

Samples		No. of Samples	Range	Average
Tilapia lidole	F	12	0.02–0.21	0.07
	G	12	0.04–1.87	0.54
Sarotherodon shiranus	F	9	0.02–0.11	0.05
	G	8	0.02–1.37	0.50
Tilapia rendalli	F	3	0.02–0.05	0.03
	G	3	0.07–0.38	0.18
Barbus paludinasus	F	3	0.01–0.25	0.09
	G	3	0.02–2.33	0.88
Clarias gariepinus	F	2	0.04–0.12	0.08
	G	2	0.17–0.22	0.17–0.20

Source: Annual Report (1977) Makoka Cotton Research Station.

A more detailed analysis of the organs of the catfish, *Clarias gariepinus*, is given in Table 17.9. The highest levels were found in the liver and ovaries. It is speculated that 'this might affect the fertility of the fish' (Annual Report, Makoka 1975). This instance of a few years' absorption of DDT residues by one large female catfish in a small 'closed basin', very close to the cotton fields, is certainly an exaggeration of the present danger to the lake, although 1 p.p.m. in the liver is pretty close to the maximum permissible (1.25 p.p.m. in fat for human consumption) laid down by WHO/FAO (1972). It is, nevertheless, indicative of what might happen with large-scale application of DDT on large areas of cotton on the Chilwa plains, if this crop becomes popular. The danger

Table 17.9 Insecticide Residues in individual organs – *Clarias gariepinus*, 1 sample mg kg^{-1}

Sample	DDE	DDD	op.DDT	p.pDDT	Total 'DDT'
C. gariepinus (3833gm)					
Flesh	0.10	0.18	0.01	0.06	0.35
Heart	0.19	0.31	0.02	0.09	0.61
Liver	0.30	0.62	0.01	0.07	1.00
Kidneys	0.26	0.45	0.02	0.14	0.85
Ovaries	0.82	1.38	0.03	0.51	2.74
Gut	0.17	0.33	0.01	0.14	0.65

Source: Annual Report (1977) Makoka Cotton Research Station.

is high-lighted by an analysis of the soils alongside the Lower Shire River, where there is a very extensive Cotton Development Project, which is shown in Table 17.10.

Table 17.10 'DDT' Residues in Cotton Gardens Means in mg kg^{-1} dry weight of soil from 6 sprayed and 2 unsprayed gardens.

Soil Type	Months after stopping spraying		6 or more season's spraying	1–5 season's spraying	Unsprayed
	0	T	2.33	1.86	0.01
			(4.75)	(2.86)	(1.00)
		B	0.49	0.65	0.01
Clay	3	T	1.90	1.73	0.09
			(2.18)	(4.11)	(0.90)
		B	0.87	0.42	0.10
	6	T	1.31	1.98	0.04
			(2.01)	(4.40)	(4.00)
		B	0.65	0.45	0.01
	0	T	0.49	0.19	0.05
			(2.45)	(2.37)	(0.83)
		B	0.20	0.08	0.06
Sandy	3	T	0.78	0.36	0.31
			(2.10)	(2.25)	(3.44)
		B	0.37	0.16	0.09
	6	T	0.89	0.73	0.25
			(1.09)	(2.21)	(3.57)
		B	0.81	0.33	0.07

Figures in parenthesis are ratios of top to bottom residues.
T = Top 0–50 mm.   B = Bottom 150–200 mm.
Source: Annual Report (1975) Makoka Cotton Research Station.

Clay soil retains more DDT residue in the top layer than 15–20 cm below the surface. The opposite occurs in sandy soil, in which more residue is leached to the bottom. This distinction may be important for Lake Chilwa where the soil is largely clay with pockets of sand. In the case of the rich fishing grounds of the Elephant Marshes of the Lower Shire River, near the cotton farms which prompted the investigation, 'the mud levels of DDT residues were very low' (G. K. C. Nyirenda pers. comm.). This suggested that very little may be reaching the river, or alternatively that the river flow washed residues out of the marshes when they are flooded in the rainy season. The difference between this situation and the closed basin of Lake Chilwa where clay sediments accumulate and where the surface layers are maintained in suspension by wind and wave action is self-evident (see Chapter 4).

Until ten years ago, DDT was recommended for annual maize storage in a domestic *nkokhwe*, i.e. a storage bin made of plaited bamboo (see Fig. 17.2b). But the maize is stored unshelled, protected by close sheathing leaves, and in fact, in the Chilwa villages almost no money is spent on insecticides and very little on fertilizers, at present. If, however, farming is modernized, with heavy applications of insecticides and fertilizers on mono-cultures of improved varieties of crops over large areas, chemicals might become a real danger to the organisms in the lake and to the people who consume them.

8. Necessity for an integrated economy

It is evident that irrigated rice cultivation enables most participants to earn more cash than was possible with floodplain rice farming. In the villages studied in the Kawinga plain, only half the households entered the money economy at all through rice cultivation, and then to a very minor extent. In the irrigated rice scheme in the central plains, the sales varied in the four schemes from 560 kg per head at Njala, the newest scheme, to 1716 kg per head in the oldest, Likangala scheme. If the rice was grade A, sold at 8 tambala per kg, the highest average income would have been about K137 per annum for a farmer in charge of an average of 3 plots amounting to 0.45 ha. This should be compared with the various incomes from fishing in the next chapter, which range from K200 to K600 **before** expenses are paid.

The progress towards a cash economy through the sales of surplus crops in the three districts of the Chilwa area is shown in Table 17.11. It is of interest to compare the twelve-fold increase in ADMARC purchases in the Zomba District where the Development schemes have been started with the six-fold increase in Kasupe District, where there was little more inducement than the creation of a trading centre where surplus rice could be sold. Crop production varies from year to year considerably because of variable rainfall, but there has been an overall increase in both rice and cotton from 1969–77 (Ministry of Agriculture Report 1977). Future planning should bear in mind the dangers of the use of persistent insecticides to the food chain in an endorheic lake such as Chilwa. In regard to maize, the staple food, it is aimed to produce a surplus of about 7 per cent per annum (when possible) in order to export in good years to cover the costs of import in poor years. Irrigation from the permanent rivers is possible in the southern Phalombe plain and the central Chilwa plain of the Zomba District. But it remains to be seen how a balance will be maintained between the demands of the fisheries for water to maintain the lake level and the need for irrigation. It appears from the experience of 1966 and 1973 that in

Table 17.11 Purchase of crops in tonnes on the Chilwa plains by the Agricultural Development and Marketing Corporation (ADMARC).

Crop	Date	Zomba District	Kasupe District	Mulanje District
Maize	1964	x	x	x
	1975	235	862	261
Paddy rice	1964	122	329	377
	1975	1431	1918	216
Seed cotton	1964	8	14	–
	1975	56	1470	–

x Figures not available, probably negligible.
Source: National Statistical Office, Zomba 1978.

the minor recessions in lake level, the needs of agriculture did not contribute significantly to the fall of lake level. In the major recession, the river flow was itself reduced considerably. Plans are proceeding to increase the production of both rice and cotton with the assistance of international aid. These schemes will consume far more of the inflowing water and should allow for the almost certain occurrence of further periodic minor and major decreases in river flow, such as occurred in 1966–68 (see Chapter 3).

Pauline Phipps (1973) underlines the need to treat the economy of the Chilwa area as a whole, considering the impact of lake recessions and the low rainfall which precede them. 'The repercussions of recession in Lake Chilwa waters and consequent decline of fishing are much wider (than on fishing alone). The whole of the Chilwa plains and lake must be seen as an economic network. Not only are there links between fishing and various ancillary services, but also between fishing and farming. Investigation has shown that there is a complementary flow of income between fishing and farming activities. The successful fishermen have larger gardens and produce more cash crops than other fishermen.' This is especially true of the more southern area on the Phalombe plain. But there are also farmers who do not fish for sale, nor trade in fish for cash. They constitute the more stable communities a little further inland, where farming alone is their way of life.

One of the main features of the inhabitants of the Chilwa plains is the overlapping of roles, such as farmer and cattle-owner, farmer and trader, farmer and fisherman, and farmer and fish-trader, which has emerged historically. There is also the shared threat to the livelihood of all by the periodic low rainfall, reduced river flow, and recession of the lake every six years or so.

The prerequisites for meeting the lean year exist in the mixed economy, but the yields from farms must be increased. It has been proved by Development Schemes that the soil will yield more, provided there is an input of mechanical energy in the preparation of the ground, although cultivation remains labour intensive and the people own the land. Similarly, the use of oxen by the single farmer has been shown to lead to the ability to farm more land and to produce surpluses. For this to become widespread, a pool of oxen might be created by those who already own cattle, and farmers might borrow them, initially on credit. In good years, a few cattle might be bought from surplus sales of maize or rice and be kept to sell before times of food shortage.

With foreknowledge that years of water shortage will occur crops should be diversified and cultivation intensified so as to have surplus for storage. The planned integration of the economy in each Chief's area, coupled with the well-knit organization and co-operative efforts already exhibited by plotholders in Development Schemes, should do much to alleviate the stress of the climatic fluctuations.

References

Agnew, S. 1970. The People of the Chilwa Plains. In: M. Kalk (ed.) Decline and Recovery of a lake. Government Printer, Zomba, Malawi, pp. 41–44.

Agnew, S. 1972. Environment and History: The Malawian Setting. In: B. Pachai (ed.). The Early History of Malawi. Longman, pp. 28–48.

Agnew, S. 1973. The indigenous growth towards a cash economy in Africa: a case study of a lacustrine economy in Malawi. Geoforum 16:16–28.

Agnew, S. & Stubbs, M. 1972. Malawi in Maps. University of London Press. 143 pp.

Agro-Economic Survey, 1972. Lake Chilwa. No. 9. Ministry of Agriculture and Natural Resources (Planning Unit). Zomba, Malawi. 63 pp.

Annual Report, 1975. Makoka Agricultural Research Station, Thondwe, Malawi, pp. 74–75.

Annual Report, 1977. Makoka Agricultural Research Station, Thondwe, Malawi, pp. 91–95.

Annual Report of the Department of Veterinary Services and Animal Industry (1973). 1975. Government Printer, Zomba, Malawi. 54 pp.

Bandawe, L. M. 1971. Memoirs of a Malawian. B. Pachai (ed.), CLAIM, Blantyre.

Chikwapulo, A. 1971. Malawi: Cotton growing in the southern Lake Chilwa Plain. Dziko, 13. Geog. Soc. University of Malawi, Zomba.

Chilivumbo, A. 1969. The response to planned change: a study of the Rice Scheme in Chief Mwambo's Area, Lake Chilwa, Zomba, Malawi. Soc. Malawi J. 22:38–57.

Chilivumbo, A. 1972. The fishermen of Lake Chilwa: a study in adaptability. In: D. J. M. Vorster (ed.). Human Biology of Environmental Change. I.B.P. (HA) Conference, Blantyre 1971. International Biological Programme, Marylebone Road, London, pp. 15–20.

Coleman, G. 1976. Economic Development and Population Change in Malawi. School of Development Studies, Seminar. Univ. East Anglia, Norwich.

Fagan, B. 1969. Early trade and raw materials in South Central Africa. J. Afr. Hist. 10:1–13.

Feijen, H. R. 1977. Research on the rice stem-borer *Diopsis macrophthalma* Westwood and other Diopsids in Malawi. In: Overdr. WOTRO-Jaarboek. Amsterdam, The Netherlands, pp. 38–41.

Howard-Williams, C. 1977. A check-list of the vascular plants of Lake Chilwa, Malawi, with special reference to the influence of environmental factors on the distribution of taxa. Kirkia, 10:563–579.

Ministry of Agriculture and Natural Resources Report 1977. Guide to Agricultural Production in Malawi. Government Printer, Zomba.

Mowlam, M. D. 1974. New developments in ultra-low-volume cotton spraying. Malawi J. Sci. 2:108–109.

National Statistical Office. Cattle numbers; cattle sales; rice crops; population: data supplied 1977 & 1978. Zomba, Malawi.

Phipps, P. 1973. The 'Big' Fishermen of Lake Chilwa: a preliminary survey of entrepreneurs in rural Malawi. In: M. E. Page (ed.). Land and Labour in Rural Malawi. Rural Africana 21:39–48.

Population Census Report, 1972. The census of 1966. National Statistical Office, Zomba.

Population Census, 1977. Preliminary Results of 1977. Statistical Office, Zomba, Malawi. 4 pp.

Schmidt, R. 1969. Cattle in Malawi's Southern Region. Soc. Malawi J. 22:57–72.

Schulten, G. G. M. 1971. The use of DDT at Lake Chilwa. University of Malawi. (Cyclostyled) 5 pp.

Statement of Development Policies in Malawi 1971–1980. 1972. Office of the President and Cabinet, Economic Planning Division. Government Printer, Zomba, Malawi.

WHO/FAO. 1972. Evaluations of some pesticide residues in food. 1973. World Hlth. Org. techn. Rep. Ser. 525. Geneva. 571 pp.

8 Fishing and fish trading: Socio-economic studies

Swanzie Agnew & C. Chipeta

Fishing and fish trading: Socio-economic studies

1. The growth of fishing and fish trading at Lake Chilwa *by* S. Agnew

In terms of economics, fish is a 'free good' to be gathered by anyone without licence in Lake Chilwa. By the simplest gear made from local fibres into basket traps or nets, by line fishing or spearing, any man has the means of supplying food to the household, of bartering fish for other goods or selling the catch for a little cash. Then with a little capital (usually obtained by earning elsewhere) a boat may be acquired which gives a man the opportunity of becoming a full-time fisherman or a specialist fish trader collecting catches from outlying parts of the lake. The full-time fisherman may extend his enterprise by hiring labour to work more sophisticated gill or seine nets and the fish trader may process larger hauls of fish and sell in bulk.

Fishing and fish trading have been steadily increasing since the Second World War in response to the doubling of the population between 1945 and 1966 and its concentration on the Shire Highlands (Population Census Report 1972). Fish traders with bicycles have increasingly used buses and lorries to transport fish, and the use of money amongst villagers has grown more common. Amongst other sources, not inconsiderable money has been remitted to Malawi every year by labour migrants in South Africa and Rhodesia, which in 1966 amounted to £2 000 000 Sterling for the whole country (Compendium of Statistics 1966). Although each individual's share was small, it may have served as capital to buy a boat or a manufactured net.

The introduction of nylon thread in 1958 to make nets by machine in a Blantyre factory (in the Shire Highlands) instead of from fibres of local plants, was the most dramatic innovation in the history of the fishing economy. In the making of fishing gear, it equalled the introduction of the iron axe and later, the steel adze, which permitted the felling of large trees and their being shaped into dug-out canoes.

Although Lake Chilwa has been considered a comparatively productive lake in Africa in terms of output per unit area (Chapter 12), the greater part of the haul was made by 'subsistence' fishermen up to the nineteen sixties. Much more is now disposed of in markets, both small village markets and those of larger towns. Simple production had recognizably changed to exploitation of the resources of the lake when, in 1965, a catch of 8800 tonnes of fish was estimated by the newly introduced recording system of the Fisheries Department. This size of fish landing high-lighted the process, which had been going on for some time, following the introduction of nylon netting, of the change from subsistence fishing to partial exploitation of the lake for cash. This became possible partly because of the greater efficiency of the nylon gill nets, *machera*, but also because of the innovation of seine nets, *makhoka*. The *chirima*, an open-water seine net, towed by two men, was also used; this appears to have been adapted from that developed in the northern waters of Lake Malawi (Phipps 1973).

In the census year 1966, 6444 persons were recorded as associated with fishing sites, which included fishermen, their relatives, ancillary helpers, women and children and fish traders and their families. There was a major decline in the population during the recessions of 1967–68 and in the minor recession of 1973, but in the recent census in 1977, the population in the fishing villages in a belt within 8 km of the water's edge, totalled 25 347 persons (Population census 1977). Barely ten years after the drying up of the lake, a Fisheries Department census in February 1977, at the height of the fishing season recorded over a thousand craft plying the lake (Fisheries Report 1977).

In the 1966 census it was reported that on Nchisi Island 'in keeping with fishing as a man's occupation, males in the villages outnumbered the females by 15 to 1. By contrast in most land settlement on the Chilwa plains, females were in excess of males' (Agnew 1970). The temporary increase in the number of men and the large fishing fleet of canoes point to a dramatic intensification of fishing in the years of high lake level. The most densely populated parts are Nchisi Island and Chinguma peninsula in Mkumbira's area, with over 5000 persons. The enumeration areas adjoining the shore in the southwest held about 250–300 persons per square kilometre. These areas of concentrated activity are among the highest rural densities in Central Africa.

The actual number of men actively engaged in fishing or fish trading, as a full-time occupation, is not yet known from the 1977 census. Most of them, if not all, have a stake in semi-subsistence farming in their wives' 'gardens'. In contrast to Ghana, for example, where women employ fishermen and own boats, at Lake Chilwa the women play little part even in activities related to fishing, such as fish processing and fish trading, and do not fish themselves. They work the land. Although the techniques of fishing are, at present, almost wholly 'traditional' and the way of life of families has not yet changed, the fishing 'industry' as a whole has entered the modern economy of Malawi.

2. Some aspects of the economics of fishing in Lake Chilwa by C. Chipeta

Between 1962 and 1974, Lake Chilwa was on average the third most important source of fish in Malawi. In 1964 it was the second most important source of fish in the country, accounting for 40 per cent of the total catch, and in 1965 and 1966 when the Lake Chilwa Research Project started, it was actually the chief source, accounting for 47 per cent and 41 per cent of the total catch, respectively. In 1976, it was 28 per cent of the total. The proportion of the national total has decreased, partly because of the major recession, 1966–69 and the minor recession of 1973. Another reason is the intensification and extension of fishing elsewhere, including trawl fishing with bulk landings at fixed points for export on Lake Malawi introduced in 1968 and the development of fisheries in the Lower Shire River and the surrounding Elephant Marshes (Economic Report 1972). Owing to the increase of fishing in these waters, Malawi became a net exporter of fish in 1969, when the water level of Lake Chilwa had just started recovering. In 1969, fish exports were K188 900 in value, while imports were K152 600, giving a net export surplus value of K36 000. In 1975, the net export surplus had grown to K585 000.

After the recovery of the fisheries in Lake Chilwa, fish landings were 19 700 tonnes in 1976. In that year, exports of fish to Rhodesia and Zambia from Lake Chilwa alone reached a value of K25 000. The value of the fish extracted from the lake and swamp may be roughly estimated from the total cost of fresh fish to traders on the beaches of Lake Chilwa, which has been recorded in connection with sales of dried or smoked fish at markets in the urban areas, which amounted to 17 per cent of the total (see Table 18.1). When fish was bought in bulk for export it was cheaper, and the average cost was K123 per tonne (Fisheries Report 1977). If the total catch recorded had been sold by the fishermen at this price, the value of fish landed would have amounted to two and a half million Kwacha (=£1.6 million Sterling). Not all the fish landed is, however, sold. A very large proportion goes to rural markets (not included in Table 18.1) and the remainder is bartered, used as part remuneration for labour or consumed by the families of fishermen and fish traders. In 1967 it was estimated by Williams (1969) that the proportion of fish that went to market was only 20 per cent. This proportion is also far too low since the fish sold at village markets and for indigenous money (barter) is not included.

The price paid by the trader at the lakeshore varies. When a basket load of 50 kg is purchased, the current price is about 9 tambala (6 pence) per kg. After processing in various ways (sun-drying, smoking or freezing) the trader disposes of his load at markets. At Zomba market in 1978, the price of fish to the consumer, after distribution by middlemen, was on average, 10 tambala per large 'plate' of dried *matemba* (*Barbus*), with a range of 5t to 15t. When transformed into weight, the price amounts to K1.00 per kg. Smoked *mlamba* (*Clarias*), which look like oily kippers, cost 10t each on average, ranging from 1t to 20t, depending on size. This is dearer, K1.40 per kg. *Makumba* (*Sarotherodon*), when dry cost 20t and when fresh cost 25t. This again amounts to about K1.00 per kg.

People buy only a few hundred grammes at a time and the fish, when cooked with a few vegetables, forms the basis of 'relish' and is served with the main dish of *nsima*, maize porridge. It would appear that fish traders and middlemen do very well from the marketing of the Chilwa fish, even allowing for a decrease of weight through preservation of over 50 per cent.

Lake Chilwa is unique in Malawi because, due to its salty waters, it produces tasty fish which the local people prefer to others. Similar fish from Lakes Malawi or Malombe are not very good substitutes. Thus even if people can afford to buy fish from elsewhere during the recessions of Lake Chilwa (and probably few can), the loss in satisfaction is not fully compensated. During such periods digging for rats and gerbils is intensified to compensate for the loss of their main source of protein.

About 31 per cent of the country's total population lives within 60 km of Lake Chilwa in the districts of Zomba, Kasupe, Mulanje, Chiradzulu and Thyolo, forming with Blantyre, the most accessible markets. In order to supply these markets the lakes and rivers thoroughout the Southern and Central Regions of Malawi form a complementary source of fish, so that shortages from one source is made good by supply from others. Although, in the case of the area near Lake Chilwa, it means that only the more expensive fish would be available in times of recession of the lake.

Cash incomes from Lake Chilwa fishing are attractive in good years. In 1965 and 1966, for example, the average sizes of gross returns for commercial fishermen were K416 and K320, respectively. These were adequate returns, considering that not much labour was employed and only intermittently during

347

Table 18.1 Distribution of fish from Lake Chilwa to Urban Markets 1976.

Markets	Barbus		Sarotheradon		Clarias		Totals	
	Tonnes	KWACHA	Tonnes	KWACHA	Tonnes	KWACHA	Tonnes	KWACHA
Mulanje	336.0	43 562	95.0	16 266	490.0	64 223	921.0	124 051
Zomba	305.0	28 363	202.0	25 882	166.0	24 010	674.0	78 255
Limbe	269.0	28 491	20.0	3 136	97.0	13 692	386.0	45 319
Thyolo	176.0	18 117	36.0	6 283	129.0	17 648	341.0	42 048
Lilongwe	273.0	30 803	2.0	238	13.0	1 972	288.0	33 013
Chiradzulu	107.0	11 756	46.0	8 095	110.0	16 065	263.0	35 916
Rhodesia	218.0	21 143	10.0	1 887	–	–	228.0	23 030
Blantyre	141.0	14 781	16.0	2 681	65.0	10 528	222.0	27 990
Zambia	16.0	1 879	0.1	27	0.2	25	16.3	1 931
Kasungu	13.0	1 537	–	–	0.5	108	13.5	1 645
Balaka	8.0	897	0.3	76	1.0	166	9.3	1 139
Mchinji	7.0	598	0.2	50	–	–	7.2	648
Ntcheu	1.0	126	–	–	0.2	19	1.2	145
Dedza	0.9	68	–	–	0.3	46	1.2	114
Kasupe	0.6	59	–	–	0.4	92	1.0	151
Dowa	0.6	80	–	–	–	–	0.6	80
Luchenza	0.6	63	–	–	–	–	0.6	63
Chikwawa	0.4	48	–	–	–	–	0.4	48
Mangochi	0.1	9	–	–	–	–	0.1	9
Total	1 873.2	202 380	428.6	64 621	1 072.6	148 594	3 374.4	415 595

Source: Fisheries Report 1977.

the year, and that the average value of the fishing equipment was then about K38 in addition to possessing a canoe. The annual average gross return for the semi-subsistence fishermen for the years 1963 to 1966 has been estimated at K220, which was eight times as much as the fire-cured tobacco grower earned and five times as much as was earned from seed cotton on customary land. For fishermen, farming was and probably still is a less attractive proposition, even if account is taken of cash income from farm produce. The fisherman in Malawi has been among the first to enter the market economy and to invest capital – quite unlike the situation in developed countries.

On Lake Chilwa, the fisheries are entirely in the hands of part-time semi-subsistence and small-scale commercial fishermen, who in the 1966 sample constituted 75 per cent and 25 per cent respectively. Since the recovery of the lake after the recession and the expansion in years of high level the number of commercial fishermen and the amount of capital required are greater. In 1976, the cost of 100 yards of seine netting was K35–45 and for gill netting, K10–20, according to the mesh size and depth of net. The smallest quantity of long line that could be bought was 25 lb (3000 yd) for K75.

It is a characteristic of the fishermen on Lake Chilwa, that whatever their incomes, the economy is a semi-subsistence one, in the sense that the fish that are caught are used both for their own consumption and for exchange for indigenous 'money' such as maize, to pay wages of helpers, as well as for cash. In addition, the capital goods they use are not all bought, but many are constructed by themselves or given to them or even inherited. All boats, fishing lines, seine nets and gill nets are bought, but almost all fish traps and baskets scoops are made by the people who use them. Net scoops are midway between home and factory manufacture, in that the raw material is bought. For canoes, the tree trunks are bought, and the labour of construction is usually paid for, although some fishermen participate. Plank boats are bought locally from skilled craftsmen.

Nevertheless there is a big difference between the 75 per cent small-scale and 25 per cent larger scale fishermen. The latter only are engaged in fishing for the greater part of the year (12 months in the good year of 1966 or 1976) and they make use of purchased capital goods, rather than home-made equipment. The type of labour they use is different. It is mostly the 25 per cent larger scale semi-subsistence fishermen who employ labour from outside the family groups. The average number employed in 1966 was six, for four months of the year only, at an average wage rate of K5.50 per month. Of these, about half combined family labour and employee labour. However, even the family labour had to be remunerated, often by surrendering part of the catch. The 75 per cent small-scale fishermen either fished alone or made use of one relative at least, a son, an uncle or a brother – who were remunerated by a share of the catch.

Regression analysis was used (Chipeta 1972) to determine the relative importance of various factors involved in the economics of fishing. These were: the number of relatives employed; the number of man/hours that labourers were employed; the time spent on fishing by the fisherman himself and the value of the fishing gear.

In 1966 the average income of the fishermen in the sample who employed labour was K608 per annum. The income of the small-scale individual fisherman averaged only K200. The gross income from fish was closely and positively correlated with the number of labourers and with the time for which they were employed; whereas the number of relatives and their work time and the fisherman's own time was not as productive. The small-scale fisherman spent more time on making or mending his equipment than did the large-scale fisherman, who had more expensive equipment. The latter spent more time fishing too, so that the use of more productive equipment and the employment of labourers accounted for the higher profits.

The shallowness of the lake and the extension of the fishery into the swamps and feeder streams account for the popularity of basket scoops, fish traps and spears; and there is also the large reward for comparatively little effort. The use of such capital goods would not be very effective in Lakes Malawi or Malombe where the water is deeper and reliance must be placed on gill and seine nets. Living is cheap on Lake Chilwa. Fishing camps consisting of temporary huts may be built on shore or even on water far from the shore and in the swamp (Fig. 17.2a). *Typha* shoots are cut and piled high until the mound rises above the water. Then more is added so as to keep the platform firm, poles are sunk into the mud to erect the walls of the hut and tied together with ropes or fibre. The walls and roof are composed of grass and *Typha*, woven together as thatch. But the life of these assets is short and some weeks must be spent every year in constructing replacements. Fish traps and basket scoops last only a year and nets, one to two years.

The response of the Lake Chilwa Fishery to lake level fluctuations is shown in Table 18.2. The data commences after a minor recession in 1960–61 and includes the major recession of 1966–68, followed by recovery and another minor recession in 1973–74, up to the years of high level 1976. The first peak was in 1965 before the decline towards the dry period, but it will be seen that the later years of high level show about twice the quantity of fish landed in the earlier peak period.

Table 18.2 The Estimated Quantity of Lake Chilwa Fish Output 1962–76. Figures in metric tonnes.

Year	Quantity	Change in Quantity
1962	544	—
1963	3262	+ 2718
1964	5255	+ 1993
1965	8820	+ 3565
1966	7100	− 1620
1967	3139	− 4061
1968	97	− 3042
1969	3326	+ 3229
1970	4166	+ 840
1971	3595	− 571
1972	5246	+ 1651
1973	1903	− 3343
1974	3171	+ 1268
1975	2808	− 362
1976	19 746	+16 938

Source: Annual Report (1965). Dept. of Agriculture and Region, and Fisheries Report 1977.

The main causes for the decline in the fisheries were first of all the high mortalities of *Sarotherodon shiranus chilwae* in the dry seasons of 1966 and 1967, and secondly the exit of the fish to spawn in the swamp in 1967, from which they could not return, since the swamp was cut off from the lake by a vast expanse of mud and the rivers were barred by sand at their mouths. The natural reservoirs in the lagoons and streams were still fished in 1967 and 1968, and in fact, probably overfished. When the water level rose again and refilled in 1969, the fishes returned to the lake. The migrating catfish, *Clarias*, were caught in transit to the lake on long lines, but the juveniles grew and bred within one year. Numbers of *Barbus* swelled the catch in 1970 and 1971, but *Sarotherodon* did not appear significantly until 1972. The minor recession of 1973 again caused a decline in fishing, from which complete recovery took three years to attain, although the drop in actual tonnes landed was not as severe as in the major recession (see Chapters 11 and 12).

Long term and short term adjustments in fishing during the decline and recovery of the lake

Although a number of alternatives, most of which are less remunerative than fishing, are available to Lake Chilwa fishermen, it took time to adjust during the local depression of the late 1960s, partly because of lack of knowledge concerning the recovery of the lake. It was widely expected that the water level would start rising again and that fishing would resume at the beginning of 1968. This hope was not fulfilled.

In 1967 fishermen had the choice of the following: (1) fishing on a very much reduced scale in the swamps, streams and Lake Mpoto (a lagoon of the River Sombani) or in lagoons in the northern swamps; (2) transfer to Lakes Malombe, Malawi or Chiuta; (3) increasing the cultivation of rice, cotton, cassava and vegetables; (4) a switch over to commercial handicrafts such as plaiting carpets; (5) spending considerable time trapping birds and digging for rodents; or (6) seeking employment elsewhere.

The fishing that remained during the major recession was limited to small areas of the swamps in the south at the mouths of rivers or in lagoons for a few months in 1968, and in 1969. The commercial fishermen suffered most, but the subsistence fishermen were also severely handicapped.

Two hundred fishermen are known to have gone to Lake Malombe and eight to Lake Malawi (Min. Natural Resources Report 1970). They constituted as little as 10 per cent of the fishermen and those who were among the 25 per cent larger-scale fishermen, who owned seine nets, boats or canoes and who had enough money for transport of goods. Such fishermen obviously must have had enough capital to enable them to employ labour, since to set up a new fishing beach and to pull an offshore seine net requires ten or more men. Among the fishermen whom I visited, the range of the number of labourers employed was from 10 to 30. These fishermen were probably the last to return to Lake Chilwa fishing because of restrictions on the use of seine nets (Chapter 12). Nevertheless, their catch elsewhere must have contributed in no small way towards the continued flow of fish even to the remotest rural markets.

Out of the 25 fishermen interviewed in the area of Mwambo, in the west of the Chilwa plain around the Likangala river, 9 had continued fishing in rivers, three continued fishing in Lake Chilwa swamps and thirteen (50 per cent) turned to farming. The proportion of fishermen who turned to farming is higher in the areas of Nazombe (south Phalombe plain) (61 per cent), Kumtumanji (northwest Chilwa plain) (95 per cent) and Kawinga plain in the north (62 per cent), because of the greater opportunities for growing vegetables, rice, cassava and cotton and less opportunity for fishing at that time. In Kawinga some continued fishing in the swamps while others turned to carpet mat plaiting.

Amongst businessmen, some shop-owners added itinerant hawking to their normal trading practices. Some hawkers extended the area covered without finding more custom. Others migrated to other parts of the district or turned to farming. It is clear that an economic depression resulted from the loss of fishing revenue.

Conclusion

Cyclical fluctuations are part and parcel of the life of the economy of Lake Chilwa. Prevention of such fluctuations is impossible since they are caused by natural not economic factors. Their ill effects could be reduced only if agriculture in the region were developed to a stage where it became a greater source of cash than fishing or if any other industry were developed to such a level. This kind of development is becoming very desirable since periodic recessions of the lake are likely to re-occur. But care should be taken that people should not lose all if the lake level should rise and inundate the floodplain for some years.

The use of more efficient gear should be encouraged when the rate of return warrants it, since it would facilitate the movement from Lake Chilwa to other shallow lakes. Borrowing of large sums by fishermen is unwarranted as even prosperous fishermen can do with little capital equipment. The employment of labour appears to be an important factor even in the adjustment, but in this respect fishing is self-financing.

Further research is needed into the character of cyclical fluctuations so that better prediction can be made regarding the decline, the length of the dry time and recovery phases (see Chapter 3). The past situation, where little was known, led to waste of resources when fishermen stayed idle, hoping that the lake would recover instead of switching to alternative activities straightaway. More work is also needed on the problems of expanding agricultural and pastoral pursuits for which the potential is there.

3. Fishing craft *by* Swanzie Agnew

The success of the water economy so largely depends on fishing craft (Fig. 18.1) that it is worth considering the process of manufacture, the costs involved and the attendant problems. Buoyancy, durability and size are the qualities sought in the wood from which canoes are made. A poor canoe in the hard conditions prevailing may last for as little as five years, but canoes made from *Chonya* (*Adina microcephala*) or *Mtondo* (*Cordyla africana*) might last for thirty years. In general, the life expectancy of a canoe is between four and ten years. The cost of a canoe is considerable in the mini-economics of the village and in relation to earnings.

The Forestry Department has had control of all trees in the Forest Reserves, since British colonial times. In 1966, trees were valued between £3 and £7 (Sterling); but a forest giant from which two boats might be made, then cost £15. The cost was about double in 1976. Because credit was not available, a fisherman might purchase a tree from the Forestry Department, then wait a year or more to collect funds before ordering the felling and excavation of the heartwood and fashioning of the canoe. Both felling and shaping are skilled crafts and specially experienced men are employed. First the feller exposes the

Fig. 18.1   Canoes with the Fisheries Jetty of Kachulu in the background (photo: G. Lenton).

tree's upper roots, then drops it on the side that is most suitable for the base of the canoe, called *diwa*. Shaping the ends to points is termed *kulenga*, and smoothing the sides *mkawoya*. The hollow is made with the least possible opening along the axis to guard against shipping water and to provide seating. A team of craftsmen may take two or three weeks to complete operations, which are all done with hand tools, nowadays made of steel, and formerly of locally smelted iron.

A great many taboos and obligations are associated with the making of a canoe. Men on the job must remain celibate; women are engaged during the transport of the canoe overland to the lake to sing haulage songs and to call out imprecations at the difficulties encountered. Villagers coming to help in haulage expect beer and provender to be offered. The launching culminates in a celebration with food, drink and dancing (Chirwa et al. 1966). All this cost the owner in 1966 about £30. This cost may be compared with the price of a western-type standard plank boat, made by local carpenters or by the Fishermen's Institute at Lake Chilwa, which at that time was £50 Sterling. They were sold in 1977 for K180 (now = £120) and a few only are being made at present (Fig. 18.2a and b). The new type of plank boat, made with the minimum of machinery, should be more economic, since less wood is wasted in chippings of the heartwood. It should be more durable in storms, if properly caulked, and its stability lends itself to the attachment of an outboard petrol engine. But petrol and oil are probably beyond the pocket of the fisherman still. Plank boats could be driven by oars as well as by poling (the method of propulsion adopted from dug-out canoes) and might take small sails. Such adaptations would suit the users of gill nets and long lines, although efficient trawling would require an engine to reach the central deeper parts of the lake and to work there. During the depression caused by the recession in 1966, canoes were constantly changing hands at half the original cost. Thieving of nets sometimes was the cause of a man deciding to sell up and cut his losses. The 1966 minor recession of the lake became the major recession of 1967–68 and many canoes were abandoned at the lake edge for several years. Some were in good enough condition to refloat when the level rose in 1969 and 1970.

Trees suitable for canoes are becoming very scarce. The plains around Lake Chilwa are now devoid of trees and the gallery forests of the hills have been decimated of timber. Boats are now recorded as coming from the less densely populated adjacent areas of Mozambique and some from as far afield as Chiradzulu, Mulanje or Zomba mountains. The Forestry Department now offers a service for transporting canoes, which has meant that canoes from even more distant forests can be brought to the lake.

Not all fishermen own canoes. Chilivumbo (1972) found that under half of the 411 men questioned from various shores of Lake Chilwa owned a canoe; the rest borrowed or hired canoes not in use by the owner. In the southern sector of the lake more people own their own craft (Agnew 1973) and the greatest density of canoes is found there (Fisheries Report 1977). Fig 18.3 shows the distribution of boat owners who fish in the southern waters of the lake, by village, that is to say, the registered village of the fisherman, where his family and wife's land is situated (Agnew 1973). It is apparent that a line may be drawn separating seasonally inundated land where most boat owners live from dry-land farming areas. Nevertheless, there were a few boat owners

(a)

(b)

Fig. 18.2 (a) The 'modern' type of plank boat (photo: A. J. McLachlan); (b) Plank boat under construction by local craftsman (photo: A. J. McLachlan).

355

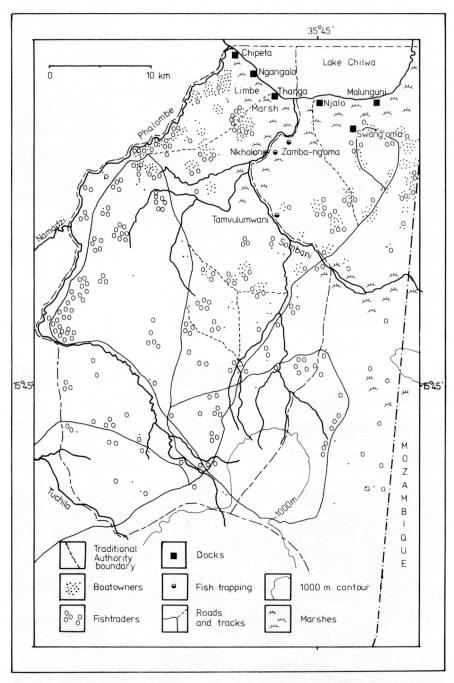

Fig. 18.3 The distribution of fishermen owning dug-out canoes or fish traders using bicycles for fish marketing, and the location of landing beaches in the (southern) Phalombe plain (after Agnew 1973).

356

working on the lake, who were resident in villages as far away from the lake as the foothills of Mount Mulanje, 25 km distant. The heaviest concentrations of boat owners are along the deltaic stretch of the Phalombe River and the area surrounding the Sombani embayment. When the plains are flooded, canoes are also used to travel about when rising levels of water impede movement on foot.

There has been a continuous increase in the number of fishing craft on the lake since its recovery from the dry period of 1967–68. In 1967 there were 981 fishing craft on the whole lake, and in 1977 the number had increased by 27 per cent to 1243. In the Northern Marshes, which were little known to the Fisheries Department before the major recession of the lake, there were 349 fishing craft (including bark canoes) and 34 per cent of the total fish catch was landed there (Fisheries Report 1977).

4. Fish distribution centres

The main landing sites for fishing have increased in the last seven years. The biggest change has been seen in the northern marshes where fishermen may spend several days on the water and land their fish catch, when some of it is already dried, on certain days at special sites, where villages have grown up. The main collecting points are at ten villages around the periphery of the marshes beyond the *Typha* swamp on the sand bar, which extends from the northwest to the northeast and which separates Lake Chilwa from Lake Chiuta. These villages have some access to the railway opened in 1970, which connects the area with the Central Region to the northwest and with Mozambique to the east.

Fig. 18.4   Ferries unloading at a landing beach (photo: A. J. McLachlan).

There are also fifteen landing beaches on the west shore, seven in the south, three in the southwest and three in the northeast on a sandy promontory, making a total of 36 in all. Three main dispersion centres handle the fish catches. The main one is at Kachulu Harbour on the west (with 40 per cent of the landings), which is connected by an all-weather road to Zomba and thence to the Shire Highland villages and the small industrial town of Blantyre. Narrow bicycle tracks lead from the main roads to all the villages on the plains and Highlands. Kachulu receives the catches from the northeast grounds of Chinguma by daily ferry (Fig. 18.4).

In the south, the Swang'oma 'docks' or beaches are the centre of distribution to the southern villages and by road to Mulanje, the town centre of the tea plantations. In the northwest, Mposa village has an outlet by road to the northwest villages and to Liwonde on the Shire River. The fish landings at different centres of distribution around Lake Chilwa for the months of 1976 are given in Table 12.2.

5. Fishermen

Fishing is a skilled occupation. To the outsider it may appear to be inefficient, consisting of a number of alternative operations, some more fruitful than others, with no mechanization. In fact, it is a very complex, highly integrated economic effort in the lake, rivers, swamps and marshes. The variations in the amounts of the various fish species at the various centres shown in Tables 12.1 and 12.2 (p. 222) are at first sight inexplicable, since different centres have peak periods at different times for the same species of fishes. This is partly due to the migrations of the fishes to and from the swamp or sandy nesting places (see Chapters 11 and 12), but it is also largely due to the migratory habits of the fishermen, who follow the fish and set up a new base at any site which proves profitable. They may remain there for a few days or weeks and then return to their own villages to fish. The monthly pattern also depends on other activities of the fishermen, sometimes connected with repairs to fishing nets or hut building and sometimes involving farming. In addition, different methods are used by the same fisherman where and when they are necessary. Labour may be employed temporarily, when it is available from other work, and so the intensity of fishing varies according to the demands of village life as well as to the availability of fish.

Chilivumbo (1972) submitted a questionnaire in 1969–70 to a large sample (13.7 per cent of 3000 registered fishermen) to ascertain the degree of adaptability of this sector of the population to changes after the recovery of the lake from the recession of 1966–68. The social changes in these circumstances were the following: (1) a steadily increasing population; (2) the opportunity offered by the growth of the monetary factor in commercial fishing; and (3) the possibility of participation in the rice schemes, which offered more reliable cash earnings than could be derived from fishing.

The response to changes was measured against the three ethnic groups, resident in the Chilwa basin: the indigenous Nyanja section, the earlier Yao invaders and the Lomwe, who had recently penetrated from Mozambique. Chilivumbo postulated that there would be a greater likelihood that the group least integrated into social values, tribal attachments and ancestral holdings would be those that would seek support from non-traditional ventures, such as fishing for profit. This assumption was strongly borne out by the large number, i.e. 52 per cent of the fishermen interviewed, who were recent Lomwe settlers. The Nyanja came second with 36 per cent of those interviewed. The Yao, with 10 per cent, were least represented as fishermen, as they had owned land by conquest and infiltration since the mid-nineteenth century. They are not indigenous to a water economy as are the Nyanja and are not socially insecure as are the Lomwe immigrants. The remainder were Khokola from Mozambique, and Ngoni and Tonga from other parts of Malawi. The last two groups undertook fishing specifically for the cash earned by fishing during the boom period after 1961. Most fishermen were born either in the same Chief's area as they fish (26 per cent) or in an area of another Chief, but in the same district (50 per cent). 10 per cent were born in Mozambique and the remainder were born in other parts of Malawi. The district of birth of males in the Chilwa population is illustrated in Fig. 18.5 from Agnew (1970).

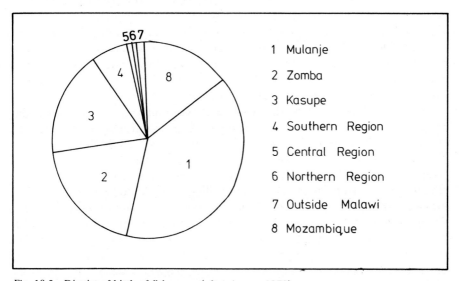

Fig. 18.5   District of birth of fishermen (after Agnew 1970).

For the Lake Chilwa fishermen, 25–39 years was the modal age group; 50 per cent fell into this category, followed by 29 per cent for the 40–59 age group. There were very few older men (3 per cent) and none under the age of 20 years.

The level of education declined with age, because there were few schools in the area before Independence. Men under 24 years of age had received some education; only 9 per cent completed primary education while 23 per cent had been to primary school for the first few years. Formal education however bears little relationship to skill and experience in fishing; but it does have a bearing on the ability of the individual to acquire capital by semi-skilled work elsewhere, and hence the means to acquire boats and nets.

In the domestic sphere, it was ascertained that 92.5 per cent of the sample had only one wife, which may be explained by the fact that 90 per cent were Christians. The size of the family tended to be small; 25.5 per cent had no child or one child only (surviving) and only 7 per cent had six or more children.

Fishermen do not appear to be much involved in village affairs, since they are away from home so much. Only 9.5 per cent claimed the status of headman, acting headman, elder or *mwalimu*, church official or member of a political or educational committee.

Since fishing gear (nets, lines and traps) is an important aspect in commercialization of fishing, it is of interest that 30 per cent possessed equipment valued at under £1; and only 35 per cent had equipment worth between £5 and £20. Among the big-scale fishermen in this sample, values jumped to over £100 owned by 5 per cent of the fishermen; and four fishermen in the sample had *makhoka* or shore seines, worth £250 each. The findings of the survey showed that most fishermen started their careers by working for a time with an older member of the family. The simple equipment of home-made traps and grass scoops were at first used by those without initial capital. When savings had been accumulated from paid employment, either at the lake or elsewhere, scoop nets or gill nets were the first investment. As Chilivumbo observed, it would be a mistake to presume that there is a linear progression from simple gear to more sophisticated equipment, since many kinds of fishing techniques are used by those who own nets, and most fishermen never acquire enough capital to buy expensive nets. The income of fishermen fluctuates widely according to the season, the water level and weather conditions and the time lost to repair or replace depreciating equipment. When high winds do not permit the use of boats, gill nets cannot be used on the open lake.

In 1976, 82 per cent of the total catch was analysed to determine the proportion of fishing techniques employed (Fisheries Report 1977), and the results give a fair indication of the effort devoted to catching different species of fish at different places. 58 per cent of the total analysed comprised *Barbus* (*matemba*) caught by shore seine net, which can only be used where there are firm beaches on which to land the catch. Gill nets caught 20 per cent of the fish, mainly *Sarotherodon* (*makumba*) and *Clarias* (*mlamba*) with fewer *Barbus*. Long lines catch *Clarias* and they brought in 9 per cent of the total fish. Fish traps in this analysis caught 13 per cent of the total with 0.6 kg per trap. The number of times fish traps were set in a year in Lake Chilwa runs into millions, and fishermen usually set many at a time, to form barriers across channels from lake to swamp. Every household probably uses fish traps for unrecorded subsistence fishing.

The term 'big fishermen' was used by Pauline Phipps (1973) to identify those fishermen in the sample, who appeared to earn over £1 a day from their fishing in the survey conducted by Chilivumbo (1972). This dividing line distinguished 63 fishermen in the 411 interviewed, and of these 12 (or 3 per cent) claimed to earn more than £2 daily from the sale of the catch. It seemed that these successful fishermen had gradually accumulated capital and gear over the years. Those in possession of shore seine nets and gill nets were in a position to command a steady commercial market for their catch.

The successful fishermen have more land in their home village: 50 per cent of those earnings more than £2 a day have farms of over 1.5 hectares in area and 10 per cent of these owned 4.5 hectares or more. It is amongst these larger landowners that further capital could be derived by hiring labourers to work at producing cash crops, particularly cotton, as well as tobacco, rice, beans, pineapples and various staple grains.

360

These big fishermen also owned more cattle than others; 50 per cent had more than 6 head of cattle. Lastly investment was made in other enterprises such as stores and canteens, from which an income varying between £1 and £9 a week was mentioned by those willing to reveal profits. The big fishermen are able to diversify their income by depending on other interests during the lake's recession and so can minimize their risks from fishing. Some had sufficient capital in hand to transport their boats and gear to Lakes Malombe, Malawi and Chiuta further north, to continue fishing when Lake Chilwa dried up. In answer to the question as to what they would do in the event of the lake drying up again, the majority of big fishermen answered that they would transfer to other fisheries. These reactions probably indicate a reluctance to allow the larger amounts of capital to lie idle, whereas the small fisherman's reaction to adversity is to return to his village or seek a job outside until productivity returned to Lake Chilwa (see section 18.2).

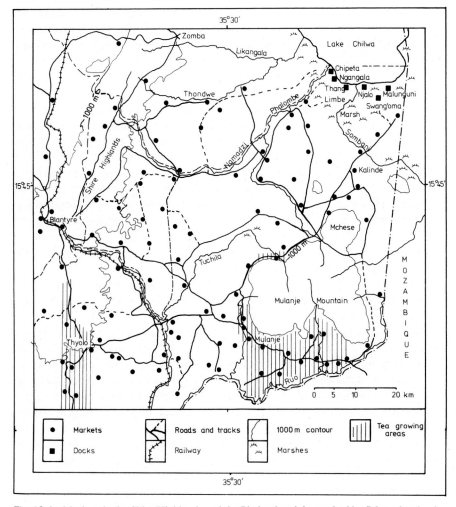

Fig. 18.6 Markets in the Shire Highlands and the Phalombe plain reached by fish traders buying fish at the landing beaches (after Agnew 1973).

361

6. Fish traders

Traders use bicycles to transport a basket of fish on the rear carrier, as the main means of disposing of dried or fresh fish to the rural markets. Of a sample of fishermen in the western Chiefs' areas in 1970–71, only 10 per cent sold to traders buying in bulk to transport fish to the urban markets (Chilivumbo 1972). Fig. 18.6 shows the distribution of markets served by fish traders on bicycles who bought the fish at the landing beaches, south of the Phalombe River in the Swang'oma area, and carried the fish inland for distances up to 150 km from the lakeside (Agnew 1973). It is clear that the Thondwe River, tributary to the Likangala, forms the accepted boundary between traders hauling fish from the lakeside beaches to the north of the Phalombe River (including Kachulu harbour) and those obtaining fish to the south of the river.

The distribution of traders in the southern part of the basin is in contrast to that of boat owners (Fig. 18.7). Though every fishing community has one or more fish traders resident in it, the distribution of fish traders is to be found in all parts of the plain. The network of roads and tracks linking the landing sites with the retail markets has the fish traders as its hub, for it is there that he recuperates, as he needs to do, since the average weight carried on a bicycle is 50 kg.

Information, collected by Chilivumbo and Phipps on the manner of selling by the fishermen, confirms the importance of bicycle traders in the distribution of fish. More than 80 per cent of the fishermen in the southern half of the basin sold their fish to 'bicycle' traders, while only 8 per cent sold to lorry drivers. The opinion elicited from the fishermen was that there is a proper separation between fishing and trading in fish. 'A fisherman cannot do two jobs' was frequently heard. This finding is in contrast to that by Klausen (1968) who advised that fishermen in Kerala (India) would achieve much higher returns by selling their catches themselves. The price paid by fish traders to fishermen at the lakeshore for fish was, in 1976, about 12 tambala or 8 pence per kg. In the urban markets after processing and delivery by bicycle, fish costs many times as much (Fisheries Report 1977), but the fish traders also make a living and the fishermen can devote their whole time to fishing.

Williams (1969) in a wider survey of fish trading in Malawi was able to establish the source of fish in various districts of Malawi in 1966–67. These are shown in the map, Fig. 18.8. Lake Chilwa before the major recession of 1967–68 accounted for 20 per cent of the fish offered in the main markets of Ncheu, Dedza, and Lilongwe Districts, up to 150 km away, because of easy accessibility by the main road from Zomba. Fish from Chilwa was in competition with that from Lake Malawi and the Lower Shire River, wherever bituminized roads or rail impinged on the bicycle traders from Chilwa. Two districts in particular were predominantly served by Chilwa: Mulanje and Chiradzulu (each 90 per cent of the fish sold), where distance and poor accessibility were in favour of direct use by bicycle distributors. The 34 000 workers on the Mulanje and Thyolo Tea Estates constitute the most reliable wage-earning clients within the economic province of Lake Chilwa.

Williams categorizes traders into: (a) Seasonal – those who are farmers or have other seasonal occupations, such as tea-plucking or tobacco grading, who trade in fish during the slack months; (b) part-time traders who visit markets

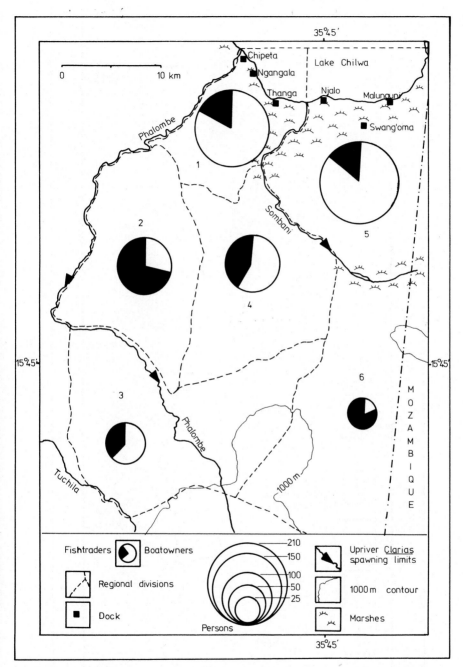

Fig. 18.7 Comparison between the distribution and numbers of boat owners and fish traders on the Phalombe plain (after Agnew 1973).

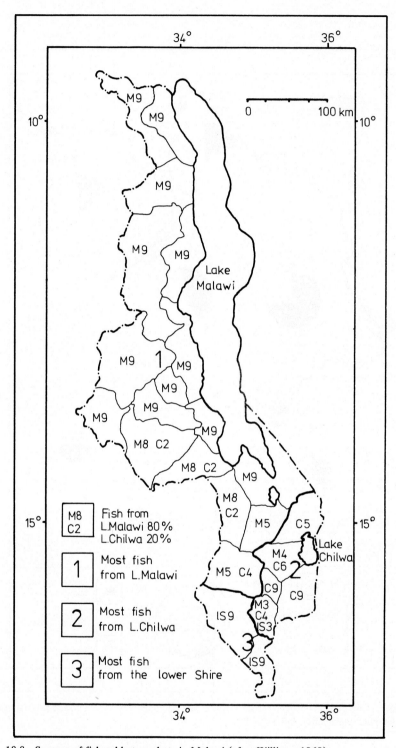

Fig. 18.8 Sources of fish sold at markets in Malawi (after Williams 1969).

throughout the year, but only for 4–5 days for one or two periods only each month. The intervening time is spent in getting fresh supplies; (c) fringe traders who, becoming unemployed, invest in a fish load to sell at markets. Such men have usually acquired a bicycle during their period of employment; (d) fishermen, who take their own catches to market because they are unable to find traders in that part of the lakeshore or in the marshes of the northern sector of the basin, where they operate; (e) lastly, entrepreneurial traders with capital gained from wage employment of at least £20 (in 1967) who work full-time and who may own a boat to collect fish from outlying shores.

There are four distinct stages in fish distribution to markets and some stages are handled by the same man. These may be described as: (1) Processing fish bought at the lakeshore; (2) transport to a larger market in one of the main urban centres; (3) there the stock of fish is sold to smaller traders or to other wholesalers, who again sell to the small trader; (4) the last stage is that of the small trader who will distribute the fish to the remotest selling point within his range of operation. This complexity may account for the price charged to the consumer which is so very much higher than that paid by the trader to the fisherman at the lakeshore.

Many traders in Lake Chilwa own boats by which they visit marsh and island sites to gather fish, which they smoke and dry and transport for sale at the lakeshore landing beach. Many 'bicycle' traders will purchase fresh fish daily and process them by smoking in crude kilns until a sufficient load of 50 kg has been acquired, before setting off on the market round. In the northern marshes many fishermen smoke their catches and take them out of the marshes to outlets where meetings between fishermen and traders are arranged on selected days or at the month's end (Fisheries Report 1977).

The urban markets are under the control of the municipal authority and licence to trade in them is statutory. These urban areas have daily markets, except Sundays; the smaller trading centres are controlled by District Councils, and two to three days a week, including Sundays, are scheduled in such a way that neighbouring markets do not coincide. The fish trader will move from market to market to take advantage of the gathering of customers on the special market day, but he may remain to dispose of his fish to casual purchasers on non-market days. Markets on tea estates are outside the District Council control, as are those set up casually at cross-roads, with the fish displayed on the grass verge.

Many small traders reach a wider market area by carrying their bicycle and load of fish by bus or by lorry (Fig. 18.9a and b). The bus routes are scheduled, while the lorries prefer to keep to the better roads and transport the largest amount of fish to the urban markets, which are thus serviced more cheaply than remoter areas, reached by bicycle or on foot. Williams calculated that 50 per cent of all traders attended markets with 10 traders or less, and that 50 per cent of all markets had four traders or less. The expansion of sales in years of high level on Lake Chilwa now embraces foreign exports as well as the main towns in the central region (Table 18.1).

It is of interest that at full production fish from Lake Chilwa fisheries competes successfully with the other major fisheries. In the recession of 1967–68, a substantial flow of fish came from Lake Malawi even to as far away as Mulanje (160 km). Traders shifted from Lake Chilwa and sought supplies

(a)

(b)

Fig. 18.9 (a) Bicycle traders loading their baskets (photo: A. J. McLachlan); (b) Bicycle traders setting off to market (photo: A. J. McLachlan).

366

from Lakes Chiuta and Amaramba, where this fish supply was recorded on the markets in Chiradzulu, Mulanje, Zomba and Blantyre, having travelled distances of up to 120–160 km. Table 18.1 also demonstrates how far afield fish from Lake Chilwa are sold (Fisheries Report 1977).

The trading organization was summed up by Williams (1969) by observing that the distribution pattern may be compared to one developed in a well-organized economic system. The wholesale and retail functions appeared (in 1967) to be sensitive to a variety of specific needs in so far as the densest population attracted the greatest number of traders, even if there was no urban centre. There was also a rapid readjustment in source of supply when, as in the case of the recession of Lake Chilwa, the nearest source failed. The traders at Lake Chilwa immediately reasserted their hegemony on the districts bordering the lake when the fishing started again in 1969, which is an indication of its favourable location near the most ample and profitable market in Malawi. (However the price to the consumer rises by ten times between the lakeshore and urban market).

7. The economic environment

The fisheries of Lake Chilwa offer an economically unstable environment, determined by the seasonal and long-term fluctuations in lake level (Chipeta 1972). Yet at high productions, the fisheries have permitted comparatively readily earned cash, while a substantial number of men gained an income five or more times greater than that prevailing for unskilled or agricultural labour. For a few entrepreneurs, fishing has been a means of acquiring cash with which expensive nets have been bought, hired labour employed on farm holdings or in bigger fishing units, besides investments in cattle, canteens and vehicles for transport.

The uncertainties in the behaviour of the lake have hindered the popularity of trawling or the establishment of viable commercial fisheries supported by refrigeration and a transport fleet. In a sense there is compensation in safeguarding the lake for small-scale fishermen and a few full-time fishermen and traders indigenous to the area, thus maintaining their interest in the land and supporting their families.

In the last decade subtle changes in the economic climate of the basin have been taking place. To be more than a 'subsistence' fisherman, greater capital is needed than in 1965 when £30 would suffice to become an independent fisherman or trader. Gill nets and shore seines are very costly today, motorized plank boats and brick kilns which are far more efficient for processing fish, need the additional cost of fuel. Bicycles are three times as expensive and transport costs have risen steadily, while the Government has attempted to control the price of fish in markets to protect the lower income groups, who depend on fish for protein in their diet.

During the time that the surveys in this chapter were undertaken (1966–72) the fishermen and fish traders established on Lake Chilwa were there by their own initiative and enterprise. In reply to a questionnaire as to the reasons for fishing, 31 per cent had been attracted by the quick cash earnings; 34 per cent replied in a general kind of statement that 'fishing is their way of life'. The remaining number had been fishing part of their working lives but had been in paid employment for shorter or longer periods elsewhere (Chilivumbo 1972). Yet only about 3 per cent of them had become moderately 'prosperous' (Phipps 1973).

Four factors which have emerged in the last three years, 1976–78, may have changed the economic outlook: (1) many more men are now engaged in fishing for the quick cash return; (2) the total annual yields from the lake have increased spectacularly and there seems no indication of overfishing; (3) simple shallow-lake trawling with plank boat and outboard engine has experimentally demonstrated (Chapter 12) that it is profitable to fish the centre of the lake in the cooler months by this means. This was an area which was virtually untouched by traditional fishing methods; (4) In this monograph in Chapter 3 an analysis of the 25-year record of lake level fluctuations is shown, which gives a clearer picture of the periodicity of the behaviour of the lake. Briefly, minor recessions sufficient to reduce fishing for one or two years can be expected every six years or so and major recessions which will interfere with fishing in the open lake for three to five years can be expected every 60–70 years, with a possibility of an intermediate recession in 30–40 years. It may be possible to determine which kind of recession it is going to be, by analysis of the lake level spectrum of periodicities, combined with rainfall figures and chemical analysis of the water (Chapters 2, 3 and 4). Continued planning by the Fisheries Department to meet the special exploitation problems of the lake over the long term is now possible.

References

Agnew, S. 1970. The People of Lake Chilwa. In: M. Kalk (ed.). Decline and Recovery of a Lake. Government Printer, Zomba, Malawi, pp. 40–41.

Agnew, S. 1973. The Indigenous Growth towards a Cash economy in Africa. Geoforum 16:26–35.

Annual Report, 1965. Department of Agriculture & Fisheries 1964 Part I, Appendix. Government Printer, Zomba, Malawi.

Annual Report 1970. Ministry of Agriculture and Natural Resources. Government Printer, Zomba, Malawi.

Chilivumbo, A. 1972. The Fishermen of Lake Chilwa: A study in Adaptibility. In: D. J. M. Vorster (ed.). Human Biology of Environmental Change IBP (HA) Conference Proceedings 1971, Blantyre, Malawi. International Biological Programme, Marylebone Road, London. pp. 15–20.

Chipeta, W. 1972. A note on Lake Chilwa Cyclic Changes. J. Interdisc. Cyclic Res. 3(1):87–90.

Chirwa, G., Kango, C. & Munthali, M. 1966. Dugout canoes in Lake Chilwa. Soc. Malawi J. 19:58–61.

Compendium of Statistics for Malawi (1966). Government Printer, Zomba, Malawi.

Economic Report, 1972. Office of the President and Cabinet Economic Planning Division, Government Printer, Zomba, Malawi.

Fisheries Report, 1977. Lakes Chilwa and Chiuta Statistics. Department of Fisheries, Malawi.

Klausen, A. M. 1968. Kerala Fishermen and the Indo-Norwegian Plot Project. Allen & Unwin, London. 201 pp.

Phipps, P. 1973. The 'Big' fishermen of Lake Chilwa. A preliminary study in entrepreneurism in rural Malawi. In: M. E. Page, Land and Labour in Rural Malawi. Rural Africana 21:39–48.

Population Census Report 1972. The census of 1966. National Statistical Office, Zomba, Malawi.

Population Census 1977. A Preliminary Report. National Statistical Office, Zomba, Malawi.

Williams, R. D. 1969. Fish Trading in Malawi. In: I. G. Stewart, Economic Development and Structural Change. Edinburgh Univ. Press, pp. 83–104.

M. Kalk, C. Chipeta & P. R. Morgan

The range of diseases prevalent in developing tropical countries of Africa are found in Malawi, and the people inhabiting low-lying areas near swamps are more prone to infection than those of the drier Highlands (Stubbs 1972). Because of its location, the people of the Chilwa plain suffer from vector-borne diseases, such as malaria and bilharzia, and from diseases spread through the agency of water and bad sanitation, such as enteritis, cholera and hookworm. Leprosy has been a problem, and, in addition, there are the diseases spread through contact with white populations either within the country, such as measles and whooping cough, or tuberculosis spread through migrant mine workers. Famines occurred periodically through the ravages of the red locust until about thirty years ago, and in the lake itself, Nchisi Island has a background radiation in certain parts which is thirty times normal.

In spite of this daunting range of possible disease, the Chilwa plains have been inhabited for many centuries, and many ethnic groups have arrived and intermingled. The population has grown from a very small number after the slave trade (Drummond 1903), more especially in the last thirty years, and increasingly during the last ten years, until some parts have the densest population amongst rural Africans south of the Sahara. They are at present completely self-supporting (see Chapters 16, 17 and 18).

The health and education of the people of the Chilwa plains, features which are very much inter-related, may be viewed against the historical, geographical and economic background already described and in special relation to the changes of the last ten years. These are predominantly: (1) Independence, development and the resultant small advances to a cash economy; (2) the effects of the drying-up and recovery of the large lake which, although economically harmful, had ironically some temporarily useful impact on disease.

These days the people resort to both African medicine and 'western' medicine and prefer whichever they think might be more efficacious at the time. It is not surprising to find African herbal medicines useful, since many western pharmaceutical preparations had their origin in tropical plants. *Strophanthes* sp., a climbing shrub, for example, which still occurs on Nchisi Island, had until recently been exported to Europe from southern Malawi, where it grows on wooded hillsides. It is a source of a glycoside which stimulates the heart. In Malawi, in olden days, extracts of the seeds made a potent arrow poison (Williamson 1974).

1. Traditional medicine

Remedies for infectious diseases and psychosomatic or psychological disorders are dispensed in villages by African herbalists or *sing'anga*, whose skill in 'diagnosis' and preparation of medicines is passed on from father to son. African medicine is, by and large, a secret practice and the nature of the diseases treated are known only from symptoms and their relative importance. Nevertheless there is a trend in Central Africa today towards co-operation by the western doctor with the African medicine man. Psychological disturbances are often treated very well by the *sing'anga* (Gelfand 1973). The Medical Association of Malawi held a joint conference with the African Herbalists' Association at the Polytechnic, Blantyre in 1971. One result has been access to the ingredients and recipes of some traditional medicines to science so that their effects can be quantified and the active principles identified. Attention was drawn to the harmful effects of some eye 'medicines' suffered by patients before they attended the hospital for treatment. Mitchell et al. (1975) investigated one of the extracts, an infusion of the leaves of *Steganotaenia araliacea* Hochst. (*mpoloni*) which is used to treat 'sore eyes'. Experiments showed a general bacteriocidal effect on bacterial cultures *in vitro*, but attempts to infect animal eyes to test the extract *in vivo* were unsuccessful. It seems as though the extract by itself, while having a marked irritant effect, did not cause permanent damage to rabbit eyes. There is, however, ample circumstantial medical evidence that permanent damage is caused when eyes already inflamed by bacterial infection are subjected to 'treatment' with these extracts. The same herbal extract is used in conjunction with others to treat all urinary and sexually transmitted infections. Experiments to demonstrate bacteriocidal effects after metabolism and secretion via the urine were not conclusive. Nor were attempts to demonstrate a diuretic effect of these extracts proved.

A few common plants with possible remedial properties will suffice to illustrate their range and the mixture of empirical usefulness and belief in their unproven magic qualities. These have been described by Williamson (1975), who listed 121 species in Malawi. They are common in *Brachystegia* woodland, which once covered the higher ground of the plains around Lake Chilwa and many of the species are found on the hillsides of Nchisi Island (Howard-Williams 1977).

Several species of Leguminosæ are used; an infusion of the bark of *Acacia albida* Del., a thorn tree, is drunk to stop diarrhoea. *Acacia* species are widely used for this purpose in Africa, even by the early white settlers in South Africa (Watt & Breyer-Brandwijk 1962). The roots of *Albizzia versicolor*, a tall tree, are used as a cure for intestinal worms in strong infusion and in weaker doses as a purgative. Another *Albizzia* species is actually called *A. anthelmintica* (A. Rich.) Brogn. and its medicine is said to have a moderate efficiency against tape-worm (Watt & Breyer-Brandwijk 1962). *Cassia petersiana* Bolle is the source of a variety of medicines for stomach-ache, coughs and venereal diseases. Some *Cassia* spp. contain quinones which have the strong purgative effect of 'senna' leaves. *Cassia* is also one of the components of a medicine for snake bite. But belief goes even further, for when rubbed into artificial cuts on ankles and hands, the mixture is said to 'inoculate' a person against snake bite. Snake-bites are greatly feared, but they are not common – so the belief is strengthened.

A preparation of the bark of the sausage tree, *Kigelia africana* (Lom.) Benth. (Bignonaceae) found along the banks of inflowing rivers, is the basis of a balm for sores of many kinds and for syphilis. *Lippia javanica* (Burm. f.) Spreng., (Verbenaceae) found on the lake islands, has many uses, which one might imagine may be related physiologically, with active principles in some way associated with the biochemistry of sex hormones, because of their related applications. An infusion of the roots may be used as a contraceptive. With other roots an infusion is made, which when inhaled from a boiling pot is believed to cure anti-social psychological states connected with rules for sexual conduct, or 'madness' of some kind. Girls at first menstruation are given a mixture of this and other roots dissolved in water. One of the other components is *Vernonia amygdalina* Del. (Compositae), common on the floodplain, which in strong dosage results in abortion and possibly later sterility. The leaves and stem are dried and powdered and mixed with ground-up tobacco leaves to make snuff. Brewing pots may be rubbed with the leaves to make the brew more intoxicating (Williamson 1974).

Several common remedies for intestinal diseases have been tested for their microbiological effect. Mwanza & Mwambetania (1974) found that concentrates of the leaves of the guava tree, *Psidium guajava* L. (Myrtaceae), which is abundant around villages although not indigenous, had bacteriostatic effects. The concentrate was compared with the effect of chloramphenicol, a broad spectrum antibiotic, used for typhoid and other bacterial infections of the intestine. Agar cultures of the bacteria *Serratia marcescens*, *Shigella sonnei*, *Salmonella paratyphi* and *S. typhimurium* and pathogenic *E. coli* were inhibited, but not cultures of *Proteus vulgaris* or *Proteus morgani*. The active principle of the guava leaves was probably composed of more than one component and first results suggested that they were phenolic compounds which are known to be present in many tropical plants and to be bacteriostatic (Watt & Breyer-Brandwijk 1962). The effect of the guava leaf infusion was directly proportional to concentration and in villages it is used in a very strong dose.

2. Modern medicine on the Chilwa plains *by* C. Chipeta

Between the year of Independence in 1964 and 1968, there was a 30 per cent increase in hospital beds in the whole country. Mission facilities tend to fill the gaps between Government Medical Centres. On the average most areas (in 1968) were within 24 km of a hospital facility, although on the lakeshores (including Lake Chilwa) with relatively dense populations, the distance was greater than this (Stubbs 1972). Since then, the number of Health Centres, Dispensaries and Clinics, scattered through the Chilwa plains according to the concentrations of the population, has increased to some extent.

More importantly, the nature of these centres has changed in the last ten years. Although anti-malarial tablets, antibiotics and other medicines are dispensed when patients need them, the emphasis has changed from curative medicines to preventive medicine. Existing dispensaries as well as new centres now serve as 'Under-Five Clinics', where mothers are taught how to look after the nutrition and health of the children, who may be immunized against various diseases. This change of policy after Independence was greatly stimulated through the work of Dr. Sue Cole-King who started the first clinic of this kind at Namitambo on the Shire Highlands. She studied the health patterns and nutritional status of the children who were brought to the clinic and concluded, among other things, that the one constant factor associated with the absence of malnutrition among the children was the former attendance of the mother at school for two or more years (Cole-King 1971). A similar factor was observed in a small follow up study at the same clinic (Masiku & Kawonga 1974).

373

The hospitals serving the people of the Chilwa plains are situated on the periphery of the area, at Mpiri in the North, Zomba and Malosa in the west, Phalombe and Mulanje in the south. These are used by outpatients who live near enough, but the majority of people go to the nearest Health Centre in their Chief's area.

2.1   A Case Study August 1977

A sociological study of the problems of women and children in Malawi was organized by UNICEF in 1977. It is based on a random sample survey of 554 primary family units on the Phalombe plain and as there was an estimated 36 580 primary family units in the area, the sample size was 1.4 per cent. This is believed to be large enough to give a reliable picture of some aspects of the quality of life in an area closely dependent on the fluctuations of rainfall and lake level. The changing attitudes of the people towards social problems are an integral part of the human ecology, without which the management of the environment is not possible.

2.1.1   Child mortality and disease

The average family on the plain consists of four to five people of whom two are adults and two to three are children. On the average, each women gives birth to four or five children, but two to three of them die. 22 per cent of the women in the area of Mkhumba to the west of the Sombani River and 26 per cent in Nazombe to the east, had experienced miscarriages or still births, or their babies had died immediately after birth.

Over 90 per cent of the families in both areas reported that the cause of death of their children was illness and not accidents nor witchcraft, which were mentioned only by insignificant numbers of those people interviewed. The chief child-killer in Mkhumba was reported to be diarrhoea, the second, measles and the third, malaria. In Nazombe area, the chief cause reported was measles, followed by diarrhoea and the third was whooping cough. Other worrisome diseases were respiratory tuberculosis, 'fever' and 'stomach-ache'. Worms and bilharzia were seldom mentioned because the families have no way of detecting these diseases unless the children have been to hospital for examination.

Water-borne diseases are contracted through contact with infected water (bilharzia and hookworm) or drinking dirty water (gastro-enteritis, cholera, typhoid). With an increase in the use of piped water the dread of such diseases is decreasing.

In order to shield children against communicable diseases many, in some cases all families reported that their children has been vaccinated against smallpox, measles, whooping cough, tetanus, tuberculosis and typhoid. Most of them also reported that they fed babies more than twice a day, which is the normal custom for older children and adults, and they wash them at least twice a day. But not many of them boiled equipment used in feeding babies nor knew of the medicines that can be used to cure diseases.

374

Most of the miscarriages were said to be caused by illness or by accidents. Witchcraft was mentioned in Mkumba, but not in Nazombe, as a cause of miscarriage. About half the total number of births were at home in Mkhumba and 42 per cent in Nazombe. These numbers are bound to decline with the increasing number of trained maternity nurses at nearby clinics to help deliver babies. The long-term plan in Malawi is to have maternity/delivery units stationed at 16 km intervals and eventually at 8 km intervals.

Another problem is diet. There are taboos, such as the prohibition of eggs for women and children, although poultry are kept (see section 17.7). For fear of illness or that the child will be born without hair or that delivery might be difficult, many pregnant women stop eating fish and meat, although fish is plentiful, thereby denying themselves protein when they most need it. It is hoped that health education of the type received at clinics and in Home Economics courses at schools will have a beneficial effect in this regard. Since the quantity of food available depends entirely on local production, periodic fluctuations of the climate make considerable impact.

2.1.3   Hygiene and sanitation

Owing to the relative 'prosperity' of this area in terms of cash incomes (see Chapter 17), more than half the families have houses made with mud bricks, which are rated as modern or as of 'mixed' type (with both traditional and modern features). Almost 20 per cent of the houses are roofed with iron sheets, and the average number of rooms is three. The condition of most of the houses is average or good. Traditional housing provides a hut separate from the parents for children of the 'extended family' from the age of six years (Mwanza 1972).

In Mkhumba and Nazombe 60 per cent of the families possess a pit-latrine. In some areas, especially near the lake where the water table is close to the surface, the walls of the pits fall in and the pits fill with water in the wet season. This becomes a nuisance to health; enteritis, hookworm and cholera are spread by contamination of water and soil with faeces. A possible solution is to strengthen the pit walls with metal drums or with wood. A water-borne sewerage system is too expensive for the village community at this stage. A slightly modified pit-latrine with mudded walls and a tall, wide, black ventilation pipe has been designed by Morgan (1977). Experiments showed that it is almost completely fly-proof since the ventilation pipe covered with fly-proof netting takes advantage of the physical principles of ventilation and the behaviour patterns of flies, so that the entry of flies and mosquitoes is minimal. The pit is deep but collection of water at the bottom hastens the decomposition processes.

2.2   *Water on the Phalombe plain*

Paradoxically, although villages are near the lake or streams, clean water is not available. An opportunity to compare two areas which differed in the state of the provision of a clean piped water supply from the mountain streams of

Mount Mulanje was offered by the Mkhumba and Nazombe areas. A contrast was presented between villages in the Mkhumba area, of which seven out of eight already had a piped water supply and five villages in Nazombe area where no village yet had access to piped water, although pipes were then being laid for that purpose. In Mkumba the piped water has been available for only two years, but a comparison between the two areas already shows its impact.

The main sources of water in both areas are tabulated in Table 19.1. It will be seen that 80 per cent of the people in Mkhumba used clean water, but in Nazombe, wells, boreholes, streams and even the brackish lake water were used. These depend on the fluctuating annual rainfall. In the former, women draw water 3–4 times a day and in the latter 2–3 times a day. Women have to walk a fairly long distance before they can draw water. In Nazombe the average distance to water is about 1 km, while in Mkhumba with piped water, they walk 0.5 km each time. The closer proximity of water encourages more use of water and saves time. They keep water in calabashes standing in their houses for a day. In Mkhumba most of the families stated that their water was sufficient in the dry season as well as the wet season. In Nazombe only 45 per cent had sufficient water in the dry season and most of them had insufficient clean water in the wet season as well.

Table 19.1 A comparison of water supplies used in two Chief's areas on the Phalombe plain.

| Source of Water | Number of Families | |
	Mkhumba	Nazombe
Dug out wells	41	175
Boreholes	2	81
Rivers	1	18
Springs	2	3
Swamps	1	2
Water taps	197	0
Lake Chilwa	0	50

For those who depend entirely on Lake Chilwa for their water supply, the periodic decline of the water level inevitably removes their supply completely.

The overwhelming majority of the families in both areas believed that clean water was necessary for good health. However, it is only in the piped water areas of Mkhumba that clean water is available. In Nazombe the water was dirty and salty. In spite of this, not many families boiled their water before use, either because they do not have adequate pots for doing this, or because the fuel is inadequate. They may think it unnecessary, unless there is an outbreak of a serious disease like cholera. In the face of a standard of living without sufficient individual family means to improve hygienic measures, the provision of free clean water to the community as a whole is of paramount importance. The introduction of the piped water system in Mkhumba area has already created an appreciation of this for health in both villages, but further community education is clearly needed to ensure proper maintenance.

It is difficult to give precise causal relationships between the provision of piped water and health benefits, but it seems that the overall impact has been good. In all areas where piped water is available, cholera has been eliminated. The incidence of gastro-enteritis, eye diseases and skin diseases appears to be reduced, but there are no figures available. Potential health benefits are upset when the system breaks down and recourse to traditional sources of water has to be made. Contact with impure water also occurs in bathing and when washing clothes in water often shared with domestic animals. Bovine bilharzia has been found to infect man (see section 19.5 and Gear 1968).

Plans are under way to cover the remaining area of the Phalombe plain with piped water. The limited availability of funds and skilled staff may not allow the provision of piped water to all villages at first, but eventually it is hoped that this will be possible.

The community is involved in every component of the provision of piped water. It was observed that the people take pride in the responsibility for maintaining water points and that they are very glad to have safe water. The majority of the women in Mkhumba area have given help to the water project on more than five occasions, in contrast to Nazombe area where the gravity piped water project is only just beginning. Maintenance of the water points is largely the responsibility of the community with local leaders and committees which supervise maintenance and repair and also the collecting of money for repair and the cleaning of the areas around the taps.

The following rules for the use of piped water points have been agreed: the water is free to everyone; no personal bathing or washing of cooking pots to be done near the site; children are not allowed to play with the tap; women must clean the point regularly and dig up the drainage pit and re-place the stones; users of the point must contribute labour and money for repairs.

The water is used primarily for domestic consumption, but it is also used for brewing beer, making bricks and mudding houses. A notable indirect benefit for agriculture is that its provision will allow the relocation of families from densely populated areas into new settlement schemes. Provision of clean drinking water has made greater amounts of unsafe water available for gardening and irrigation, but without hygienic practices health will still be impaired.

In addition, intangible benefits gained from 'self-help' have been valuable in that community 'infrastructures' have been formed, which assist with the implementation of other services.

With the broader spread of the piped water service, there will certainly be a reduction in the scourge of water-borne diseases in other areas as well; pure water will be available throughout the year and the time spent on fetching water by women will be reduced. The water supply will be independent of the decline and recovery of the level of Lake Chilwa since it is derived from the perennial mountain streams nearby. In the long run improvements in the health of the community will be derived from health education and the provision of clean piped water.

3. Common Diseases *by* M. Kalk

The Agro Economic Survey (1972) of four villages on the Kawinga plain northwest of Lake Chilwa found that 5.5 per cent of the working hours of adults was lost for 'illness or tending the sick'. This is not dissimilar to the amount of sick leave (2 weeks per year) allowed in western society in certain employment. On the other hand, life expectancy is lower in the tropics.

3.1 Malaria

Malaria used to be the most important disease of young children. But, as seen above, first place now goes to imported diseases: measles and whooping cough. Until the age of three months, babies have a relative immunity transferred by the mother. From that age to about 12 months babies are very susceptible to malaria and mortality may be high. Should the child recover a certain degree of immunity may develop which is maintained into adult life, provided there is re-infection from time to time. Mortality from malaria is low in adults with treatment but sickness is common (McGregor 1970). *Plasmodium falciparum* is the most frequent cause of malaria in Malawi, which is transmitted by *Anopheles funestus* and *A. gambiae*. The latter breeds in water standing in household utensils, temporary puddles, garden plants, seepages and rice fields, while both occur in ponds, streams and canals, but neither will breed in brackish water (like the swamps and water of Lake Chilwa) (Wilcocks & Manson-Blair 1972).

3.2 Hookworm

Hookworm disease comes second in causes of morbidity amongst parasitic diseases, being exceeded only by malaria in Malawi (Min. Health 1971). It causes anaemia from loss of blood if the infection is heavy. It is transmitted directly from faeces in the soil to the skin in contact with it.

A study of the prevalence of hookworm in a village of 204 people near Kachulu Harbour (Gadabu & Magendantz 1975) showed that children from the ages of two to ten years were the most infected (70 per cent), and the age group 26–30 years were next (64 per cent). There is a drop in prevalence between 10 years and 20 years, possibly because immunity had built up (Goldsmid 1972) and the worms already in the gut have died. The increase in prevalence after 20 years of age may perhaps be explained in this case by the large number of immigrants during the successful fishing in years of high lake level (Chapter 18). Of the two species of hookworm found *Ancylostoma duodenale* was rare, and the so-called American hookworm, *Necator americanus* was the more important parasite. Both are widespread in the tropics. The false hookworm *Ternidens diminutus*, was not detected in this study. Hookworm anaemia induces iron deficiency which varies with the worm load, the age of infection, the nutrition of the patient and the natural or acquired resistance of the patient to the worms (Chandler & Reed 1961). A comparison of this study (Gadabu & Magendantz 1975) and that of Msiska & Magendantz (1974) suggests that the prevalence of hookworm may vary from year to year with the amount of rainfall and with the season of the year, that is, with variation of the amount of exposure of bare feet to wet ground around the houses where there is no sanitation system.

3.3 *Leprosy**

Leprosy was fairly common in the Chilwa area, estimated to have a prevalence of over 2 per cent (Molesworth 1967). The LEPRA Control Project, partly supported by international aid, has been operating in the Chilwa area since 1966, and aimed at its inception to eradicate leprosy in ten years. Four thousand cases were registered in the first year and treated with weekly doses of dapsone. 3500 uninfected contacts, under 20 years of age, received BCG vaccinations in the first year, and the whole area has now been covered. The method which was used, was the dissemination of information to every village by trained personnel, explaining that complete cures were possible if a weekly medication of dapsone was taken regularly. This was followed up by the registration of all patients and the delivery of tablets once a week at treatment points to each person. If the patient did not attend the mobile clinic, then he was sought out by bicycle or on foot into the remotest parts.

Up to about 8 km from the lakeshore, this plan of action has worked well and in 1977 over 50 per cent of the patients had been cured and discharged.

The exodus of people from the lakeshore villages in the time of the lake recession (1967–68) and the even greater influx of temporary fishermen and fish traders in the years of high level since 1974 has put a large strain on the programme. Patients do not turn up regularly for their medicines and defaulters cannot be found because it is necessary for the men to travel across the lake or to the north, when fishing is reported better in various parts. In the hope of intercepting patients a clinic was set up on the eastern shore of the lake, but it was not successful in finding old patients.

Ironically, the economic disaster of the dry years made it easier to find patients suffering from leprosy and to treat them, and successful fishing and trading has interfered with treatment.

In 1975 the administration of the LEPRA Control Project was taken over by the Ministry of Health. One advantage is that education about the successful treatment of leprosy will continue at all the health centres, which are being established on the plains. But a disadvantage is that the medicines are issued monthly instead of weekly at the existing health centres or hospitals. This involves a longer distance to walk for medicine than previously; attendance has decreased and a whole month's supply of medicine may be forfeited.

At present the LEPRA Control Project is discovering new cases in those areas where every former patient had already been registered. These may be new arrivals at the lake, or might even be those who had been infected by the patients who became defaulters (in regard to their medicine supply) and could not be traced.

It has been shown that leprosy can be controlled with the co-operation of the patients, and an intensification of the education of the people in the villages would lead to its elimination, if the whole community was convinced of the need to treat every patient regularly.

* The editor is indebted to Mr. S. Lipenga, supervisor of LEPRA Control in the Chilwa area for this short report on the effect of lake levels on the leprosy campaign.

In response to a question about the prevalence of cancer in the Chilwa area, Dr. Paul Keen, a specialist in cancer for south and eastern Africa has written the following:

'It is now recognised that more than 80 per cent of cancers are caused, at least in part, by environmental factors. In western countries, industrialisation has produced environmental carcinogens and changes in diet and habits have produced precipitating factors.

Among Africans, many "trigger factors" have been noted in local diets and customs. For example the role of mycotoxins (poisonous fungi such as afflotoxin, a black blemish on badly stored groundnuts) has been invoked in the genesis of primary liver cancer.

The two commonest cancers in African males are primary liver cancer and oesophageal cancer, but they have unexpected ethnic or geographical distribution. The people of the northern part of South Africa and Mozambique have a high incidence of primary liver cancer, whereas in the Transkei in the south-east, they are more prone to oesophageal cancer.

Malawi is unique in Southern Africa in the sense that these cancers are equally present, but with a different geographical distribution in the country itself. In the Chilwa region, both are present. An ecological study including diets, habits and customs should prove of great benefit in elucidating some of the possible causes of the disease.'

4. The influence of the fluctuations of the lake level on bilharzia transmission *by* P. R. Morgan

Bilharzia, which is caused by blood flukes, is wide-spread among the people of the Chilwa plains, because there is close contact between the human population, the water and the abundant vector snails *Bulinus* (*Physopsis*) *globosus* and the less abundant *Biomphalaria pfeifferi* Krauss.

Areas of human infection are associated with freshwater swamps, especially the cleared channels and lagoons, with streams and water holes. Around the lower reaches of the Domasi River, all the schoolchildren were infected with either urinary bilharzia (*Schistosoma haematobium*) carried by *Bulinus*, or intestinal bilharzia (*Schistosoma mansoni*), carried by *Biomphalaria*. A similar picture was reported by the dispensaries of Chamba and Makwapala near the Likangala River (Morgan 1972a). During the heavy rains roadside ditches fill with water for a few months and children wash and swim in it. Snails multiply and are infected by the urine or faeces of the sufferers. Rice cultivation in flooded fields particularly favours the increase of snail populations and their infection, and paddy rice cultivation is extending every year.

In the early years of the Lake Chilwa Research Project, before, during and immediately after the major recession of the lake (1967–68), no snails infected with human schistosomes could be found on the lake edge, although a relatively high proportion of infected snails might have been expected in waters which drain areas where bilharzia is common. Morgan (1972a), for example, found no human cercariae released from a collection of 1300 *Bulinus*, taken from the northern shores of Kachulu Bay. Snail populations were then low compared with that reported in 1976 (see Chapter 10). They had apparently died out when the lake bed dried and numbers grew slowly until the lake became more dilute after 1974, when numbers were recruited from snails which had aestivated in the swamp or which were brought down from permanent streams.

It was not until May 1974 and May 1975 that infected snails were found at the lake edge. This was the month in each of those years when the lake was at its highest level and infection of the snails appeared to have taken place a few weeks earlier when the water was most dilute, at conductivity 530 μS cm^{-1} or about 0.5 parts of salt per litre (Mkolesia & Magendantz 1975). The percentage patent infection of the snails was determined by the number of cercaria larvae released under a strong light.

The apparent low level of infection recorded by earlier workers, followed by a completely different picture of high infection in the following years of high lake level (M. Magendantz pers. comm.) may indicate that the variable chemical quality of the waters of Lake Chilwa may influence the transmission cycle of bilharzia. The lake water is very turbid and alkaline at times of major and minor recessions, unlike the streams and lagoons which act as potential sources of human infection. A study of the effects of Lake Chilwa waters upon the ova and of miracidia (larvae) of *S. haematobium* was undertaken to ascertain whether seasonal and periodic fluctuations in water quality might influence the transmission of the disease (Morgan 1972a). Eggs were obtained from patients' urine and miracidia hatched from them. The percentage hatching rate of *Schistosoma* ova, their mean hatching time after immersion in water and the life span of the larval miracidia were determined in a series of water samples with increasing concentration at constant temperatures. Natural lake water was either diluted or concentrated by slow evaporation to simulate the concentrations of river water and lake water up to conductivities similar to those of the 1966–67 period of the lake recession.

The results are shown in Table 19.2 and Fig. 19.1. They indicate that alkaline waters can have a pronounced effect on the ova and miracidia of *S. haematobium*, if the concentration of salts is high. In experimental conditions miracidia lived up to 15 hours in dilute salt water and for less than one hour at the highest concentration. They become dumb-bell shaped as a result of dehydration. Ova became less viable at higher concentrations of salts and took longer to hatch. All these factors may influence the success of transmission of the schistosome to the snails.

Table 19.2 Effects of Lake Chilwa water on ova and miracidia of *Schistosoma haematobium*.

Conductivity μS cm^{-1}	Alkalinity meq l^{-1}	pH	Mean life span of miracidia minutes	Ova % hatching	Ova; mean hatching time minutes
100	1.0	8.0	638.9 (36)	87.2 (447)	18.2 (15)
2000	13.0	8.4	579.9 (38)	85.9 (264)	19.5 (15)
4000	23.0	9.0	415.3 (42)	69.1 (777)	24.7 (15)
6000	34.0	9.2	169.9 (40)	30.7 (862)	40.6 (15)
8000	46.0	9.4	55.6 (31)	18.0 (874)	67.6 (15)

Figures in brackets are numbers of observations.
Source: P. R. Morgan 1972a.

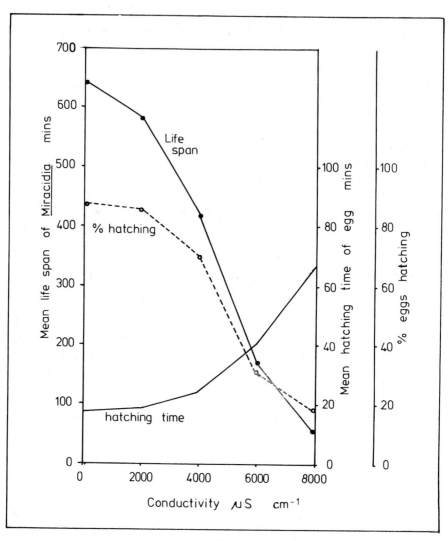

Fig. 19.1   Schistosoma haematobium: Hatching time, percentage hatching of ova and life span of miracidia at various concentrations of Chilwa water, shown as conductivity μS cm⁻¹.

In the context of the variability of Lake Chilwa as a habitat for the snails, the experimental results suggest that when conductivity rises above 4000 μS cm⁻¹, which happens only during marked recessions of the lake such as in 1960, 1966–68 and 1973, the life cycle of the blood fluke will be broken. During 'normal' dry seasons, when the conductivity may exceed 2000 μS cm⁻¹, the life cycle may be more difficult to complete in the lake and swamp. But in years of high lake level such as 1975–78 which equals that of the 1950's, no such inhibition will occur because the water is dilute (see Chapter 4). The channels have ideal conditions for snails and disease transmission (see Fig. 10.2).

Nevertheless, the conductivity in February 1971 when over 1300 snails were examined by Morgan (1972a), two years after the filling of the lake after dryness, no snails were found to be infected with human schistosomes. There was a lag of six years before the parasites were found in the snails. The alkalinity (or salinity) alone does not offer a complete explanation for the apparently low level of bilharzia transmission, and another factor was considered.

When the lake fills after dryness, it is remarkably turbid and during the succeeding years the turbidity decreases (see Chapter 4). Hira (1967) has noted that reduced light levels suppress the hatching of schistosome ova. In addition, the muddy lake bottom, on which the ova fall is low in oxygen (Morgan 1972b), which also inhibits the hatching of the fluke ova. Nowhere are the conditions for hatching better than in the newly cleared swamp channels cut by fishermen in years of high lake level between the lake and the floodplain where they live. Here the bottom is often sandy, the water is clear, the snail population is extremely high (see Chapter 10), and there is intensive human contact.

Although the open lake does not support snails (for lack of submerged vegetation), it is known that small highly frequented infected areas influence a considerable number of people. This is the case described by Webbe & Jordan (1966) for Lake Victoria. The high prevalence of bilharzia among the people of the Chilwa plain probably results from a number of ideal transmission points in marginal streams, swamp channels and marshes rather than from the lake itself.

The distribution of infection amongst 245 adults and children at Kachulu village was analysed according to occupation (Mkolesia & Magendantz 1975). 100 per cent of the 'herd boys' (aged about 10–16) were infected, but the Fisheries Officer, the boat builders and teachers were not infected at all. Only 24 per cent of the fishermen as a whole group appeared to be infected; among older fishermen who had resided at the lakeshore for many years the incidence of eggs in urine was low. This study took place, however, at a time of high lake level when successful fishing had attracted a very large number of 'male immigrants' to the area (see Chapter 17). The incidence among the 'immigrants' (probably from inland areas) was 29 per cent in residents of six weeks standing. When this group had lived in Kachulu for a year the incidence was 100 per cent. These comparisons may be explained by the fact that absence of voided ova in these fishermen is not indicative of acquired immunity to bilharzia; and it does not exclude serious chronic inflammatory changes due to schistosomes in the liver, spleen, urinary tract, colon or rectum. Obstruction of these organs and of the cardio-pulmonary circulation is common (Gear 1961).

Amongst children the highest incidence was in the age group, 10–14 years, when it was in general over 50 per cent in boys and girls. This reflects the greater contact of this group with infected water for play as well as for cleanliness.

Nkomo & Magendantz (1975) estimated the degree of haematuria (i.e. blood in urine) in a small sample of 21 infected children at Kachulu village. An average of 811 eggs per 10 ml urine from a single sample (per day) was counted. The red blood cell count in the urine was from 4–7.5×10^6 rbc ml^{-1} urine. These children were treated with Ambilhar and five weeks later there were no eggs in the urine and the rbc ml^{-1} urine count had diminished to an average of 1.5×10^5 ml^{-1}. Unfortunately no follow-up study of these children was possible to see whether re-infection had occurred. It is usual for re-infection to occur in such areas.

In these small studies, an attempt was made to find a correlation between prevalence of infection and the number of years schooling. There was none, although all primary schools teach hygiene. The conclusion was reached that 'Education about schistosomiasis and about the use of latrines is of little use unless an entire community received such education and unless a clean water supply is provided, so that villagers no longer have to use the naturally infected water for bathing.'

It is of interest that the hazard of bathing in water used by cattle mentioned above by Chipeta (section 19.3) was expressed as bovine bilharzia in a child who also suffered from human bilharzia in one case in this small sample. Bovine schistosome cercariae, *Schistosoma matheei*, are often found in *Bulinus* snails, and it is expected that these may infect a significant number of people. *S. matheei* is common in children near dams in South Africa (Gear 1968).

Urinary bilharzia is much more common than intestinal bilharzia, *S. mansoni*, in Malawi (Min. Health, 1938, 1950, 1969). The vector *Biomphalaria* has not been found in the Chilwa swamp, but does occur in fish ponds and dams nearby, but these were found not to be infected (Nankhonya & Magendantz 1976). The absence of infected snails in the study site used as a source of snails (Makoka Cotton Research Station) was considered to be due to the high standard of sanitation there, since the dam was protected from human faecal contamination. Laboratory bred *Biomphalaria* snails from this source were infected by exposure to schistosome miracidia and found to be susceptible. Hamsters exposed to their cercaria became infected as shown by eggs in their faecal pellets. A circum-ova precipitin serological test on hamster and human sera with the disease showed the presence of anti-body properties, which according to Smithers & Terry (1969) protect the human host against new infections, but are unable to reach established fluke populations in the tissues through their protective sheaths.

5. Nutrition and malnutrition *by* M. Kalk

In the face of so many environmental hazards to health, nutrition is of importance. The normal diet consists of *nsima* and *ndiwo*, that is, a share of thick porridge and a 'side-dish' or 'relish', made of vegetables and a protein-rich food such as beans and groundnuts or, near Lake Chilwa, mainly fish. The staple cereal for *nsima* is maize, supplemented by cassava. The former contains about 8–10 per cent protein, the latter almost none. Vegetable growing is encouraged in the whole country, since vitamin deficiencies are very common towards the end of the dry season before the new crops are ready (Cole-King 1973).

The preparation of the flour for *nsima* is done at home. First the grain is removed from the cob by hand, by rubbing one cob against another. About 15 kg of grain, enough to last a household for four to five days, is prepared at one time. The first pounding and winnowing of the grain takes one woman about six hours. Bran and germ are separated and usually kept. Coarse bran may be fed to the fowls and the rest may be made into a cooked snack for the women and children, and a proportion is added back to the flour in proportions that vary with the amount of maize available for the family. After the pounded grain has been soaked in water for a few days, it is pounded again for about seven hours. Then the flour is dried on a mat in the sun and later stored in a home-made earthenware pot. There is a variable amount of wastage during pounding and drying, estimated to be between 15 per cent and 20 per cent (Williamson 1975).

This method of preparation is very time-consuming and advantage is now being taken of the local mechanical mills which are being set up in many villages on the plains by men with a little capital earned by employment in towns. The product of the mechanical mill is not as finely separated into flour and bran as the manual product and includes a little more protein than the village flour. Women may walk up to 10 km to these mills (Williamson 1975).

The nutritional status of the people on the Chilwa plain has not been studied directly, as far as is known. Nevertheless, many sociological field workers have remarked that, when there is a lake recession, the low water level and consequent fall in the fish supply and the relative failure of crops for two or three years have far-reaching effects on the availability of foods (see Chapters 17 and 18). In 1975, a good year for crops, small surpluses of maize, groundnuts and pulses were sold by the few more well-to-do villagers to the trading centres of the Agricultural Development and Marketing Corporation (see Chapter 17). But the margin of excess is small and in a dry year there is no surplus. Supplementary protein is then sought from termites which swarm after heavy rains, from red locusts which roost on the *Typha* plants on the landward swamp margin in the north after the rainy season, from hunting gerbils and rats on the plains in the dry season and from snaring birds. Seeds, fruits and roots of wild plants which are normally welcome famine foods (Williamson 1975) are not abundant in the Chilwa area because the natural vegetation has been cleared to prepare agricultural land. Trees with edible fruits now occur only on islands in the lake or on hills and are not easily accessible. Of the 80 species of plants listed as useful as vegetables, very few are available; ten species only occur in the checklist of the floodplain (Howard-Williams 1977).

In general, the men are usually considered to be better-nourished than women and children; Nurse (1968a) found in his study of well-nourished men of the Central Region during two years of good harvests, that

(i) the mean height was 154 cm and the mean weight was 54.5 kg;

(ii) the weight for height was lower than that of the standards for East Africans, which themselves are lower than that for Americans (Latham 1965). Nurse discusses the possibility of the smaller stature and lighter weight being due to nutritional, climatic or racial factors.

Villagers in Malawi frequently refer to the month of January as the 'time of hunger'. 'This appeared to be the case even when the previous harvest has produced a good crop' (Nurse 1968b). January is the month when little maize is left in store in the family *nkokhwe* (maize crib) and the new crop is not yet ripe. The mean weight of his sample of 7000 men from the Central Region fell sharply by 2 kg from maxima in August–October to minima in November–January, and then the mean weight increased to the April level. Nurse attributes the months of maximum weight to success in hunting gerbils (*Tatera afra*) and the fat mouse (*Steatomys pratensis*). Similarly, gerbils are a supplementary source of protein in years of low level at Chilwa (Chapter 18).

The people at Chilwa are a fish-eating community and the annual fluctuation in weight found by Nurse may not occur here, but the rainfall is unpredictable (see Chapter 3) and the possibility of a reduction of intake in staple food (maize) should be investigated. The Agro-economic Survey (1972) recommended that more cassava should be grown on the Chilwa plain as an 'insurance' against maize crop failure, since cassava requires little labour input and can be stored in the ground.

Schulten (1975) in a four-year study of losses in stored maize in Malawi, mentions that 'most of the maize is stored by the farmers for home consumption for less than a year.' Only about 1 per cent of unimproved village maize, stored unshelled in village *nkokhwe* is lost from insect damage by the maize weevil *Sitophilus zeamais* and the grain moth, *Sitotraga cerealella*. The close sheathing leaves on the local variety of maize prevents the entry of pests in the field.

Loss during storage is therefore not a reason for the expected lower availability of maize in the 'month of hunger'. It seems as though it is necessary to increase the yield of village 'gardens' to prevent the annual occurrence of shortages.

In addition, Nurse (1968b) observed that the 'hunger months' coincided with the time when women who work all day in the fields had not the opportunity to prepare the usual two meals a day and this too aggravated hunger, as well as the diminishing store of food. On the Chilwa plains 'famine' years occur periodically in years of lower rainfall than average (see Chapter 3). Shortage of maize thus occurs in years of low fish catch so that it is imperative that crops be diversified, with greater emphasis on foods that can be stored. This requires a change of attitude in the adults, and less emphasis on cash sales of surplus crops in hand in order to gain some ready money, and greater attention to storage.

Traditionally men eat the meals prepared by women before the rest of the family. The women and children do not eat with them and the children share the remains in a common family pot, according to their ability.

Mothers are often unaware that lack of good food has precipitated illness in their children. Cole-King (1969) emphasized, as a result of a diet survey in Malawi, that adverse effects of an unbalanced or insufficient diet tended to show more in young children below five years of age. This underlines again the need for education of the parents. (Data on the possible incidence of protein and calorie-malnutrition are not available for the Chilwa area.)

6. Natural radio-activity on an island in Lake Chilwa *by* M. Kalk

Nchisi Island, near the western shore of Lake Chilwa, lies 3 km due east of the biggest fishing harbour at Kachulu (Fig. 19.2). It is roughly square and about 9 km² in area, with closely forested peaks rising to 400 m above the level of the lake. The island is considered to be of volcanic origin (Chapter 2).

For tens of thousands of years, waters at a much higher level than today in the Proto-Lake Chilwa (Chapter 3), caused erosion of the west side of the volcanic cone and left a semi-circle of hills which form a natural amphitheatre around a small plain, with an area of 1 km², known as the 'arena'. It slopes gently down to the shore from 30 m higher at the foot of the steep hills. Low-lying land also

surrounds the rest of the island which is densely populated, intensively cultivated and with good beaches for fishing.

The 'arena' is distinguished from other plains on the island by a background
radiation thirty times normal (Garson & Campbell-Smith 1958). The soil is red
in colour and 10 m deep; it has been derived from carbonatite rocks on the hills
and has adsorbed material from weathered radio-active dykes on the hills,
whose material has been washed down a steep gulley on to the plain. The dykes
vary in thickness from 1–10 m and one was 500 m long.

The richest rock samples analysed contained '2.2 per cent thoria and 2.13
per cent rare earth oxides'. The radio-active material occurred in small patches
of yellow-brown material in the samples, believed to be 'goethite', a mineral
known to be capable of adsorbing radio-active substances from solution in
times of intense heat. The chemical content of radio-active material increased
after crushing and leaching with water, which suggested that the thorium
complex was relatively insoluble.

Fig. 19.2 Kachulu Harbour and Nchisi Island, showing the radio-active 'arena' (photo: N.
Lancaster).

The radiometric and chemical analyses of the parent rocks in the dykes were
in close agreement. This indicated that thorium is approximately in equilibrium
with its decay products and that there was no appreciable uranium in the
samples. Low radio-activity was also exhibited by samples of pyrochlore.

On the other hand, chemical and radiometric analyses of the red soils in the
arena did not agree. The chemical analysis showed that the content of thorium
was about one third less than that expected from radiation values, although it
was higher than the uranium content. It was concluded that both uranium and
thorium had been adsorbed on to ferruginous material in the soil, but that
uranium had been preferentially dissolved out. Mainly decay products were
left, chiefly radium. The radiation from the soil was highest near the foot of the

gulley and decreased towards the lakeshore. In the centre of the plain it was equivalent to 0.025 per cent U_3O_8, one twentieth of that of the rocks.

The question arises whether the radiation is of a level which constitutes a health hazard to the people who live in the 'arena', who cultivate the soil and eat the produce. Attention had been drawn to this possibility by Hammond (1958) who reported a 'high incidence of skin cancer' on the Island. Later, Weir (1962) undertook medical examinations of samples of people from the 'arena' and from another village on the island for comparison. In a very small pilot study it was found that, on the whole, the health of the people in both samples was above the average for the mainland, but it seemed as though more people in whom the white blood cell count was lower than normal lived in the 'arena' village than in the control village. Radiation exposure was suggested as a possible reason for the few cases of leucopaenia among the people of the 'arena' area. It was considered that there was a case for further investigation.

This medical study recommended that samples of the population should wear radiation-protection badges for a time to detect the level of exposure; and the amount of radiation absorbed by animals and plants might also be tested.

The Lake Chilwa Co-ordinated Research Project organized various tests. The distribution pattern of the radiation levels was first confirmed with a geiger counter by the University's physicists (B. Whitmore, J. Blackburn & A. Pepper pers. comm.). Twenty villagers wore radiation-protection badges for 25 days in each month for 16 months. They lived in the 'arena' or beyond it. The badges were assayed by the Radiation-Protection Service, Pretoria, South Africa. Of the 320 exposed badges, only 12 positive exposures were noted, i.e. 3.7 per cent. All occurred in farmers in the south and the west parts of the 'arena', and no fishermen, school teachers nor farmers elsewhere were affected. These positive exposures were recorded only in November and July, months when women and men might be working in the fields. The exposure varied from 0.2 to 0.02 R gamma radiation, which is considered low. It was probably the result of more prolonged contact of the people's clothing, arms and legs with soil dust. Even if this exposure had proved to be continuous throughout the year at these levels, the exposure for the year would amount to only 3.6 rem per year absorption of energy by the bare skin. The internationally accepted standard of Maximum Permissable Exposure for arms and legs for the general public is 7.5 rem per year. On the evidence of the badges alone, it would thus appear that exposure to gamma radiation in the 'arena' is not a health hazard.

Samples of maize stalks, baobab fruit and bark, drinking water from a clay-lined pool, lake water and red soils from different parts of the 'arena' were assayed for radiation activity by the S.A. Atomic Energy Agency. Soils showed a variation from 118–658 pCi/gm dry weight for alpha-radiation and 192–372 pCi gm^{-1} for beta-radiation. Although the beta-radiation was about 100 times higher than in the soils around Pelindaba, where the Atomic Energy Board is sited, the values are very small and insignificant from the point of view of health unless the radio-active material is taken internally.

The values for maize stalks (no seed was available at the time of collection) were 71 pCi gm^{-1} dry weight (alpha-radiation) and 37 pCi gm^{-1} (beta-radiation), which shows very little uptake from the soil. Baobab fruit from a very old tree gave only 0.5 pCi gm^{-1} dry weight and 35 pCi gm respectively. The bark was only slightly higher, 77 and 55 pCi gm^{-1} respectively. In fact, uptake of radio-active substances by plants was of the same order in the 'arena' as around Pelindaba, which is considered safe. The water samples had an even lower contamination and was also similar to the ground water near Pelindaba (D. Van Uys 1966, pers. comm.). This supports the suggestion from the chemical analyses, mentioned above, that the radio-active substances in the rocks are relatively insoluble.

It would seem that the fears of Bowie (pers. comm. 1962) are not justified. He considered that the conditions at Chilwa (Nchisi) Island differ from those at Kerala in that radio-active minerals are broken down and the radio-elements are adsorbed on the soil, so that both thorium and uranium together with their decay products would be more readily available for uptake by plants.

The insignificance of the effect of the radiation on crops was confirmed to some extent by a short intensive breeding programme using maize from the 'arena' village. The number of lethal mutants was high in the first two generations, but not significantly higher than in other isolated populations of local maize in Malawi (T. Pinney pers. comm.).

Since radiation effects are cumulative, it might be important to know how long people live in the 'arena'. Although the population on Nchisi Island is dense at times, it is not permanent. 62 per cent of the people on the Island were found to have left when fishing declined from 1967–69, but many returned later, or others migrated there. The members of the younger generation usually leave the Island to find employment elsewhere. The majority of the inhabitants are fishermen whose daily drenching with water would remove contaminating soil. The lake water is usually accessible for domestic use, which presupposes a high level of hygiene among the women who work in the 'gardens'. When the lake level and rainfall are low for a few years, the people who remain to farm are temporarily more exposed to radiation from the increased dust.

The number of measurements taken in this study was few, since our work was interrupted by the falling lake level and the Island became inaccessible at that time. There are indications, however, that although the thorium deposits and, more particularly, the rare earth minerals may become of interest economically later, the health of the people living in the 'arena' is at a very slight hazard from radiation alone. Sickness which might occasionally arise through inhalation or swallowing the dust from the soil would merit investigation. The occurrence of skin cancer first noticed by Hammond (1958) should immediately be reported at the nearest Health Clinic.

7. Education *by* M. Kalk

Traditional education involves learning by doing. Young girls at an early age (8–10 years) take charge of younger children and help in cooking and cleaning, fetching water and gathering wood for fuel, while their mothers work in the 'gardens'. Boys help their uncles and father to repair and build houses and are responsible for looking after the cattle. The Agro-Economic Survey (1972) of the Kawinga plain villages did not find that children contributed very much to growing crops, although they do participate in weeding.

At puberty, 'schools' for initiation into adult responsibilities, usually directed by the leaders and elders of a village are held separately for boys and girls. With a great deal of ceremony, co-operation with adults, conformity to strict rules of conduct and loyalty to the village is taught and a code of behaviour is assimilated, so that the young do not as a rule make innovations, but carry on their elders' practices.

Mission schools were started in the late nineteenth century on some parts of the Chilwa plain, but it is only in the last ten years that opportunities have been created for many children to attend schools run by the Government, which are aligned towards the need for literacy and numeracy and the understanding of the environment, which are so necessary for agriculture and for health.

Table 19.3 shows the attendance at Primary Schools on the Chilwa plain in 1977. If the age of the children attending school is assumed to be between eight and seventeen years and that this is 25 per cent of the population, then about 30 per cent of the children eligible on the Chilwa plains attend school. The assumption may be wrong but it is necessary because the details of the 1977 census have not yet been published. The figure does have some basis in the composition of villages studied in the Agro-economic Survey (1972) where 25 per cent of the population was under 5 years and 25 per cent between the ages of 6 and 14 years.

Table 19.3 Attendance at Primary School on the Chilwa plains.

District	Traditional Authority	Total Population	No. Schools Total (enrolment)	No. Full Primary		No. Junior Schools		%[x]
Kasupe	Mposa & Mlomba	29 082	7 (1276)	4	(855)	3	(234)	17
	Kawinga	54 052	11 (2154)	5	(1750)	6	(444)	16
Zomba	Kumtumanje	40 829	7 (2850)	6	(2645)	11	(205)	16
	Mwambo & Mkumbira	77 988	21 (6087)	14	(5145)	7	(946)	31
Mulanje	Nazombe	57 996	16 (5765)	122	(4965)	4	(800)[xx]	39
	Mkhumba	122 048	28 (10 307)	1	(433)	27	(9874)[xx]	34
	Total	382 895	90 (28 439)	42 (15 793)		38 (12 503)		29.5

[x] Per cent enrolment of children in the estimated school-going age group.
[xx] The enrolment at 4 schools in each case has been assumed to be 200 in the absence of an exact figure. Source: National Statistical Office, Zomba.

Since school fees have to be paid, and schooling is one of the first 'big' expenditures undertaken by parents, the percentage of children attending school is an indication of the local wealth in a traditional area. The proportion of children in school is higher in the south (Phalombe plain) and centre (Chilwa plain) than in the Kawinga plain in the north. This confirms the greater prosperity noticed in the housing of the Phalombe plain, observed above by Chipeta section (19.2) and the contrast in the economies, described in Chapters 17 and 18. During years of lake 'decline' and movement of the population, attendance at schools falls.

The total 'Full Primary Schools' (standards 1–8) have a larger number of children than 'Junior Primary Schools' (standards 1–7 or less) and have usually been open longer. Schools are built in villages by parents by 'self-help' and, when completed, the Ministry of Education provides one or more teachers according to the number of children enrolled (Fig. 19.3). The number of small new schools almost equals the number of the larger older schools. There is clearly a demand for new and better equipped schools. The number of children who might be attending secondary schools was not included above. This would probably amount to less than 1 per cent. Secondary Schools are available on the periphery of the Chilwa basin at Domasi, Malosa, Zomba, Chiradzulu and Mulanje, and at other smaller Day Secondary Schools which provide two or three years of classes, instead of the usual four years. There were fifty full Secondary Schools (often boarding schools) and ten Junior Day Secondary Schools in the whole country in 1977, compared with eight full Secondary Schools before Independence in 1964.

Fig. 19.3   A new primary school on the Chilwa plain (photo: N. Lancaster).

Since 1967, the school syllabuses have been completely rewritten in every subject, for both primary and secondary schools. A great effort has been made to make education relevant to Malawi and to eliminate the hangover from colonial times of the British-orientated schooling of the early twentieth century. Hygiene and elementary Science, Nutrition and Agriculture are being introduced to primary schools. But to follow up these new ideas in the villages it will be necessary to educate the parents, who have not had the opportunity for schooling. Some progress is being made in this direction through clinics, extension services in farming and fishing, community development and political organizations such as the Women's League, Youth League and Young Pioneers. Contact with the small white population and their industrial or domestic skills is also a means of education for those in employment, more especially since Independence, because the focus of many of the remaining white employees is on development rather than on private gain from farming or administration. General literacy has not yet been achieved, but many skills are handed down from one generation to the next.

Unlike primary and secondary education, tertiary education is free. Intermediate training for Agriculture, Public Health, Medical Assistants, Nursing, Home Economics and especially training for Primary School teachers is given at a number of special colleges. The injection of higher education is very necessary to nurture primary, secondary and adult education.

The national target is 50 per cent enrolment of pupils at schools by 1980 (Developing Malawi 1971), which might well be achieved in some Chiefs' areas in the Chilwa plains.

The University of Malawi is also free to suitably qualified secondary school leavers and was established in 1965. It offers degrees and diplomas which give students a general level of education and an introduction to various professional fields. Further specialist training is received while graduates are understudies in their posts and many later obtain further qualifications in Europe, Africa or America. The research at the University is oriented towards development and the present study in this monograph is an example of its policy.

8. Summary of responses in health and education to changes in lake level and to development

8.1   Drying phase (1966–68)

The overall effects of three years of lower than average rainfall and the drying of the lake were poor food crops and a smaller supply of fish protein. In the absence of statistics on mortality and morbidity, a qualitative picture only can be drawn. The food intake of the people suffered in quantity and quality, except in the case of about 10 per cent of the fishermen who had gone to fish elsewhere. It is possible that among infants the poverty might have been expressed as malnutrition. Fewer children attended school because there was less money available from sales of fish.

On the other hand, older children might have gained from the dry years since bilharzia vectors were fewer in the swamps (although still plentiful in the streams) and conditions which facilitate the spread of hookworm were favourable for fewer months in the year. Leprosy was relatively easy to treat regularly, since the population was static. Stagnant water did not remain around so long for malaria mosquitoes to breed. The dreaded impending red locust plague did not materialize.

During the middle of the major recession, the first of the Rice Development Schemes, using the permanent waters of the Likangala and Domasi Rivers, were commenced. These were successful, and although they improved the lot of only a small number of people, they signified a promise of a better life for many more.

8.2 Recovery years 1970–73

During the post-filling years of the lake, fishing and fish trading increased slowly again, crops improved and the 'years of suffering' came to an end. The people of the southern plains expanded their farming economy to include cotton, while the people of the central Chilwa plains joined the rice schemes in greater numbers. Purchasing power increased a little. But in the Kawinga plain, although rice growing was encouraged, only about 10 per cent of the people sold surpluses and crossed over to a cash economy. Gradual improvements in the standard of living, to which health and educational status are closely related, occurred. But in 1973, rainfall was again low, the lake waters receded and fishing was reduced. The economy took two to three years to recover from this minor recession. From 1973 cholera appeared throughout the area of the lakeshore villages to add to the endemic diseases, and a vigorous hygiene campaign was mounted.

8.3 Years of high lake level, 1976

Although the economy is improving, the fisheries boom has brought a large influx of migrants seeking work on the lake, who do not have the immunity to various diseases such as malaria and hookworm which had been built up by continuous exposure in the local inhabitants. Parasitic diseases from malaria to worms have increased in adults and children. Leprosy appears on the increase, since patients cannot be traced to receive their regular medications. The concentration of the population and the longer times of exposure to standing water to mosquitoes and contaminated soil or water on the floodplain make transmission of diseases easier in years of high level.

On the other hand, progress in the management of health and nutrition has commenced with the increase and reorganization of clinics towards preventive medicine and maternal and child care. Attitudes towards ill-health are changing from a hopeless belief in witchcraft as a cause of sickness to a wide-spread understanding that there are material causative agents of disease that can be controlled by simple preventive measures and the timely use of medicine. The implementation of the preventive measures still depends on that small progress towards a cash economy which is being urged upon the community and which can be achieved by more intensive and diversified farming, so as to be able to sell surpluses for cash, and by fishing and fish trading. The completed construction of a gravity-piped water supply in one Chief's area has brought about the elimination of cholera there and a diminution in the amount of water-borne and fly-borne diseases. A piped water supply is under construction in the area of Nazombe and also in Kumtumanje's area in the west.

The number of primary schools has doubled in recent years and many more children are attending school. More pupils are reaching the top classes of primary schools and some will proceed to secondary schools on the fringe of the Chilwa area, or to technical training in fishing or agriculture. Roads are being improved so that access to hospitals, clinics and markets is being made easier. Housing is improving in the more prosperous southern Phalombe plain. Rice schemes have grown and a Phalombe Cotton Development Scheme is projected.

References

Agro-Economic Survey, Lake Chilwa. 1972. Ministry of Agriculture and Natural Resources. Government Printer, Zomba, Malawi. 63 pp.

Chandler, A. C. & Reed, C. P. 1961. Introduction to Parasitology. Wiley, N.Y. 822 pp.

Cole-King, S. M. 1969. Problems of malnutrition in Malawi. In: N. P. Mwanza & M. Kalk, Problems of Natural Resources in Malawi. IBP Symposium, Blantyre, Malawi, pp. 25–27.

Cole-King, S. M. 1971. Analysis of the nutritional status of children at Namitambo clinic, Malawi. Ministry of Health (unpublished). 25 pp.

Cole-King, S. M. 1973. Malnutrition in Malawi. City of Blantyre Public Health Seminar Medical Bulletin 3:3–146.

Drummond, H. 1903. Tropical Africa. 11th edition. Hodder & Stoughton, London. 228 pp.

Gadabu, A. D. & Magendantz, M. 1975. Studies on hookworm species and the prevalence of disease in the people of Kachulu village on the shore of Lake Chilwa. Biol. Dept. Report, Univ. Malawi, Zomba (unpublished). 10 pp.

Garson, M. S. & Campbell-Smith, W. 1958. Chilwa Island. Geol. Surv. Dept. Memoir 1. Zomba, Nyasaland. 127 pp.

Gear, J. H. S. 1961. Schistosomiasis – a major problem. Industry and Tropical Health 4:1–16.

Gear, J. H. S. 1968. A balanced view of the diagnosis of bilharziasis. Central. Afr. J. Med. 14:89–93.

Gelfand, M. 1973. The Shona Ng'ana as I know him. Soc. Malawi J. 26:563–579.

Goldsmid, J. N. 1972. T. deminutus and hookworm in Rhodesia. C. Afr. J. Med. 18: Suppl. 1–10.

Hammond, J. 1958. Medical Report, Ministry of Health, Nyasaland.

Hira, P. R. 1967. Studies in the hatching of Schistosoma haematobium ova and some factors influencing the process. 3. West Afr. Sci. Assoc. 12:95–102.

Howard-Williams, C. 1977. A check-list of the vascular plants of Lake Chilwa, Malawi, with special reference to the influence of environmental factors on the distribution of taxa. Kirkia 10:563–579.

Latham, M. 1965. Human Nutrition in Tropical Africa F.A.O. Rome. 268 pp.

Masiku, M. & Kawonga, R. 1973. An evaluation of the Under-Fives Clinics programme at Namitambo clinic in Malawi. Biol. Dept. Univ. Malawi, Zomba (unpublished). 20 pp.

McGregor, L. A. 1970. Immunity to plasmodial infections, considerations of factors relevant to malaria in man. Int. Rev. Trop. Med. Part I.

Ministry of Health Reports, Nyasaland (Malawi). 1938, 1950, 1969, 1970 and 1971. Blantyre, Malawi.

Mitchell, M. P., Ahluwalia, R., Ndovi, E. D., Ngulube, N. N. & Sobratee, H. 1977. Investigations of extracts of Steganotaenia sp. used as an 'eye medicine' by African herbalists. Biology Department, University of Malawi (unpublished).

Mkolesia, P. G. & Magendantz, M. 1975. Urinary schistomiasis at Kachulu Village on the shore of Lake Chilwa. Biol. Dept. Univ. Malawi. Zomba, Malawi (unpublished). 5 pp.

Molesworth, B. D. 1967. Malawi Leprosy Control Project. Soc. Malawi J. 21:58–69.

Morgan, P. R. 1972a. The effect of natural alkaline waters upon the ova and miracidia of Schistosoma haematobium. Centr. Afr. J. Med. 18:182–186.

Morgan, P. R. 1972b. Studies of mortality in the endemic Tilapia (Sarotherodon) of Lake Chilwa (Malawi). Hydrobiologia, 40:101–119.

Morgan, P. R. 1977. The Pit Latrine – revived. Central African J. Med. 23:1–4.

Msiska, M. D. & Magendantz, M. 1974. Studies on hookworm species and the prevalence of hookworm disease in children at Mtala School in Zomba District. Biol. Dept. Univ. Malawi, Zomba (unpublished). 10 pp.

Mwanza, N. P. 1972. Human Environment in Malawi. Report to U.N. Human Environment Conference, Stockholm. 20 pp.

Mwanza, N. P. & Mwambetania, F. 1974. A preliminary report on the effect of herbal extracts on medically important bacteria. Malawi J. Sci. 2:52–58.

Nankonya, J. M. & Magendantz, M. 1976. Laboratory studies on susceptibility of snails and hamsters to intestinal schistosomiasis and a simple serological test. Biol. Dept. Univ. Malawi, Zomba (unpublished). 17 pp.

Nkoma, E. C. & Magendantz, M. 1975. Intensity of urinary schistosomiasis, the severity of associated haematuria and the effects of chemotherapy. Biol. Dept. Univ. of Malawi, Zomba (unpublished). 7 pp.

Nurse, G. T. 1968a. The body weight of African village men, Central Afr. J. Med. 14:94–96.

Nurse, G. T. 1968b. Seasonal Fluctuations in the body weight of African villagers. Central Afr. J. Mid. 14:122–132.

Schulten, G. G. M. 1975. Losses in stored maize in Malawi (C. Africa) and work undertaken to prevent them. EPPO. Bull. 5:113–120.

Smithers, S. R. & Terry, R. J. 1969. The infection of laboratory hosts with cercaria of *Schistosoma mansoni* and the recovery of adult worms. Parasitology 55:695–700.

Statistical Office, Malawi, 1978. Data on population, schools and school enrolment in the Chief's areas adjoining Lake Chilwa. Tables (unpublished).

Stubbs, M. 1972. Health Services in Malawi 1969. In: S. Agnew & M. Stubbs (eds.) Malawi in Maps. Univ. London Press. London, pp. 76–77.

Watt, J. M. & Breyer-Brandwijk, M. G. 1962. Medicinal and Poisonous Plants of Southern and Eastern Africa. 2nd edn. Livingstone, London. 789 pp.

Webbe, G. P. & Jordan, P. 1966. Recent advances in knowledge of Schistosomiasis in East Africa. Trans. Roy. Soc. Trop. Med. Hyg. 60:279–306.

Weir, R. N. 1962. High background radiation on a small island. Nature Lond. 194:265–267.

Wilcocks, C. & Manson-Blair, P. E. C. 1972. Manson's Tropical Diseases. 17th edn. Whitefriars Press. London. 1164 pp.

Williamson, J. 1975. Useful Plants of Malawi. Univ. of Malawi Press, Zomba. 336 pp.

Part 4. Conclusions

) The Lake Chilwa ecosystem – A limnological overview

Brian Moss

1. Introduction

Limnologists have been studying lakes in earnest since before the turn of the 20th century, and like all scientists trying to cope with a great deal of new information, their initial instinct was to classify lakes into types or categories. The classifications have been based on many features – water chemistry, species of bottom-living animals, production of the plankton, types of fish. In the end all such schemes have failed for every lake is different from the next, and when sufficient lakes were examined exceptions which did not fit into the schemes of classification were inevitably found. Any limnologist wants to know how the lake which he has studied is related to other lakes, and that is the purpose of this chapter. Classification schemes used to be a convenient way of doing this, but since lakes do not conveniently fit into them, they are only a limited way of comparing information. Lake Chilwa provided an interesting example of this during its cycle of drying. In 1965, Talling and Talling ordered a vast amount of chemical information on the water of African lakes by dividing it into three categories of increasing total salinity, I, II and III. Chliwa would be classified in II normally but was for a time in III as it evaporated away, and has now returned to II. Another general classification has compared 'temperate' with 'tropical' lakes. Here the contrast between Lake Chilwa and the similarly shallow, but entirely different Lake George in Uganda (Greenwood 1976) is so great that to attempt any comparison of the two with a temperate lake is meaningless.

If classifications are limited, how is a limnologist to place a lake in perspective? A valid way seems to be to recognize that for any measurable characteristic – sodium content, fish yield, percentage of the basin covered by swamp, for example, a lake will fall somewhere along a series of continuous variation. The lake can be described by its position in that range of variation or continuum. There is an almost unlimited number of characteristics by which a lake can be described, and they may be highly correlated with one another, or they may be quite independent of one another. If there were only two of them, one could describe a lake by the intersection of their co-ordinates on a simple two-dimensional graph. If there were only three, then points on a three-dimensional graph would serve to relate one lake to the next. But there are many characteristics, so many continua, and the graph must be one of a similar number of dimensions – not a three-dimensional graph enclosing a three-dimensional space, but a multi-dimensional graph enclosing a multi-dimensional hyper-volume. This is not something with which the human mind can readily cope! Some continua seem more important in determining the major features of a lake than others, however.

This may be a human rationalization, but it is impossible mentally to cope with variation in more than a few dimensions at once. The total energy input as photosynthetically useful radiation, the rate of supply of phosphorus and nitrogen and the depth of the lake basin seem together to explain much of the variance in primary productivity (and hence of secondary productivity and fish yield) as well as its distribution between planktonic and littoral, algal and plant communities in whole series of lakes (Brylinsky & Mann 1973, Schindler 1978, Moss 1980, Vollenweider 1975, Ryder et al. 1974, Oglesby 1977). All of this work has been based on research in exorheic lakes and a fourth axis, water balance, is needed to make the framework closer to a universal one.

One role of an overview such as this is to test the validity of such a framework by examining where a particular lake fits into it. Do the features of Lake Chilwa extend the continua on which the frame is based and does the lake form a link in the hyperspace (in the above sense of Hutchinson 1957), which furthers our understanding of processes in other lakes close to it in the hyperspace? This seems to be so in respect of its phosphorus dynamics and in the relative roles of the littoral and open water.

A second major principle that has emerged in limnology is that not the lake ecosystem alone but the catchment area plus basin is the fundamental unit of study. Almost everything that happens in a lake is a function of what happens in its catchment area, for this determines what dissolved and suspended materials reach the basin. The lake basin is a rubbish bin for the exports of the land in its catchment area, and itself either exports these materials downstream, perhaps elaborated by metabolic processes or stores them in its sediments. There is thus a net one-way flow of materials from the catchment to the outflowing rivers or to the sediment, via the intermediaries first of the littoral zone, then of the open lake water. There may be temporary back-flows from the open water to the littoral, or from the sediments to the open water, but the overall movement is inexorably unidirectional. Again these ideas have been developed mainly from studies of temperate and exorheic lakes. What does an endorheic lake, particularly one which is less anciently a closed basin than those of the eastern rift valley in Kenya and hence has less extreme water chemistry, contribute to our understanding of this principle of net one-way flow of materials? An overview of the limnological work done on Lake Chilwa is perhaps best made firstly in the context of the second principle and considered in terms of catchment, littoral and open water, and secondly in terms of the whole system and the first principle, that of the general significance of Lake Chilwa to the science of limnology.

2. The Lake Chilwa basin as a whole

2.1 The catchment area

The water which flows to Lake Chilwa is not particularly fertile. Measurements, which are not subject to error through delay in analysis, are available for the key elements which determine lake fertility, nitrogen and phosphorus, only from the Likangala River. Total nitrogen and total phosphorus data even then are lacking, but in a flowing system when discharge is high (also the period when most nutrients reach a lake from the catchment) nitrate and soluble reactive phosphorus levels are probably at worst within the correct order of magnitude of the level of total N and total P. In the upper reaches of a tributary of the river, on Zomba mountain, phosphate was almost indetectable and nitrate-N never rose above $68 \mu g \, l^{-1}$ (Moss 1970), and on the lake plain, even after addition of some sewage effluent from the town of Zomba, the mean

PO_4–P level was only 5 μg l^{-1}, and mean NO_3–N 49 μg l^{-1}. Maxima recorded were 24 μg l^{-1} and 171 μg l^{-1} respectively. In an open lake system mean concentrations of a substance in the lake water cannot exceed the mean level in the inflow waters, except temporarily, when intense sediment release may occur in hypereutrophic waters (Osborne & Phillips 1978).

As an open basin prior to about 9000 years ago, Lake Chilwa would have been relatively infertile, with planktonic chlorophyll a levels perhaps around 10 μg l^{-1}. Even this figure is an upper estimate since it depends on current inflow nutrient levels, influenced as they are today by Zomba sewage effluent. Presently, as an endorheic lake, Chilwa has concentrations of phosphorus and nitrogen sometimes very much greater than those predictable from the river inflows, and the sources of these will be examined below.

A second point regarding the influence of the catchment area is the question of how increasing clearance of the natural vegetation might have changed both nutrient and silt inputs. Tillage almost always leads to increased nutrient loading (Likens & Bormann 1974) and any modern increase serves to emphasize again the potential natural infertility of the lake as an open basin. The silt question is more difficult. The lake has an extremely high turbidity from suspended clay, which has wide ranging effects from limitation of light availability for phytoplankton photosynthesis to provision of an unstructured bottom which seems unfavourable to benthic invertebrate colonization. Has the lake always been so turbid, perhaps increasingly so, as normal infilling has brought the lake bed close to the wind-disturbed water surface, or have silt levels increased markedly following increase in human populations and disturbance of the natural vegetation?

The exact extent of agricultural disturbance of the catchment is unknown but the plains are heavily cultivated with only fragments of the original *Brachystegia* woodland remaining. The Likangala River is heavily silt-stained in the wet season. Reports from elsewhere in the world suggest that natural ecosystems have developed in such a way as to conserve fine soil particles and the nutrients associated with them (Likens et al. 1977). It is a speculation, though a reasonable one, that silt loads on Lake Chilwa have increased greatly in the last century. The implication is that the water would have been less turbid previously, permitting higher phytoplankton productivity and that when the basin was open, presumably coinciding with a greater inflow to lake volume ratio and lower salinities than now, areas of submerged aquatic macrophytes would have been a permanent feature, increasing the overall structural diversity of the system.

In an open system much of even an increased silt load would be washed downstream; with closure of the basin the clay and silt must accumulate giving a large reserve available for re-circulation in the water column, more rapid filling-in of the basin and the present turbid water. Emergent vegetation, though extensive in clear, shallow waters, is much less affected by turbidity than planktonic and submerged macrophyte communities (Wetzel 1975, Wetzel & Hough 1973), and its proportional importance in lake productivity and provision of habitat increases. It is logical, therefore, that the Chilwa *Typha* swamps have a key role in the Chilwa ecosystem.

2.2 The swamps

Many features of the Lake Chilwa swamps confirm their general importance in lake ecosystems. Howard-Williams has listed and comprehensively discussed these in Chapters 5 and 13. Unquestionably they provide the structure, or part of it, for the niches of many organisms, including commercially important fish, that could not persist otherwise. This is reflected in the greater physico-chemical variability of the swamp habitat compared with that of the open water (Howard-Williams & Lenton 1975), in greater diversity of invertebrate and fish communities found in the swamps, in the aggregation of benthic inverte-brates of the lake bed close to the swamp fringes, where pieces of detritus and litter washed out of the swamps modify the texture of an otherwise amorphous mud (McLachlan 1974, 1975), and in the spawning movements of *Clarias* and perhaps *Barbus* from the lake to the swamp.

The swamps also export organic matter to the lake. Rates of supply are difficult to establish but the greater populations of zooplankton and catches of fish close to the swamp fringes are indicative. It is interesting therefore that the emergent swamp plants are not directly eaten to any great extent by herbivores. This is a general feature of emergent swamp plants (Hutchinson 1975, Scul-thorpe 1967) and perhaps reflects the importance of the habitat structure they provide. Evolution of the swamp ecosystem has moved in such a way as to eliminate any species making significant inroads on the habitat structure. Hutchinson (1975) has speculated that the tolerance of algal epiphytes, which cause significant shading (Phillips et al. 1978), by submerged macrophytes, may be a device to divert herbivore activity from the host plant. For emergent plants like *Phragmites* and *Typha* this is not important, for their shoots rapidly reach the high light intensities above water, supported during their initial growth by energy reserves in the rhizome. There seems even to have been no tendency for swamp emergents to evolve toxic secondary metabolites as anti-herbivore devices (Hutchinson 1975) as many land plants have done (Levin 1976). Indeed *Typha* is highly palatable to geese which, at high population levels, will eliminate the plant by grazing the shoots (R. VanDeusen, pers. comm.). That extensive stands of *Typha* persist in Lake Chilwa and elsewhere is perhaps a reflection of the strong selection pressure against species likely to destroy the main framework of the habitat.

The question of how swamps might regulate nutrient levels in the open water of a lake is a general one of great significance. Swamps generally stand between the inflow water and the open water. They may therefore absorb dissolved substances from the inflow and deliver a changed water supply to the lake. Generally this change is one of reduced scarce key nutrients like phosphorus and nitrogen, for where peat or organic mud is laid down in a swamp then some finite storage of these nutrients must be occurring. This principle has been proposed as a means of reducing phosphorus levels of lakes receiving sewage effluent (Toth 1972), and has been shown to occur in papyrus and other sorts of swamps (Gaudet 1977, Brinson 1977). It is invoked as a major agent in the natural development of less fertile conditions in small lakes of mid-western North America (Wetzel 1975). This seems logical – a steady-state community cannot export more lithospherically supplied nutrients (i.e. derived from the land surface and not from the atmosphere) than it receives from the catchment.

It can, of course, export greater quantities of atmospherically derived nutrients such as carbon and nitrogen (through photosynthesis and prokaryote nitrogen fixation) than it receives in its inflow water, but this cannot be true of phosphorus which is derived only from the lithosphere.

A complication ensues because swamps also receive inorganic nutrients as particulate matter which is incorporated into the swamp soil. The nutrients adsorbed or combined in this material are relatively insoluble in simple physico-chemical systems but might be made available by the activity of the swamp plant root hairs. It is conceivable therefore that swamps could mobilize nutrients like phosphorus from this 'insoluble' fraction and export them in soluble form to the water. The waters of Lake Chilwa at times have phosphate levels vastly in excess of inflow levels – sometimes of the order of several mg PO_4–P l^{-1} in the lake compared with only several μg l^{-1} in the inflows (Moss & Moss 1969, and Chapter 4). Evaporative concentration could be involved to explain this, but since phosphorus is an element readily precipitated in mud, particularly where clay levels are high and iron is present, this is most unlikely to be the explanation.

The Chilwa swamps are not laying down peat, so a net export of soil-derived nutrients cannot be discounted. Some organic matter, inevitably containing nutrients, is exported, but phosphorus in these pathways is likely to remain largely in combined form and to find its way as such into the lake sediments; it is unlikely to explain the milligram levels of inorganic phosphate often present in the water. Howard-Williams & Howard-Williams (1978) found rapid release of various ions from decomposing *Typha* and because of the annual dry down and improved oxygen penetration to the swamp soils, decomposition of *Typha* in the Chilwa swamps is undoubtedly rapid. Large amounts of Na, K, Ca and Mg were lost but these elements are present in the lake (as in all waters yet studied) at levels vastly in excess of these likely to influence plant or algal production. Phosphate was also released, but rapidly taken up again, probably by associated microflora. From the point of view of the swamp community this again seems logical. Steady state communities tend to conserve scarce nutrients by maintaining them as much as possible in their biomass and there is evidence that external enrichment increases the production of swamp communities (Onuf et al. 1977) suggesting that they are nutrient-limited. Waters deep in swamps and emerging from them often have very low phosphate levels (Sioli 1975) and the phosphorus levels of the open water of a reed and alder-swamp-lake system in England were highly predictable on the bases of events in the inflow, open water and lake mud, the surrounding swamps acting as a self-contained system (Osborne 1978).

The swamp waters of Lake Chilwa nonetheless were at times rather rich in phosphate (Chapter 5), particularly close to the lake itself, and there was some evidence that these levels declined with distance from the lake. A simple mobilization of phosphorus from swamp to lake (the nutrient pump hypothesis) seems unlikely in its simplest form, but an indirect mechanism, discussed below, involving both swamp and lake may be invoked to explain the high levels in the lake water.

There is one further matter concerning the swamps which should be noted first however – their role in infilling of lakes. As Lake Chilwa dried down the vigorous *Typha* swamps did not encroach on the area that had been open water.

Though abundant and fertile, *Typha* seeds were unable to germinate on the saline mud (Howard-Williams 1975), and there was no obvious vegetative invasion. It is commonly believed that marginal vegetation inevitably encroaches on open water, despite some evidence to the contrary (Spence 1964). The Chilwa swamp has clearly progressively colonized a very large area but may now have reached a state where further movement is unlikely. Prediction is difficult but certainly an extra insight has been added to a common limnological belief (or misapprehension).

2.3   *The open water*

Limnology has developed primarily in studies of the open water and through investigation of the planktonic system in particular. Even the voluminous studies of benthic invertebrates have rested on the apparent dependence of these animals on sedimented organic matter suppled from the plankton. The important roles of the swamp littoral zone, (its contribution of detritus, for example) have only recently been emphasized and a limnologist of the littoral will seize upon the data in this book with avidity. A replacement of much of the open water by swamp might have little effect on the Chilwa ecosystem as a whole; a converse major increase in open water probably would.

As matters stand, the open water away from the swamp fringes seems to harbour very much a dependent community. Phytoplankton populations do develop there, but the major activity of zooplankton and fish seems to be inshore. The mud surface has been shown experimentally and in the field (McLachlan 1974, 1975) to be physically unsuitable for benthic invertebrate colonization again except at the swamp fringe where structure is given to it by debris washed out of the swamp.

The only positive inputs the open water seems to make are in the provision of space which conceivably influences the degree of piscivore activity and perhaps the ultimate yield of fish, and in the provision of sufficient fetch for some inshore areas to be swept clear of fine particles, thus providing the sandy bottom essential for spawning of *Sarotherodon shiranus chilwae*. As the volume of the lake increases during filling and the silt and clay, provided ultimately from the catchment, are allowed to settle out somewhat, the open water plankton seems to develop some autonomy but the benthos does not. If the loading of silt and clay has increased greatly, following human disturbance of the catchment, it seems likely that previously the open water community would have had a much greater development.

Sarotherodon seems a fish more associated with the open water than *Barbus* and *Clarias*, which move into the swamps to breed, and its relatively lower resistance to the disturbance of drying out of the lake, reflected in the delayed recovery of its populations, might reflect its evolution during a period when the lake was much less turbid and in which open water conditions were normally less extreme.

As in many African lake waters available nitrogen was relatively scarce and available phosphorus relatively abundant in Lake Chilwa. Both nutrients are required to support phytoplankton growth, but as the biomass increases it seems that nitrogen supplies are often exhausted before those of phosphate (Moss 1969, Talling & Talling 1965). The nitrate levels of Lake Chilwa were moderate, always less than 1 mg NO_3–N l^{-1}, and sometimes very low; bioassays showed a likelihood of nitrogen limitation, and at high water levels potentially nitrogen-fixing Cyanophyta (*Anabaena*, *Anabaenopsis*) were prominent in the phytoplankton. Nitrogen limitation theoretically can only be a temporary phenomenon in lake waters (Schindler 1977), provided the water mass persists (hydrologically) long enough for nitrogen fixers to develop populations, because there is effectively an infinite supply of nitrogen available to be fed into the system from the atmosphere. The low dissolved inorganic nitrogen levels in African waters may reflect high bacterial denitrification rates at tropical temperatures particularly in wet soils and sediments where nitrate is used as an oxidant under anaerobic conditions. The very conditions which lead to low combined inorganic nitrogen levels may also favour increases in phosphate levels, however.

Phosphate is readily adsorbed by clays and precipitated with oxidized iron or aluminium, both elements very common in soils and sediments. Under aerobic conditions sediments are usually a sink for phosphate underneath a surface crust of oxidized ferric compounds which effectively prevents much release of phosphate from the sediment. If labile organic matter is added to sediment surfaces, usually by development of large overlying plankton crops, then the sediment surface may become reduced, through bacterial activity and Fe^{3+} is reduced to Fe^{2+}. Under these conditions both Fe^{2+} and phosphate may diffuse into the overlying water, usually to become immediately reprecipitated in the aerobic conditions above the mud surface. However, if bacterial activity in the surface sediments is very intense the redox potential of the sediment may fall so low that sulphate-reducing bacteria produce sulphide ions which may then form a very insoluble precipitate with Fe^{2+}. Phosphate may then diffuse into the overlying water without immediate reprecipitation as iron complexes, and quite high concentrations – of the order of several mg PO_4–P l^{-1} – may ensue (Osborne & Phillips 1978). These may be lost in many lakes by dilution and displacement downstream during periods of high water flow. The supply of phosphorus must, of course, first be introduced to the sediment from the catchment area, and because there is loss downstream there must be continued substantial input from the catchment to maintain the process indefinitely.

In Lake Chilwa there is the interesting variant that the inflow concentration is not substantial yet the phosphate levels in the lake water may approach those expected in, for example, sewage effluent and are one to two orders of magnitude greater than those predictable from simple exchange between water and clay-absorbed phosphate. There is no loss of phosphate by outflow from Lake Chilwa yet normal precipitation processes must be operating to remove phosphate from the water just as they remove calcium and magnesium ions (Moss & Moss 1969, Morgan & Kalk 1970). Indeed calcium and magnesium carbonates are very effective complexers of phosphate (Wetzel 1975). Evaporative concentration cannot thus maintain the high phosphate levels.

A reasonable mechanism for the production of the high phosphate levels is release from the sediments powered by decomposition of organic matter originating **mainly** from the swamps. Decomposition rates were high and the relatively low organic content of the sediments (only a few per cent compared with 50 per cent or more in sediments offshore from temperate reed swamps) seems to suggest a rapid turnover of organic matter and a revision of our current classification of detrital organic matter into labile and refractory. Material which is refractory under temperate conditions may not be so at tropical temperatures.

The process will, of course, go on in the swamp waters as well as those offshore, but there seems a possibility that released phosphate may be rapidly reabsorbed there in contrast to the open lake, where higher plant uptake is absent. Much of the dissolved phosphate in the swamp water near the open lake may be introduced by wind surges. As the lake dries down in major drying phases such as that of 1967/68 the crust of accumulating salts on the mud surface is rich in phosphates, whilst the underlying mud is dark and anaerobic. Under these conditions other mechanisms may also operate as large numbers of invertebrates stranded by the retreating water were decomposed. Nitrate levels were high in the salt crust as a consequence. Under the more usual conditions where the mud surface is still overlain by water, even during the dry season, the sediment release mechanism proposed should be dominant. Support is lent to the significance of this mechanism in Lake Chilwa by the variation in iron content of the waters. It is significant that deep ground water in the area was not phosphate rich (Moss & Moss 1969) but had PO_4–P levels of the same order as that of the surface river water.

Other endorheic lakes also have high phosphate levels, sometimes spectacular ones (Talling & Talling 1965, Millbrink 1977) despite low influent concentrations. A similar mechanism can be adduced for these, with the variants that the organic matter needed to drive the process need not come from surrounding swamps. It can come from phytoplankton production and dense blooms of *Spirulina* and other blue green algae found in such lakes. In Lake Chilwa the process seems driven by the swamps to the extent that planktonic sources seem **minor**, owing to silt turbidity.

3. The significance of the Chilwa ecosystem

All organisms that persist are adapted to their habitat and all communities are adapted to the physico-chemical regime which forms the background to their ecosystem. Adaptation is not fixed – continual adjustment is essential to meet changes in the environment and need not involve very specialist features – a lack of them may be most adaptive in certain environments. In an environment of high physico-chemical predictability, expressed in regular seasons of climate relatively constant from year to year, a high degree of specialization may evolve among a diverse collection of species. The rock-fish communities of the huge and old Lake Malawi are excellent examples (Fryer & Iles 1972), where high environmental predictability has led to very fine partitioning of food resources among a lot of species, each with narrow diets. Thus, quite apart from food specialization on different groups of invertebrates or on other fish, there has been specialization among feeders on the epilithic algae coating the shoreline rock into those that feed largely from vertical or horizontal surfaces or from loosely attached algal species or those that are firmly anchored. Among the piscivores are specialists on fin or scale feeding. Such environments have a very high species diversity which is likely to be greatly changed by any disturbance of the environment, any flaw in its predictability.

At another extreme are highly predictable ancient environments which pose very rigorous conditions for living organisms. The deep sea bed, with its constant high pressure, low temperature and complete darkness is an example (Sanders & Hessler 1969, Dayton & Hessler 1972, Grassle & Sanders 1973). Here too the predictability seems to have allowed evolution of a much wider variety of species than might be anticipated, some with very specialist adaptations such as huge mouths to take advantage of rare but large meals, when carcasses of fish falling from higher in the water column become available. On a smaller scale the bacterial community of anaerobic mud is also diverse. Accumulations of organic matter under water have been a continuous habitat for a long time and evolution has produced a wide variety of heterotrophs, chemotrophs, and autotrophs capable of dividing effectively the inorganic and organic resources provided. This habitat itself may reflect a great microbial diversity in the 2×10^9 or so years of anaerobic constancy that constituted the middle Precambrian era. Now that suitable methods are available it is notable that new micro-organisms, almost certainly originating then, are continually being discovered (Lewin 1976, Maugh 1977, Pierson & Castenholz 1971).

The common features underlying the diversity of these apparently contrasted habitats are immense age and a regular physico-chemical pattern in time, a high predictability. Other habitats have high predictability but have been short-lived. Volcanic hot springs are spatially isolated and relatively temporary but have a very predictable environment (Castenholz 1969). The bacteria and Cyanophyta that live in them are highly specialist – their enzymes and membranes can survive temperatures which would kill almost all other organisms, but the community diversity is low. There has not been time for high diversity to evolve, though there is no reason that it should not were the habitat to persist undisturbed for long enough. In less extreme cases the fish communities of north European islands like Britain are comparatively poor in species, despite the regular march of the seasons and the permanence of the water bodies. The environment is predictable, but the recent glaciation caused such disturbance that there has been little time for recovery of high diversity. In contrast, the equally seasonal and predictable regime of river flood plains in the tropics has produced, in the absence of major climatic catastrophes, if not of climatic change (Livingstone 1975) some exceptionally diverse communities (Lowe-McConnell 1975).

A third category of ecosystem is those that are ancient, but subject to unpredictable vagaries of their environment. The gut floras of omnivorous, but not herbivorous (Hungate 1975), vertebrates are perhaps examples. At irregular intervals they may be swamped with food sources of different kinds in no particular predictable sequence. The organisms surviving there, like *Escherichia coli*, must be unspecialized and able to utilize a wide range of metabolic substrates. Narrow specialists would be displaced from the gut in the periods between supplies of their particular food.

The fourth category – of ecosystems which have neither time nor predictability on their side – is exemplified by the Lake Chilwa ecosystem. It underwent the major change from an exhorheic lake to a closed system only between 8000 and 10,000 years ago – the same sort of time period that has been available to northern ecosystems affected by polar glaciation, and its relatively small volume and its sensitivity to changes in precipitation make it an unpredictable environment characterized by low community diversity.

Certainly there is some periodicity in the incidence of dry-down periods, but the variability in the degree of evaporation of the approximately six-year cycle makes it useless as a regulator that species can depend upon, and to which their life histories can become adjusted, as they have, in other systems to such fluctuations as tide, lunar- and photo-period. Within this crucially unpredictable status of the system as a whole, there is a continuum of higher predictability within years and within the ecosystem. The swamp creates a certain relative temporal constancy by its own structure and by its closeness to the inflowing water supply, compared with the open lake basin. This should not be confused with the greater spatial variation in the swamp, which, together with its greater predictability in time allows a much greater diversity of organisms to persist in it, compared with the lake itself. The regular cycle in water level during overall high water periods also contributes a predictability that perhaps triggers the incidence of spawning in some of the fish.

Those organisms that have been well studied in the lake show characteristics that might be described as r-strategies for survival. r is the conventional symbol for the intrinsic rate of population increase in the well-known logistic equation of population growth $\frac{dN}{dt} = rN(\frac{K-N}{K})$ where N is the number of organisms in a population), and r-strategies are those devoted to maintaining populations in environments where mortality is likely to be high, usually owing to the action of unpredictable and density-independent physical or chemical changes. r-strategies contrast with K-strategies, where K is the symbol for maximum sustainable population in the logistic equation, and which characterize strategies for maximum utilization of resources in a predictable environment such as that of Lake Malawi, or a tropical rain forest. Of course there is a continuum of combinations of r and K strategies in a range of organisms (Southwood 1976) and it would be best to refer to r-orientated and K-orientated combinations to acknowledge this.

r-orientation is a usual feature of all plankton communities because the open water in any lake is a relatively unpredictable habitat, subject to disturbance by mixing, flushing and the activities of planktivorous fish. Phytoplankton have high reproductive rates through simple cell division, and the parthenogenesis common in Cladocera and Rotifera is a similar r-stratagem. The plankton community of Lake Chilwa would not be expected to be particularly remarkable therefore in this respect, and it is not. The benthos, also, though living in a more structured habitat is disturbed by many random factors of sediment mixing and of changes imposed by these in the overlying plankton. It, too, is not fundamentally dissimilar from that in other lakes.

Vertebrates, on the other hand, are able to regulate their habitat by seasonal movements and their longer life-spans allow some accommodation to temporary habitat disturbance. A rather greater degree of K-orientation might thus be expected, in a given ecosystem, for vertebrates compared with smaller animals. Some specialization is usually possible, but although it is difficult to assign quantitatively degrees of r- or K-orientation, the major Lake Chilwa fishes certainly display a very high level of r-orientation.

This is reflected in their very great flexibility. They are highly fecund, reproduce at a relatively early age, are able to persist in the swamp and streams as well as in the open lake, have very broad diets with considerable overlap and show much opportunism in feeding, have unspecialized spawning habits and

410

demands, so far as is known, and are highly tolerant of a wide range of habitat factors, though these ranges are inevitably exceeded as the lake dries down. These fishes exhibit to a high degree one side of the contrast which Fryer & Iles (1969) drew between the *'Tilapia'* group of r-strategists in African lakes and contrasted with the species flocks of *Haplochromis* and other genera which exhibit extreme K-orientated specialization in the littoral of Lake Malawi and other large lakes in East Africa.

Though the Chilwa fishes are clearly well fitted to persist in the unpredictable Chilwa ecosystem, provided the refugium of swamps and streams is maintained, it is interesting that the two species which survived longest and which recovered earliest as the lake passed through its drying cycle were both species of widespread distribution in eastern Africa, and, for *Clarias*, an even wider area. The zooplankton and phytoplankton taxa were also extremely wide-ranging ones. The one known endemic species, or sub-species, in the lake, *Sarotherodon shiranus chilwae*, seems the least resilient, which is surprising for an endemic species might be expected to be rather more closely adapted to the system in which it has evolved than other species which have evolved elsewhere. The key to this paradox might lie in the fact that *S. s. chilwae* is apparently still in the process of differentiation, whilst the habitat to which it is becoming adjusted may be changing at a greater rate than that with which evolution of the fish can cope. Lake Chilwa certainly is changing. Superimposed on the overall continuing climatic changes, perhaps towards still greater dryness despite the present high levels of the lake (1978), which are affecting central Africa (Livingstone 1975) and which are of much wider influence than the Lake Chilwa catchment (Chouret 1977) are the local human influences in the area, such as the suggested comparatively recent increase in silt load. Suspended silt *per se* does not harm *S. s. chilwae* in the short term at least, but deoxygenation associated with the silt does (Morgan 1972). As the lake fills with more and more sediment, conditions for temporary deoxygenation will become more frequent although the process may be temporarily halted by higher mean lake levels for a time. It may be that the probability of extinction of *S. s. chilwae* will be greater than that of its complete adaptation and that the Lake Chilwa ecosystem will eventually lack any organisms not found over a wide geographical range and capable of coping with varying conditions within broad limits. Such organisms are called generalists.

4. The role of human populations in the ecology of Lake Chilwa

Culture, in human beings, is the ultimate adaptation, for it allows modification of the environment to the capabilities of the organism and in a limnological context is expressed in the establishment of fish hatcheries, river engineering and dam building among other enterprises. Culture is subject ultimately to natural selection in that changes made to the environment may not prove to be favourable in the long term, no matter how desirable or humanitarian they may seem in the short term. An overview of the ecology of any ecosystem cannot avoid considering the human element and for Lake Chilwa it falls into two phases.

Firstly there is the role that human populations in the pre-technological period (for Lake Chilwa, until quite recently) have played and secondly there is the role that in future they might play. This latter is considered more fully in the next chapter.

The indigenous peoples of Lake Chilwa seem to have a cultural organization which is well adapted to the ecology of an unpredictable ecosystem. Like that of the fish it has been a generalist strategy. There has been no exclusive dependence on one food source, but an informal organization which has coped variously with fishing and farming as the opportunities became available. Even the fishing methods used have been unspecialized compared with the sometimes elaborate devices and techniques used on more permanent African lakes (Worthington & Worthington 1933). Various nets, traps and lines have been employed where they have proved effective. The materials were simple, often locally obtained and readily replaced. Capital-intensive equipment is less likely to pay when there may be long periods of disuse when it deteriorates.

There has also been an acceptance of migration during the periods when the lake has dried down. This has not been without hardship and undoubtedly has meant adjustment in the areas to which the migrants move for food and work when fishing is impossible, but it has been acknowledged as inevitable. This contrasts greatly with the problems and resistance experienced among peoples of more predictable habitats when some major engineering work of the technological period – a new dam for example (Scudder 1973) has meant an enforced move.

It has been suggested that compared with the rigours imposed by the environment on the Lake Chilwa fish populations the impact of fishing during periods of high water level has an insignificant effect on the ultimate fish populations recovering after one of the periodic dry phases. There is probably truth in this, though, if the population of fishermen is increasing a complete abandonment of fishery regulations, without more information on the fish population dynamics, would probably be premature. Even resilient ecosystems must have some limits to recovery. Much more danger of the impact of human populations, however, may come from subtle influences – the dangers of persistent pesticides, which have been discussed elsewhere, and habitat destruction.

In 'habitat destruction' I include firstly possible threats to the swamps, through 'reclamation' for agriculture or perhaps as irrigation reservoirs, or through increased burning, a phenomenon which seems to occur extensively but for which the reasons or significance are unknown. Secondly, I include the effects of siltation on the lake through changes in land management in the catchment. An urgent need is for thorough palaeolimnological studies to put the modern limnology of Lake Chilwa into historical perspective. Has there been a marked increase in siltation, has the nature of the surface sediments in the lake changed greatly in consequence, and was the open lake so dominated by inorganic turbidity in the early part of this century and previously as it is now?

Lake Chilwa was classified in 'Project Aqua' (Luther & Rzóska 1971) in category A, a site in its natural state or only slightly modified. Much hinges on the meaning of 'slightly', but it seems certain that the lake is coming under increasing anthropogenic influence. The history of applied limnology in temperate regions is the history of how the consequences of one man-made change to a waterway have been countered with further correctives, each with its own consequences so that ultimately the ecosystems have become not only less diverse but also costly to maintain. A river canalized so that development could take place on its floodplain must continue to have expensive flood banks kept in order, whilst further expense is incurred in artificial irrigation of once naturally wet areas.

There might be arguments put in favour of stabilizing the water levels of Lake Chilwa, say by reconnection of it to Lake Chiuta via a canal through the northern sandbar, which presently separates the lakes. A further positive consequence of this might be flush through of silt and restoration of a more productive open water community. On the other hand it might simply create a larger, turbid, unpredictable and still endorheic lake! It will always be better to avoid trying to modify the consequences of large-scale climatic trends, and to counteract anthropogenic changes not by treating the symptoms, thus perhaps creating new, often worse problems, but by treating the root cause of the problem. A thorough soil conservation policy in the catchment area could have the widest ranging benefits for the Lake Chilwa ecosystem and its peoples.

References

Brinson, M. M. 1977. Decomposition and nutrient exchange of litter in an alluvial swamp forest. Ecology 58:601–609.
Brylinsky, M. & Mann, K. H. 1973. An analysis of factors governing productivity in lakes and reservoirs. Limnol. Oceanogr. 18:1–14.
Castenholz, R. W. 1969. Thermophilic blue-green algae and the thermal environment. Bact. Rev. 33:476–504.
Chouret, A. 1977. La persistance des effets de la secheresse sur le lac Tchad. Proceedings of UNFAO Symposium 28; Burundi; 1977. Cyclostyled. 16 pp.
Dayton, R. R. & Hessler, R. R. 1972. Role of biological disturbance in maintaining diversity in the deep sea. Deep Sea Res. 19:199–212.
Fryer, G. & Iles, T. D. 1969. Alternative routes to evolutionary success as exhibited by African cichlid fishes of the genus *Tilapia* and the species flocks of the Great Lakes. Evolution 23:359–369.
Fryer, G. & Iles, T. D. 1972. The Cichlid fishes of the Great Lakes of Africa. Oliver and Boyd, Edinburgh. 641 pp.
Gaudet, J. J. 1977. Uptake, accumulation and loss of nutrients by papyrus in tropical swamps. Ecology 58: 415–422.
Grassle, J. F. & Sanders, H. L. 1973. Life histories and the role of disturbance. Deep Sea Res. 20:643–659.
Greenwood, P. H. 1976. Lake George, Uganda. Phil. Trans. R. Soc. B. 274:375–391.
Howard-Williams, C. 1975. Vegetation changes in a shallow African lake: response of the vegetation to a recent dry period. Hydrobiologia 47:281–398.
Howard-Williams, C. & Howard-Williams, W. 1978. Nutrient leaching from the swamp vegetation of Lake Chilwa, a shallow African lake. Aquat. Bot. 4:257–268.
Howard-Williams, C. & Lenton, G. M. 1975. The role of the littoral zone in the functioning of a shallow tropical lake ecosystem. Freshwat. Biol. 5:445–449.
Hungate, R. E. 1975. The rumen microbial ecosystem. Ann. Rev. Ecol. Syst. 6:39–66.
Hutchinson, G. E. 1957. Population Studies: Animal Ecology and Demography. Concluding remarks. Cold Spring Harbour Symposium. 22:415–427.
Hutchinson, G. E. 1975. A treatise on limnology III. Limnological botany. Wiley, N.Y.
Levin, D. 1976. The chemical defenses of plants to pathogens and herbivores. Ann. Rev. Ecol. Syst. 7:121–160.
Lewin, R. A. 1976. Prochlorophyta as a proposed new division of algae. Nature, Lond. 261:697–698.
Likens, G. E. & Bormann, F. H. 1974. Linkages between terrestrial and aquatic ecosystems. Bioscience 24:447–456.

413

Likens, G. E., Bormann, F. H., Pierce, R. S., Eaton, J. S. & Johnson, N. M. 1977. Biogeochemistry of a forested ecosystem. Springer-Verlag, Berlin. 146 pp.

Livingstone, D. A. 1975. Later Quaternary climatic change in Africa. Ann. Rev. Ecol. Syst. 6:249–280.

Lowe-McConnell, R. H. 1975. Fish communities in tropical freshwaters. Longman, London. 337 pp.

Luther, H. & Rzóśka, J. 1971. Project Aqua. Blackwell, Oxford.

Maugh, T. M. 1977. Phylogeny: are methanogens a third class of life. Science, N.Y. 198:812.

McLachlan, A. J. 1974. Recovery of the mud substrate and its associated fauna following a dry phase in a tropical lake. Limnol. Oceanogr. 19:74–83.

McLachlan, A. J. 1975. The role of aquatic macrophytes in the recovery of the benthic fauna of a tropical lake after a dry phase. Limnol. Oceanogr. 20:54–63.

Milbrink, G. 1977. On the limnology of two alkaline lakes (Nakuru & Naivasha) in the East Rift Valley system in Kenya. Int. Rev. ges. Hydrobiol. 62:1–17.

Morgan, P. R. 1972. Causes of mortality in the endemic *Tilapia* of Lake Chilwa (Malawi). Hydrobiologia 40:101–119.

Morgan, A. & Kalk, M. 1970. Seasonal changes in the waters of Lake Chilwa (Malawi) in a drying phase, 1966–68. Hydrobiologia 36:81–103.

Moss, B. 1969. Limitation of algal growth in some Central African waters. Limnol. Oceanogr. 14:591–601.

Moss, B. 1970. The algal biology of a tropical montane reservoir (Mlungusi Dam, Malawi). Br. phycol. J. 5:19–28.

Moss, B. 1980. An introduction to freshwater ecology. Blackwell Scientific Pubs. (in press).

Moss, B. & Moss, J. 1969. Aspects of the limnology of an endorheic African lake (L. Chilwa, Malawi). Ecology 50:109–118.

Oglesby, R. T. 1977. Relationships of fish yield to lake phytoplankton, standing crop, production, and morpho-edaphic factors. J. Fish. Res. Bd. Canada. 34:2271–2279.

Onuf, C. P., Teal, J. M. & Valiela, I. 1977. Interactions of nutrients, plant growth and herbivory in a mangrove ecosystem. Ecology 58:514–526.

Osborne, P. L. 1978. Relationships between the phytoplankton and nutrients in the River Ant and Barton, Sutton and Stalham Broads, Norfolk. Ph.D. Thesis, Univ. of East Anglia, Norwich. 255 pp.

Osborne, P. A. & Phillips, G. L. 1978. Evidence for nutrient release from the sediments of two shallow and productive lakes. Verh. Int. Ver. theor. angew. Limnol. 20 (in press).

Phillips, G. L., Eminson, D. R. & Moss, B. 1978. A mechanism to account for macrophyte decline in progressively eutrophicated freshwaters. Aquat. Bot. 4:103–126.

Pierson, B. K. & Castenholz, R. W. 1971. Bacteriochlorophylls in gliding filamentous prokaryotes from hot springs. Nature (New Biol.) 233:25.

Ryder, R. A., Kerr, S. R., Loftus, K. H. & Regier, H. A. 1974. The morpho-edaphic index, a fish yield estimator – review and evaluation. J. Fish. Res. Bd. Canada. 31:663–688.

Sanders, H. L. & Hessler, R. R. 1969. Ecology of the deep-sea benthos. Science 163:1419–1424.

Schindler, D. W. 1977. Evolution of phosphorus limitation in lakes. Science. N.Y. 195:260–262.

Schindler, D. W. 1978. Factors regulating phytoplankton production and standing crop in the world's freshwaters. Limnol. Oceanogr. 23:478–486.

Scudder, T. 1973. Summary: Resettlement. In: Ackerman, W. C., White, G. F. & Worthington, E. B. (eds). Man-made lakes: their problems and environmental effects. Amer. Geophys. Union. Washington, D.C., pp. 707–719.

Sculthorpe, C. D. 1967. The biology of aquatic vascular plants. Arnold, London. 610 pp.

Sioli, H. 1978. Tropical rivers as expressions of their terrestrial environments. In: F. B. Golley & E. Medina (eds). Tropical Ecological Systems. Springer-Verlag, Berlin, pp. 275–288.

Southwood, T. R. E. 1976. Bionomic strategies and population parameters. In: R. M. May (ed.) Theoretical Ecology, Blackwell, Oxford, pp. 20–48.

Spence, D. H. N. 1964. The macrophytic vegetation of lochs, swamps, and associated fens. In: J. H. Burnett (ed.) The vegetation of Scotland, Edinburgh, pp. 306–425.

Talling, J. F. & Talling, I. B. 1965. The chemical composition of African lake waters. Int. Rev. ges. Hydrobiol. 50:421–463.

Tóth, L. 1972. Reeds control eutrophication of Balaton lake. Water Research 6:1533–1539.

Vollenweider, R. A. 1975. Input-output models. Schweiz Z. Hydrol. 37:53–84.

Wetzel, R. G. 1975. Limnology. W. B. Saunders. Philadelphia. 743 pp.

Wetzel, R. G. & Hough, R. A. 1973. Productivity and the role of aquatic macrophytes in lakes: an assessment. Pol. Arch. Hydrobiol. 20:9–19.

Worthington, S. & Worthington, E. B. 1933. Inland waters of Africa. Macmillan, London. 259 pp.

21 Focus on social problems

Margaret Kalk

1. The handicaps of the Chilwa people

The geological changes, which preceded the formation of the sandbar between Lake Chilwa and Lake Chiuta about 9000 B.P., were responsible for the present form of the Lake Chilwa basin as one of closed drainage. The subsequent climatic pattern of alternating wet and dry periods annually, with heavy seasonal rainfall and high evaporation, has produced the conditions of the highly productive, shallow lake, surrounded by a *Typha* swamp, which experiences both seasonal changes in lake level and periodic low levels every six years or so. The lake is in a senescent stage, but unless there is a sudden change in the climatic trend, which has not been foreseen, the present form of the ecosystem is likely to continue. It will be subject to periodic extremes of lake level, which may be relatively minor, following only one year of lower than average rainfall, which hampers fishing and farming for one year. A range of increasingly more serious lake recessions following more years of lower than average rainfall have been experienced in the last hundred years, which may interfere with the local staple industries for two to five years. Exceptionally high levels (as in 1978) may also occur either for one year or a number of years to which the people must adapt.

The lake/swamp ecosystem contains within itself the means of conservation of its flora and fauna, even after complete dryness over the lake bed for some weeks. The *Typha* swamp plays the largest part in maintaining the *status quo* and, as has often been stressed in this book, should always be preserved. Any plan for draining the swamp or cutting off the deeper part of the lake and changing the present drainage of rivers are not only unfeasible but would be damaging to the ecosystem. The satellite photograph of the lake basin in the Frontispiece shows red patches of living swamp at the mouths of the perennial rivers, in portions of their courses and around lagoons in the marshes, as well as in four-fifths of the lake area, which remained filled with water during the minor recession of 1973. These areas were the refuges for fishes during the drought of that year.

Nevertheless, although the lake organisms recover from the periodic droughts within a number of years, depending on the severity of the lake recession, the situation generates special problems for the inhabitants of the lacustrine plains, who depend on rainfall and river flow for their crops, as well as on the swamp and lake for fish. Although agriculture in most of Africa suffers from droughts at times, the people of Chilwa are in addition doubly handicapped during periodic droughts. Firstly, as well as suffering crop failure, the main industries which are dependent on fishing, also decline at these times; and secondly, the society is in a state of transition from quasi-subsistence organization to a cash economy in a partially isolated, hitherto self-sufficient area. It is only a minority of people who rely mainly on cash earnings from fisheries, fish trading, farming and stock-keeping or depend (to a small extent) on payment for their labour elsewhere. The majority of the people continue to follow the traditional way of acquiring goods by barter exchange; they also pay in kind for assistance on farms or fishing, or receive goods as payment. Very little surplus produce is as yet produced for sale outside the settlement schemes.

It is evident that subsistence economy can, however, no longer sustain the people in years of crop failure. The 'famine foods' which grew wild in the savanna woodlands until the twentieth century are no longer available. White settlers took possession of much of the highlands (the natural reserve land) in colonial days for the cultivation of tea, tobacco and coffee, and this land still remains outside the customary control of the Chiefs. The problem is accentuated by the natural increase of the population and immigration from surrounding areas in Malawi and from Mozambique. Mounting pressure on freely drained land brought an end to the practice of shifting cultivation which allowed the soils and some of the wild plants used as food to regenerate. Williamson (1974) listed over 200 plants with edible fruits, seeds, tubers, roots or leaves which used to be gathered wild and eaten, more especially in the lean years. Just as access to wild plants is no longer possible, so is recourse to hunting. Men and boys have been reduced to digging for rodents and snaring birds because of the rapid decline of the larger mammals which were so plentiful until the wanton use of guns at the turn of the century.

Had there been consistent development during the colonial period, no doubt many more people would have been able to rely on cash savings for the purchase of cereals and fish from other areas, not affected by the low rainfall years in the Chilwa basin. The cash earners among farmers are at present for the most part limited to those who cultivate irrigated rice in organized schemes or to the larger landowners whose savings from former employment enable them to employ labour or oxen to work over 5 hectares for mixed crops. Another sector with some cash-in-hand are those who earn wages in tea, tobacco or other estates on the periphery of the Chilwa basin or who have become migrant industrial workers within Malawi, Rhodesia and South Africa.

There are, however, several positive aspects of development which augur well for a better reaction to the next lake recession. These have been the result of the surge forward in progressive measures since Independence in 1964. The tradition of 'self-help', now particularly encouraged on a community scale, has already contributed towards the building of schools and clinics and the construction of access roads and bridges as well as piped water supplies from gravity schemes centred on highland rivers. District direction of such efforts in Chiefs' areas will no doubt continue to modernize the area slowly. It would be beneficial if, in times of severe drought resulting in inevitable unemployment, schemes were organized to improve communications by road for marketing when normal rainfall returns, and concerted efforts were made at such times to improve village amenities. Such activities would obviously depend on solving the food problems arising in the periods of temporary depressions by saving either in cash or kind.

2. Development potential

For savings to be made to meet recurrent recessions in the rural economy the prerequisites are greater input of energy and the fuller exploitation of all possible resources of the basin. These two points will be examined in connection with the various interrelated activities of the lake people.

420

2.1 *Fisheries resources*

The Chilwa lake and swamp fisheries have been shown to be productive and the fishing techniques in use are effective in shallow water. A very large number of men from the villages are thus engaged and as yet no signs of overfishing have been noted. The cash earnings on the lakeshore are considerable – although much less than the prices finally paid by the consumer in the market. There is certainly no question of the majority of the fishermen changing their traditional methods of fishing in the streams, floodplain, marshes, swamp and the periphery of the lake, but three areas of potential improvement in relation to the fisheries emerge from the consideration of the periodic fluctuations in lake level.

2.1.1 Further exploitation of the lake

The experimental trawling programme of the Fisheries Research Department in the recent years of high level suggests that in the cooler months there are concentrations of fish in the deeper southeast sector of the lake, which may be caught by trawling from a plank boat, powered by an out-board engine. The traditional methods of fishing leave this potential source of wealth untouched. To take advantage of this behaviour of the fish in years of high level would require a small number of 5 m boats and trawls, which in times of lake recession could be removed to other shallow fisheries such as at Lake Malombe. Limitation of the number of trawling boats appears essential, and the confinement of trawling within certain boundaries would prevent the propellors from cutting across long-lines and gill nets used on the periphery. The area available for trawling is visible in the satellite photograph of the Frontispiece as the darker area in the southeast sector of the lake.

It is probable that fishing has been improved by the practice of cutting long channels from the lake through the swamp to the floodplain, which has been introduced since recovery of the lake from the last recession. These waterways not only facilitate the movements of boats from the lake through the swamp to the shores on the floodplain, but enrich the lake by encouraging the growth of epiphytic algae on *Typha* stems along the channel fringes, which increase dissolved oxygen in the water and provide a more diverse food source for fish.

The trees from which canoes are made are becoming scarce, and although temporary measures might be taken to bring trees from further afield, it appears that they will in time be replaced by plank boats. These are constructed locally from wood supplied by the Zomba sawmills of the Forestry Department. Plank boats do have some advantages, in that they can be more easily propelled by oars (although at present poles are used, as in canoes) and they would perhaps take sails for rapid return to beaches before a storm.

By this means, in years of high level, fishing could be spread further from the shore where fishes appear to be more numerous than they are in the years just after the refilling of the lake following a recession. Outboard engines are too expensive to be run by the average fishermen and in any case their use might have to be restricted, considering the risk of pollution in a lake with no outlet.

The Fisheries Extension Service has been concerned with improving fishing gear in the last few years. Deeper gill nets and a simpler design for scoop nets have been introduced and attention has been paid to smoking fish more effectively by modifying kilns and reducing exposure to flies so that the more distant markets can be exploited.

During a minor recession of the lake (as in 1973, 1966, 1960 and 1954) when fish still exist in that part of the lake that remains, but are inaccessible because boats cannot be pulled through soft mud, it might be possible to make the water more accessible. At such times, the area between Nchisi Island and Kachulu Harbour is the first part to be exposed, while water still reaches the outer eastern shores of the Island for many more months. In order to reach these waters it might be feasible to build a temporary causeway through 'self-help' if material could be supplied by the District Council. The causeway would extend across the 2–3 km of exposed mud to Nchisi Island so that fishermen might reach the Island. Their boats might be left on the outer shores of the Island and the catches carried back in baskets by head load after drying the fish. A causeway would extend the fishing in the dry months of a low rainfall year and would do much to stabilize the population on the Island. In minor recessions the eastern part of the lake is very easily reached on foot over the dry sandy lake bed. Fishing on this side of the lake might continue and fish be ferried across to Nchisi Island to be dried and then brought to the mainland over the causeway. In these ways the industry would not come to a premature halt, although it would obviously be reduced during the dry months of the year of drought (and possibly the next year too because the fish would not have had access to the swamp for breeding).

2.1.2 Conservation of fish

After the major recession of the lake, *Clarias* populations recovered in two years, *Barbus* populations in three years and *Sarotherodon* after four to five years. It has been advocated by Furse in this monograph that controls on fishing are inadvisable. Before a recession, the maximum possible yield should be obtained at a time when the lake becomes so shallow that mass mortalities are likely to occur. In support of this contention is the phenomenal recovery of the fishery after the year of dryness in 1968, which indicates that it is unlikely that the lake can be overfished in normal years. Maximal fishing before a major recession would probably make very little difference to the degree of recruitment from the fish that have survived.

The Fisheries Department, however, thought it wise to protect the juvenile fish after the first experience of the refilling of the lake from dryness, by prohibiting gill and seine net fishing for two breeding periods. This must have had some impact on increasing the sizes of the fish populations, but whether it was economically necessary remains an open question. No prohibition of subsistence fishing was enforced. On the other hand, the limitation of the minimum mesh size of gill nets in normal years, which was achieved by persuasion and example before the major recession, demonstrably protected the breeding stocks of *Sarotherodon*, without seriously affecting the catch of other species. This practice would also permit exploitation of larger fish which would probably evade a trawl.

An attempt to restock the lake during the refilling period of 1968–69 after the major recession, by releasing a quarter of a million fingerlings of

422

Sarotherodon shiranus chilwae and *Tilapia rendalli*, bred in the Government Fish Farm nearby at Domasi, has not been directly assessed, since marked fish were not recovered. No conclusion has been reached about the advisability of restocking the lake artificially. The small number released compared with that of the natural population of the lake probably made little difference. In addition, *T. rendalli* has been found to thrive in the lake only when the water is most dilute as in 1977–78, and this species would probably have left the lake for the swamp and streams at that time.

There are conflicting opinions about the feasibility of building a ring of large ponds near the perennial rivers or boreholes on a village scale by 'self-help'. Experience in Malawi has shown that the fish will thrive if the ponds are fed kitchen waste and garden refuse. The fish farming developed in the Lower Shire River area near rice schemes might in time serve as a model for the Chilwa area. Such ponds might contribute little to conservation of fish in the ecosystem when the lake flooded again, but would certainly supplement subsistence fishing during a year of drought. The ponds would have to be constructed a year ahead when data on the trend in the annual rainfall and salinity of the lake water would enable the Fisheries Department to forecast the impending drought.

2.2 *Increasing the input of energy and the diversity of farming*

One major finding almost self-evident and documented earlier, is the relationship between the size of the farm and the ability to grow surplus cereals for sale. The restraining factors are not yet the availability of land nor lack of ground water, but the low input of energy that a man and his wife (or a woman and her relatives, if her husband has a job elsewhere) can provide. It seems desirable to divide the remaining arable land of the lacustrine plains into larger plots per family for mixed farming to be serviced by shared boreholes, provided that extra energy can be supplied from oxen. Since cattle are already kept by the more well-to-do, a supply of oxen for part-time hire (at first on credit), to prepare the ground and cart the produce to market might readily be organized.

Exchange of labour between farmers (as has been traditional) would again be useful in raising the level of production by those who are still subsistence farmers. There is the possibility of doubling the yield of cereals by the application of fertilizer and the closer planting of seeds. Credit facilities for obtaining fertilizer and more intensive advice from the Agricultural Extension Service would achieve improvements comparable to those in other parts of Malawi if they were made more readily available in the Chilwa area, more especially since different crops could be grown throughout the year.

In addition to the use of oxen, fertilizer and expertise, a greater input of energy would also result if more time were devoted to the actual growing of crops. Women will be able to contribute more time through the use of the increasing number of mills for grinding maize into flour, since home preparation is very time-consuming. The provision of closer sources of water from piped supplies or many more boreholes will reduce the time wasted in fetching water daily. When houses become more permanent (made of sun-dried clay bricks), men will be released from the time-consuming labour of building and mending huts.

Greater diversification of crops appears to be essential, and the variable soils of the Chilwa plains lend themselves to this. The Agricultural Extension Service which freely analyses soils and advises on crops has an intricate task to perform in the Chilwa basin. The practice of mixed farming is traditional and it would take little extra impetus and organization to prepare for the periodic lean years. The increase in the intensity of labour and the concerted effort to make the maximum use of water facilities and to reduce pests (with extra input of energy from tractors to prepare the ground) is one example of increased output in the irrigated rice schemes. Such examples of highly organized labour by individual plot-holders can only gain in influence in the neighbourhood.

2.3   Extension of stock-keeping

The Department of Agriculture and the University of Malawi have been conducting experiments on the degree of grazing pressure on the Chilwa plains. The floodplain provides useful pasturage in the dry seasons and the cattle spread over the plains in the wet season and do not usually receive much supplementary feeding. Although water from some boreholes may be brackish, it is probably suitable for cattle. The herds are being increased and interbred with exotic bulls to improve the quality of meat and to enable milk to be produced.

There are, however, two important questions which confront the planners of stock-breeding. Firstly, there is the quesiton as to how many cattle the uncultivated parts of the Chilwa plains will sustain, when periodic droughts are expected. It may be advisable to advocate a higher rate of slaughter for sale before the animals lose condition in order to meet the periods of restricted grazing. The ideal of mixed farming has often been demonstrated in Malawi, so that cattle manure can fertilize 'gardens', and chicken manure and waste from crop production can be used as cattle feed. The greater use of oxen is being encouraged and education in animal husbandry is being provided. The division of land between cattle and crops is, however, a problem to be decided on ecological grounds rather than on competitive interests of rival types of farmer, bearing in mind that it is more economical to produce crops for human consumption than cattle fodder in some environments. The periodic recession of the lake which seriously limits the extent of floodplain grazing in certain years will have to be taken into account as well as the very high level of 1978, which may or may not be repeated for a few years. The accompanying heavier rainfall enables agricultural areas to be extended.

The second question concerns the desirability of introducing dairy herds to a human population which may be genetically and culturally ill-adapted to milk in the diet. It is now well-known that the non-dairying majority of the populations of the world, including some in Africa, lack adult lactase, a digestive enzyme which converts lactose (milk sugar) into absorbable sugars (Cook & Kajubi 1966, Jersky & Kinsley 1967). Malawians have not been studied in this respect, as far as is known, and it seems desirable that samples of the population should participate in the simple lactose tolerance test. Usually Africans drink fermented milk which causes no ill-effects (because the sugar is turned into lactate).

424

2.4 Trading, transport and markets

It has been shown that fish trading by bicycle is well organized and an efficient means of reaching villages off the main routes; and the combination of bicycle trader with lorry or bus transport supplies the larger markets. During the recent years of high lake level (1976–78), trading across the lake has grown, with traders plying canoes or plank boats or using the ferries to collect fish for sale and to deliver maize and other foods to migrant fishermen. During the times of lake recession, the fish traders proved adaptable and changed their routes to obtain fish from other sources which they sold further afield where the periodic drought had not affected the purchasing power and more stable conditions obtained.

The role of the fish trader will probably expand, since he is an enterprising independent agent. The social changes of the last few years, which have brought many more fishermen to the lakeshore villages and the circulation of much more money from the high yield of fish from the lake, have widened the role of the itinerant trader. There is now scope for transport of goods from towns and certainly a need to bring to the beaches supplies of wood from Mozambique, where it may still be available more readily than on the Malawi side of the lake, or from the scantlings (waste wood from cutting logs into planks) at the Zomba sawmills. The itinerant bicycle trader is the forerunner of a delivery service to and from the villages and the growing centres such as Kalinde in the south, Domasi in the west and Nayuchi in the north.

The Chilwa area is particularly well placed to dispose of its farm produce and the products of developing village industries to the densely populated Shire Highlands. Lack of all-weather roads and permanent bridges over the larger rivers makes the lot of the trader transporting goods to market, very difficult. It is as much in the interest of the industrial population of the Highlands as it is of the people dwelling on the Chilwa plains to have better means of communication and transport built by the District Councils, using local labour wherever possible.

Some trading centres for the sale of surplus crops and cattle have been established by the Agricultural and Development Corporation on the periphery of the Chilwa plains, and around these centres settlements are springing up with health centres, schools, missions and shops. These are permanent amenities in contrast to the canteens in which the more prosperous fishermen invest (in addition to putting their money into farming and cattle). These canteens are in small temporary buildings. They thrive on the migrant fishermen in times of plenty but close down completely during a major lake recession. There appears to be a social problem involved in such temporary shops, because when the fishermen do have money to spend there is little of permanent value to be obtained and money is frittered away. Until there is greater certainty of cash earnings in alternative work, such shops will not find it worthwhile to stock household goods which will attract the villager.

2.5 Village industries

One of the short-comings of the Chilwa area in this time of transition to a cash economy has been its great reliance on fishing as almost the sole means of creating wealth, although the industry suffers intermittent remissions due to the changes in the lake. In earlier times the months or years of poor fishing conditions might have been spent in constructing nets by hand from local fibre; but the sale of nylon netting since 1958 has gradually eased out this craft. Since the netting factory also supplies nylon thread of many gauges, it would be commendable to revive the craft of net-making at the lakeshore, using nylon fibre instead of the natural fibres now becoming scarce.

Similarly, the making of dug-out canoes is declining and the traditional boats are increasingly being replaced by locally-made plank boats. This is an enterprise which requires little capital and the wood is available at the Zomba sawmills. The plank boats require caulking and repair every year and the demand is growing. But these industries are tied to fishing and to the variable lake. It is therefore the more necessary to develop small scale village industries for a more assured market not affected by periodic drought.

A small leather industry based on the tanning of hides and skins may well emerge as the stock industry becomes better regulated. Leather thongs are suitable for stool and chair seats and other goods may be made using cattle hides and skins from goats and pigs. Training in the necessary skills would be required, but small scale village-based industries would require but little subsidy. This industry would have the advantage of growing when more cattle were slaughtered because of impending drought and of occupying people deprived of fishing for a period.

Furthermore, there may be a commercial basis for a pottery industry using the appropriate lacustrine clays. Other clays can be used for the making of sun-dried bricks for sale. On Nchisi Island the occurrence of limestone, which is suitable for making whitewash or coloured distempers, could probably provide a basis for an industry. The limestone is at the surface on the hills and was worked in a small way for this purpose some years ago. This too could be revived on a bigger scale, since transport by ferry to the mainland is easy in normal years. In a year of special drought, transport to Kachulu would be facilitated if the causeway, mentioned above, turned out to be feasible.

There are also potash feldspar breccias on Nchisi Island, easily mined on the surface. There was an earlier suggestion (Garson 1960) that this rock might be dug out and crushed to be used as fertilizer, without further processing. The rate at which this could be transported would determine the scale of the operation, but it would give employment to some people and the rocks could be allowed to accumulate during a temporary period of drought. The surface mining of such fertilizer locally would put it within the reach of many a Chilwa farmer who could not afford imported fertilizer, which is relatively expensive to manufacture and to buy.

Among the subsistence occupations in the past was the filling of pillows with *Typha* seed, which is covered with down, and also the weaving of *Typha* stems to make various articles. Such occupations might be expanded into small village industries. Most villages have a tailor who, with the use of a sewing machine, pursues a lively trade in dress- and shirt-making from cotton prints purchased from the nearest store. The cotton is grown locally and woven at a Blantyre factory. Small-scale furniture manufacture with the minimum of hand tools and a home-made bench is likely to be successful as the circulation of money increases, the more so since a supply of timber is available from the sawmills which use the tree plantations on Zomba Mountain that are already of an age to be felled. Subsidiary timber yards might be set up at Kachulu and other growing trading centres for the sale of wood to encourage various enterprises.

It is not unlikely, in years of high lake level that improvements in roads and canteens will bring tourists to the lake for a day's outing. Birdwatching is a pastime that is very rewarding at Chilwa in the daytime (when mosquitoes are few) and the hire of a boat or a ferry trip across the lake would be a delightfully relaxing occupation. A visit to the islands of the lake would interest the amateur naturalist. Even yachting would be an adventurous pastime, if roads were improved and fit for trailers carrying boats. Locally-made handicrafts might be sold near the jetty.

2.6   *Fuel needs*

The ordinary villager regards wood as a 'free good' and there is at present no question of using any other source of fuel for domestic cooking or for smoking fish. The severe shortage of fuel is one of the most pressing problems. The plains and the unprotected slopes of the surrounding escarpments, where until the early twentieth-century woodland flourished, are now devoid of timber. There is a severe shortage of wood even for domestic cooking. The position is the more serious in that fish smoking, the readiest means of preserving fish for long distance marketing, is at hazard because of the shortage of fuel. The fish trader is reluctant to buy wood at a great distance away to preserve the fish he has bought at the lakeshore for market. Timber is also necessary for building houses, for boat-building and for carpentry.

A small forestry project for the people of the Chilwa area is urgent and cannot be too strongly stressed. The problem centres on free access to fuel at or near the villages themselves. (There are promising aspects for building in the increase of timber for sale from the plantations on Mt. Zomba and Mt. Mulanje.) For ordinary village necessities, the problem can only be met in the short term by laying aside parts of village lands for planting of suitable species of tree which will grow quickly and produce good firewood. The aridity of the soil will have to be taken into account when advice and perhaps free seed is distributed by the Forestry Department to the Chiefs for their areas. In the meantime, there is an urgent need for an immediate fuel distribution service, through the Forestry Department, to depots sited along the network of roads, from which women and traders may fetch firewood.

Fuel is a prerequisite for boiling water from polluted sources, if health is to be improved. The storage of fish for home consumption, using brine in adequate earthenware pots needs fuel for both the preparation of the brine from natural sources and for firing the pots.

For the longer-term it will be necessary to extend the plantations, now restricted to the mountains, to other parts of the highlands and plains using suitable species for the ordinary villager's fuel consumption. Solar energy is used now (without being concentrated) for drying tonnes of the small species of fish on the Chilwa lakeshore. Perhaps the distant future will see the use of cheap solar energy batteries for cooking and small-scale industries, since sunshine is a 'free good' present in profusion (Souare 1977).

427

Output of work is immediately determined by the state of health of the worker and his family. In this sphere the efforts towards supplying clean water, immunization campaigns, the use of more and better latrines, providing Under-Five Clinics and nutritional education for mothers, and clearing domestic surroundings of the breeding places of the mosquitoes which carry malaria, as well as protecting houses and latrines from mosquitoes, will be rewarded in the long run. The prime place in the improvement of health may be the education in the rationale of preventive medicine of men as well as women.

A particular hazard for men and women engaged in rice growing and for fishermen in the swamp channels is the increase of snails which carry bilharzia. Social change has actually increased opportunities for contamination. A first approach to the problem might be the eradication of aestivating snails by hand in the dry season, followed by the provision of pit latrines or privies with buckets that can be removed at night. In time, pollution of the rice paddies might be prohibited. The swamp channels may be more difficult to control, but it is only near vegetation that the snails live. It is not possible to contaminate the open lake with bilharzia. So much has been achieved by the organization of rice culture in Development Schemes, that it may not be impossible to break the bilharzia life cycle by conscious health culture.

To prevent the under-nutrition which seems to accompany the six-year periodic cycle, storage of foods for longer than a year should receive attention on a local village basis, as well as by the established Agricultural Development and Marketing Corporation, which has already organized large district storage buildings containing sacks of shelled maize, protected against weevils and moths by the application of insecticides. This service is, however, very expensive for villagers purchasing maize in times of shortage. The traditional village *nkokhwe* (Fig. 17.2b) serves well if it has rat-proof discs around the legs. It will be necessary for notification by the District Council that more of them should be built and filled in the year before the expected periodic drought. It has long been the custom to grow 'famine crops' which resist drought, such as *Cajanus cajan* (L.) Millsp. and various other kinds of peas and *Vigna reticulata* Hook.f., and *V. unguilata* (L.) Walp. and many other plants well-known to the older generation and described by Williamson (1975). Local authorities might encourage villagers to grow such plants against the lean times. The annual seasonal dry period might well be better withstood if more cassava were grown and mixed with maize meal in the preparation of food throughout the year. The deficiency of protein in cassava is well compensated by the use of fish in the Chilwa area. The roots of cassava remain in the ground in natural storage, while the leaves are rich in protein. This recommendation has been detailed in the report on the sample villages investigated on the Kawinga plain (Chapter 18).

Short-term storage of fish has received considerable attention by the Fisheries Research Division in Malawi, to cover the periods from lakeshore to the markets and the consumers. Storage for over a year is another problem. It would appear to be worthwhile devising storage methods using brine, as has been done in Europe, deriving the salt from the lake or from plants already known to take up salt. Salt processing is traditional in the area and might be encouraged again with this end in view. The containers for the brine could be made locally from an expanded ceramics industry dependent on local pot-clay. The use of brine was reintroduced for fish storage in villages in the early 1970's by the Fisheries Department and could be more widely used in a lake recession, provided fuel is available to make the brine and the containers.

In these times when it is Government policy to provide incentives to enter the cash economy, education is perhaps the most valuable tool in nurturing human resources. The ability to read and write is required in order to understand the rationale of varied farming practices and good animal husbandry and to take full advantage of the advice of the Extension Services and the results of research. Record-keeping, book-keeping and communication are part of the adult training given at the Fishermen's Institute at Mpepwe on Lake Malawi. Such Institutes might be run at the developing centres on the Chilwa plains, where the new skills of the fully educated Malawian agriculturist or fisheries expert could be merged with the local village expertise and knowledge of plants and fish, passed on from forefathers who have lived long in the Chilwa area. The growth of schools in the area since Independence is a very hopeful sign, but the schools need assistance to get equipment to put into practice the good intentions of the new teachers. The schools might also become foci where the adult generation might learn from demonstrations and advice on how to make the best of their land and water.

3. Summary

3.1 *Research in retrospect and in the future*

The high productivity of Lake Chilwa noted when recording of fish catches was first undertaken drew attention to the lake as a valuable national resource. The recession that followed awakened interest in the cyclical changes affecting the whole ecosystem, and in the last ten years co-ordinated research in the various disciplines has brought many problems forward for consideration and evaluation. Many of the initial problems which were outlined in the introduction to this monograph have been shown to be inherent in the climatic pattern of the Chilwa basin and they may be eased in the course of development of the area, as suggested above. Other problems proved of lesser consequence and some to be due to lack of basic knowledge.

Concern was early expressed regarding the effect of the withdrawal of water for irrigation in so finely balanced a system as Lake Chilwa. Knowledge of the water budget has, however, shown that the amounts withdrawn are a small proportion of the total lost in a major recession, and in normal times the effect is insignificant. Nevertheless, irrigation itself will suffer in the years of severe drought in the future, when the river flow diminishes as a result of a few years of lower than average rainfall.

The declining fish catches, suspected at the commencement of the study to have been due to overfishing, turned out to be much more serious and were followed by large scale mortalities as well as by the inaccessibility of the lake while there were still fish there to be caught. The catastrophic recession of the lake was not irreversible, and after ten years the fishing industry has become more successful than before. But whether this is due to greater populations of fish or the increasing contribution of the swamp species to the catch or to larger numbers of fishermen (i.e. greater energy input) is not known. The stunting of fishes, which initially caused concern, proved to be an advantage in that they reproduced at a smaller size and thus helped to repopulate the lake more quickly. Stunting was later considered to be an adaptation to periodic changes.

Study of the *Typha* swamp established that it cannot advance into the lake during the years of low levels because the water has become too saline. The swamp may be retarded on the border of the floodplain in years of higher levels. At present (1978) some of the *Typha* may have been destroyed by the temporary rise of the water level to the record height of this century. In the longer term, the *Typha* swamp will probably advance in the north of the lake, where the open water is most shallow. Judging from the last hundred years, the swamp may advance about one kilometre in the next century, but this relatively slow rate depends on the curbing of the amount of silt entering the lake from the catchment, as well as on the number of years of higher level, such as 1978.

Other subjects of early concern to the Research Project were related to the red locust breeding grounds on the sandy floodplain of the northern reaches of the lake basin and the presence of radio-active dykes on Nchisi Island. Neither of these two hazards appear to be a present threat, though it is suggested that future research should be directed to both areas. The red locust did not swarm, as was expected with the exposure of so large an area of the floodplain to dryness during the major recession, but constant surveillance is still excercised by the Red Locust Service of Southern Africa. It has been suggested that the red locust potential outbreak area at Lake Chilwa should be intensively studied locally to contribute a fuller understanding of the habits and habitats of this pest for success in its wider control, since it still plagues surrounding countries. In our study the possible ill-effects of natural radiation through the long-term changes of thorium decay appeared to be less significant for the health of the people on Nchisi Island than had been feared, nevertheless the presence of irradiated soils might merit long-term monitoring of their effects on man, plants and fish.

Two greater hazards, which are increasing in importance, have been described in this monograph. The first is the rapid accumulation of silt in the lake which leads to the increasing shallowness of the lake and to its turbidity. These effects both accentuate the dangers of the periodic low levels, in leading to greater oxygen deprivation at certain times. The other is the potential threat of the accumulation of insecticides in a lake of closed drainage (Chapter 18). Attention has already been drawn (Chapter 20) to the need to pay attention to the preservation of the banks of the streams which traverse agricultural land in order to reduce washaway. The conservation of the lake itself depends on the prevention of soil erosion in the catchment. Similarly, the growing of cotton, with intensive use of insecticides as practised in most countries, must make provision for the absorption of run-off containing insecticide residues in some way, so as to prevent injury to the fish in swamp and lake.

It has been the purpose of the Lake Chilwa Co-ordinated Research Project to construct a framework for the consideration of future research workers and governmental agencies concerned with the welfare and development of this special environment and its inhabitants who are so harshly affected by the climatic periodicity. The studies indicate that still not enough is known about the lake itself and the rate of silting, as well as about the circumstances when the

430

Typha swamp encroaches on open water. A study of mud cores from the lake edge would be very rewarding. The lake budget can now be only incompletely quantified because of insufficient river gauges and the inaccuracy of the lake level gauges. The very high level of the lake in 1978 was unexpected and further analysis of more data is required to be able to forecast more accurately when the minor recessions can be expected and how serious they will be. We have now emphasized the effects of the approximately six-year recurrences of low lake level, but have pointed out that recessions are not always as serious as in 1967–68. It is currently estimated that another major recession can be expected in about 2035, but this estimate might well be modified in the light of more data.

The relative roles of algae and detritus from the swamp in the food chain as the lake changes from years of low level to years of higher level and declines again require further study to understand the food web and to quantify its links. The bacterial decay of the *Typha* vegetation and of the dominant blue-green algae should be studied in all the lake's phases. This will give a meaningful basis to a continued study of the population dynamics of the three economic species of fishes in order to determine the optimum degree of fishing in the changing lake and to determine what are the best means of conserving the fish populations.

The detailed layout of the various soils for different crops and of the various pastures for grazing needs close study to determine the optimum diverse productivity of the area. The cattle carrying capacity of the plains with the floodplain for winter grazing has still to be estimated. It seems as though the Chilwa basin has some areas of potential prosperity near perennial rivers, and other areas are difficult to work. A great deal of progress has been seen in many fields in the last ten years, more especially in organized agriculture, but much remains to be done to evaluate the potential of the ecosystem and the surrounding plains and watershed. It will be very important to communicate the results of future investigations to the people in the area so that they may realize the potential and be forewarned of impending climatic change, whether it causes drought or flooding, so that they can prepare for it.

3.2 *Planning for the Chilwa basin as a unit*

The problems which were initially recognized have prompted the wider studies on change recorded in this monograph. We have tried to gather together all the known facts and ideas about the changes in the Chilwa ecosystem which might be useful in planning for the Chilwa people. In so doing, we have been able to suggest some of the areas where further research or action is necessary. Social changes now make it possible for industries alternative to farming and fishing to be developed on a village scale. Fishing is already diversified in lake and swamp using a variety of techniques; it is also necessary to diversify farming, but this will depend on adequate planning and direction. Man must remain as 'generalized' in activity as the lake fauna in order to succeed in the Chilwa area.

It is hoped that planning authorities will be able to regard the Chilwa basin as a unit with special problems, some of which have been outlined here. A Joint Planning Board might be set up to deal with the Water Resources, Fisheries, Agriculture, Forestry, Animal Husbandry, Health, Education and Social Services, which would be responsible for directing and integrating the economy, into which safeguards to lessen the impact of the periodic changes might be built. Perhaps personnel trained at the University of Malawi in scientific disciplines and social development will help to contribute to the future well-being of the peoples of the Chilwa area, who will need guidance in their development. The Chilwa basin seems a special case for the application of Malawi's Development Policy (1971).

References

Cook, G. C. &Kajubi, S. K. 1966. Tribal incidence of lactase deficiency in Uganda. Lancet 1:725–730.

Garson, M. S. 1960. The geology of the Lake Chilwa Area. Geol. Surv. Nyasaland Bull. 12. Government Printer, Zomba. 69 pp.

Jersky, J. & Kinsley, R. H. 1967. Lactase deficiency in the Southern African Bantu. S. Afr. med. J. 41:1194–1196.

Office of the President and Cabinet, Economic Planning Division. 1971. Developing Malawi: a shortened version of Statement of Development Policies, 1971–80. Government Printer, Zomba, Malawi. 76 pp.

Souare, D. 1977. Solar energy. Bull. Unesco Region. Off. Sci. & Technol. Africa 12:22–25.

Williamson, J. 1975. Useful Plants of Malawi. Revised and extended edition. University of Malawi, Zomba, Malawi. 336 pp.

Appendix A

The plants and animals of Lake Chilwa: A provisional checklist

Part 1. Plants

1. *PROKARYOTA
   CYANOPHYTA
   Anabaena torulosa (Carm.) Lagh.
   Anabaenopsis circularis G. S. Wol. &
   Miller
   Arthrospira platensis (Nordst.)
   Oscillatoria curviceps Ag.
   O. nigra Vaucher
   O. planctonica Wol.
   O. subbrevis Scmidle
   O. terebriformis Ag.
   Oscillatoria spp.
   Phormidium sp.
   Spirulina platensis (Gom.)
   Spirulina major Kg.

2. *EUKARYOTA
   CHLOROPHYTA
   Chlorophyceae
   Eudorina elegans (Ehr.)
   Pandorina morum Bory
   Platydorina caudata Kofoid
   Scenedesmus spinosus Chod.
   Zygnemophyceae
   Mougeotia sp.
   Spirogyra sp.
   Stigeoclonium sp.
   Closterium ralfsil (Bréb.)
   C. striolatum Ehr.
   Closterium spp.
   Cosmarium supraspeciosum Wolle
   Euastrum ansatum (Ehr.)
   Microasterias denticulata Bréb.
   Pediastrum sp.
   Penium margaritaceum (Ehr.) Breb.
   Pleurotaenium trabecula (Ehr.)
   Staurastrum sp.
   EUGLENOPHYTA
   Euglena elastica Prescott
   E. sanguinea Ehr.
   E. spirogyra Ehr.
   Phacus caudata Hueb.
   Trachelomonas sp.
   BACILLARIOPHYTA
   Anomoeoneis sphaerophora (Kütz.)
   Capartogramma sp.
   Cyclotella sp.
   Cymbella cistula (Hemrich)

Eunotia pectinalis var. *minor* (Kutz.)
E. triodon Ehr.
E. veneris (Kutz.) O. Hull
Frustulia rhomboides Bréb.
Gyrosigma acuminatum (Kütz.)
Navicula cryptocephala var. *veneta* Kütz.
   Grun.
N. gracilis Ehr.
N. placentula Ehr.
N. pupula var. *minor* Kütz.
N. viridula Kütz, K. B.
Nitzschia fonticola Grun.
N. palea Kütz.
N. sigma (Kütz.) W. Sm.
Nitzschia sp.
Pinnularia brebissonii (Kütz.) Rab.
Stauroneis lauenbergiana Hust.
Synedra sp.
†BACILLARIOPHYTA (additional)
Amphora veneta Kg.   +
Anomoeonis perpusilla Grun.   +
Coscidociscus rothii (Ehr.) Grun.
Cyclotella meneghiniana Kg.
Cymbella turgida (Greg.) Cl.
C. ventricosa (Kg.) Ag.
Eunotia pectinalis (Kg.) Rabh.   +
Gomphonemia acuminatum Ehr. v. *turris*
   (Ehr.) Cl.
G. gracilis Ehr.
G. lagenula (Kg.) Freng.
G. longiceps Ehr. v. *subclavata* Grun.   +
Melosira granulata (Ehr.) Ralfs.   +
Navicula cincta (Ehr.) Ralfs.   +
N. confervacea (Kg.) Grun.   +
N. semilunoides Hust.
N. subatomoides Hust.
Nitzschia amphibia Grun.
N. clausii Hantz.   +
N. congolensis Hust.   +
N. frustulum Kg.   +
N. subacicularis Hust.   +
N. subristratoides Chol.   +
N. vanoyei Chol.   +
Pinnularia acrosphaeria Breb.
P. borealis Ehr. v. *rectangulata* Hust.   +
P. gibba (Ehr.) W. Sm.   +
P. rivularis Hust.   +
Rhoicosphenia curvata (Kg.) Grun.

*After Brian Moss, Chapter 6.
†Additional epiphytic collections of diatoms made by Dr. F. D. Hancock in 1972 and 1975.
+ indicates species on *Scirpus littoralis* in open water.

Rhopalodia gibberula (Ehr.) Mull. v. *van heurckii* O. Mull.   +
Stephanodiscus astraea (Ehr.) v. *minitula* (Kg.) Grun.   +
Synedra rumpens Kg.   +
S. ulna (Nitz.) Ehr.

*3. PTEROPHYTA
Marsiliaceae
  Marsilea sp.
Parkeriaceae
  Ceratopteris thalictroides (L.) Brongn.

*4. SPERMATOPHYTA
Monocotyledonae
Agavaceae
  Sanseveria sp.
Araceae
  Amorphophallus fischeri (Engl.) N. E. Br.
  Pistia stratiotes L.
Commelinaceae
  Commelina benghalensis L.
  C. diffusa Burm. f.
  C. nyasensis C. B. C1.
Cyperaceae
  Cyperus alopecuroides Rottb.
  C. articulatus L.
  C. dichroostachys A. Rich.
  C. difformis L.
  C. digitatus Roxb. subsp. *auricomus* (Spreng.) Kükenth.
  C. esculentis L.
  C. laevigatus L.
  C. nudicaulis Poir.
  C. papyrus L.
  C. procerus Rottb.
  Fimbristylis sp.
  Fuirena cilliaris (L.) Roxb.
  F. umbellata Rottb.
  F. chlorocarpa Ridley
  F. pubescens Kunth
  Mariscus cyperoides (Roxb.) A. Dietr.
  Pycreus flavescens (L.) Reichb.
  P. macrostachyos (Lam.) G. Raynal
  P. mundtii Nees
  P. polystachyos (R. Br.) Beauv.
  Scirpus articulatus L.
  S. littoralis Schrad.
  S. maritimus L.
  S. muricinix C.B.Cl.
Gramineae
  Acroceras macrum Stapf
  Andropogon fastigiatus Swartz
  Andropogon gayanus Kunth
  Aristida adscensionis L.
  A. diminuta (Mez.) C. E. Hubbard
  Bothriochloa glabra (Roxb.) A. Camus
  Brachiaria deflexa (Schumach.) Robyns
  Chloris gayana Kunth

C. virgata Swartz
Cleistachne sorghoides Benth.
Cynodon dactylon (L.) Pers.
Cynodon sp.
Dactyloctenium aegyptium (L.) Beauv.
D. giganteum Fisher & Schweickerdt
Dichanthium papillosum (A. Rich.) Stapf
Digitaria gazensis Rendle
Diheteropogon amplectens (Nees) W. D. Clayton
Diplachne fusca (L.) Beauv.
Echinochloa pyramidalis (Lam.) Hitchc. & Chase
Eleusine indica (L.) Gaertn.
Eragrostis aspera (Jacq.) Nees
E. chapelieri (Kunth) Nees
E. ciliaris (L.) R. Br.
E. gangetica (Roxb.) Steud.
E. inamoena K. Schum.
E. namaquensis Schrad.
Eriochloa borumensis Stapf
E. meyerana (Nees) Pilg.
Heteropogon contortus (L.) Beauv. ex Roem & Schult.
H. melanocarpus (Ell.) Benth.
Hyparrhenia bracteata (Willd.) Stapf
H. diplandra (Hack.) Stapf
H. rufa (Nees) Stapf
Ischaemum afrum (Gmel.) Dandy
Leersia hexandra Swartz
Loudetia phragmitoides (Peter) C. E. Hubbard
Oryza longistaminata Chev. & Roehr.
O. sativa L.
Odyssea jaegeri (Pilg.) Robyns & Tournay
Panicum graniflorum Stapf
P. maximum Jacq.
P. repens L.
Paspalidium geminatum (Forsk.) Stapf
Pennisetum polystachyon (L.) Schult.
Phragmites mauritianus Kunth
Phyllorhachis sagittata Trimen
Pogonarthria squarrosa (Licht.) Pilg.
Rhynchelytrum repens (Willd.) C. E. Hubbard
Sacciolepis africana Hubbard & Snowden
S. gracilis Stent & Rattray
S. huillensis (Rendle) Stapf
Setaria palustris Stapf
S. sphacelata (Schumach.) M. B. Moss
Sorghastrum rigidifolium Stapf
Sorghum versicolor Anderss.
S. verticilliflorum (Steud.) Stapf
Sporobolus pyramidalis Beauv.
Stereochlaena cameronii (Stapf) Pilg.
Tricholaena monachne (Trin.) Stapf & C. E. Hubbard
Vossia cuspidata Griff

Hydrocharitaceae
Ottelia exserta (Ridl.) Dandy
O. ulvifolia (Planch.) Walp.
Lemnaceae
Lemna perpusilla Torr.
Spirodelia polyrhiza (L.) Schleid.
Pseudowolffia hyalina (Del.) den Hartog
& van der Plas
Wolffiopsis welwitschii (Hegelm) den Hartog & van der Plas
Liliaceae
Anthericum sp.
Asparagus racemosus Willd.
Chlorophytum brachystachum Bak.
Taccaceae
Tacca leontopetaloides (L.) Kuntze
Typhaceae
Typha domingensis Pers.
Dicotyledonae
Acanthaceae
Asystasia gangetica (L.) T. Anders
Barleria spinulosa Klotzsch
Blepharis maderaspatensis (L.) Roth
Blepharis sp.
Hygrophila auriculata (Schumach.) Heine
Isoglossa floribunda C.B.Cl.
Ruspolia seticalyx (C.B.Cl) Milne-Redh.
Amaranthaceae
Achyranthes aspera L.
Alternanthera sessilis (L.) DC.
Celosia trigyna L.
Gomphrena celosioides Mart.
Pupalia lappacea (L.) Juss.
Anacardiaceae
Lannea antiscorbutica (Hiern) Engl.
Annonaceae
Annona senegalensis Pers.
Artabotrys brachypetalus Benth
Apocynaceae
Tabernaemontana elegans Stapf
Ancylobothrys petersiana (Klotzsch) Pierre
Strophanthus petersianus Klotzsch
Asclepiadaceae
Cryptolepis obtusa N. E. Br.
Stathmostelmma fornicatum (N. E. Br.) Bullock
Bignoniaceae
Kigelia africana (Lam.) Benth.
Markhamia obtusifolia (Bak.) Sprague
Stereospermum kunthianum Cham.
Bombacaceae
Adansonia digitata L.
Boraginaceae
Trichodesma physaloides (Fenzl) DC.
T. zeylanicum (Burm. f.) R. Br.
Burseraceae
Commiphora africana (A. Rich.) Engl.
C. zanzibarica (Baill.) Engl.

Campanulaceae
Lightfootia abyssinica A. Rich.
Capparaceae
Boscia angustifolia A. Rich. var. *corymbosa* (Gilg) De Wolf
Cadaba kirkii Oliv.
Capparis erythrocarpos Isert var. *rosea* (Klotzsch) De Wolf
Cladostemon kirkii (Oliv.) Pax & Gilg
Cleome gynandra Briq.
Maerua triphylla A. Rich. var. *pubescens* (Klotzsch) De Wolf
Celastraceae
Maytenus senegalensis (Lam.) Exell
Ceratophyllaceae
Ceratophyllum demersum L.
Combretaceae
Combretum imberbe Wawra
C. microphyllum Klotzsch
Terminalia sericea DC.
Compositae
Ageratum conyzoides L.
Bidens pilosa L.
Blumea aurita (L. f.) DC.
Chrysanthellum americanum (L.) Vatke
Crassocephalum rubens (Jacq.) S. Moore
Eclipta prostrata (L.) L.
Emilia sonchifolia DC.
Helichrysum sp. cf. *H. petiolatum* (L.) Dec.
Melanthera scandens (Schumach. & Thonn.) Roberty
Nidorella resedifolia DC.
Pluchea dioscoridis (L.) DC.
Tridax procumbens L.
Vernonia amygdalina Del.
V. cinerea (L.) Less.
V. petersii Oliv.
V. poskeana Hildebr. & Vatke
V. kirkii Oliv. & Hiern
Convolvulaceae
Evolvulus alsinoides (L.) L.
Ipomoea aquatica Forsk.
I. coptica (L.) Roth
I. rubens Choisy
Merremia pinnata (Choisy) Hall f.
Cucurbitaceae
Corallocarpus bainesii (Hook f.) Meeuse
Cyclantheropsis parviflora (Cogn.) Harms
Ebenaceae
Diospyros squarrosa Klotzsch
Euphorbiaceae
Acalypha ornata A. Rich.
Alchornea laxiflora (Benth.) Pax & K. Hoffm.
Bridelia cathartica Bertol. f. subsp. *melanthesoides* (Klotzsch) J. Léonard
Euphorbia ingens Boiss.
Synadenium cupulare (Boiss.) L. C. Wheeler

435

Flacourtiaceae
Flacourtia indica (Burm. f.) Merr.
Labiatae
Englerastrum schweinfurthii Briq.
Hemizygia bracteosa (Benth.) Briq.
Hyptis spicigera Lam.
Leonotis nepetifolia (L.) Ait. f.
Leguminosae
Mimosoideae
Acacia albida Del.
Acacia sp. cf. *karroo* Hayne
A. schweinfurthii Brenan & Exell
A. sieberana DC.
A. xanthophloea Benth.
Albizia tanganyicensis Bak. f.
A. versicolor Oliv.
A. zimmermannii Harms
Entada abyssinica A. Rich.
Parkia filicoidea Oliv.
Xylia torreana Brenan
Caesalpinioideae
Bauhinia petersiana Bolle
Burkea africana Hook.
Cassia mimosoides L.
C. petersiana Bolle
C. siamea Lam.
Papilionoideae
Abrus precatorius L.
Aeschynomene nilotica Taub.
A. pfundii Taub.
A. schimperi A. Rich.
Crotalaria ochroleuca Don
C. polysperma Kotschy
Dalbergia arbutifolia Bak.
Indigofera secundiflora Poir. var. *rubripilosa* De Wild.
I. spicata Forsk.
I. trita L. f. var. *subulata* (Poir.) Ali
Indigofera sp.
Mucuna pruriens (L.) DC.
Rhynchosia sublobata (Schumach.) Meikle
Sesbania goetzei Harms subsp. *goetzei*
S. rostrata Bremek. & Oberm.
S. sesban (L.) Merr.
Tephrosia rhodesica Bak. f. var. *polystachyoides* (Bak. f.) Brummitt
Lentibulariaceae
Utricularia gibba L.
U. inflexa Forsk. var. *inflexa* P. Tayl.
U. inflexa Forsk. var. *stellaris* (L. f.) P. Tayl.
U. reflexa Oliv.
U. vulgaris L.
Utricularia spp.
Lythraceae
Ammannia prieuriana Guill. & Perr.
Nesaea erecta Guill. & Perr.
N. radicans Guill. & Perr.

Malvaceae
Hibiscus cannabinus L.
H. diversifolius Jacq. subsp. *rivularis* (Bremek. & Oberm.) Exell.
Sida acuta Burm. f.
Melastomataceae
Antherotoma naudinii Hook. f.
Dissotis debilis (Sond.) Triana
D. phaeotricha (Hochst) Hook. f.
Menyanthaceae
Nymphoides indica (L.) Kuntze
Menispermaceae
Cissampelos mucronata A. Rich.
Cocculus hirsutus (L.) Diels
Tiliacora funifera (Miers) Oliv.
Moraceae
Cardiogyne africana Bureau
Ficus burkei (Miq.) Miq.
F. soldanella Warb.
F. zambesiaca Hutch.
F. sycomorus L.
Ficus sp.
Nyctaginaceae
Commicarpus plumbagineus (Cav.) Standl.
Nymphaeceae
Nymphaea caerulea Savigny
N. lotus L.
Oleaceae
Jasminum fluminense Vell.
Onagraceae
Ludwigia leptocarpa (Nutt.) Hara
L. octovalvis (Jacq.) Raven
L. perennis (L.) Brenan
L. stolonifera (Guill. & Perr.) Raven
Polygalaceae
Polygala albida Schinz var. *angustifolia* (Chod.) Exell.
P. petitiana A. Rich.
Polygonaceae
Polygonum limbatum Meisn.
P. senegalense Meisn.
Rubiaceae
Borreria subvulgata K. Schum.
Kohautia longifolia Klozsch
Oldenlandia lancifolia (K. Schum.) DC.
Paederia foetens (Hiern.) K. Schum.
Pavetta assimilis Sond.
Sapindaceae
Deinbollia xanthocarpa (Klotzsch) Radlk.
Haplocoelum foliolosum (Hiern) Bullock
Scrophulariaceae
Limnophila indica (L.) Druce
Rhamphicarpa tubulosa Benth.
Rhamphicarpa sp.
Simaroubaceae
Kirkia acuminata Oliv.
Sterculiaceae
Dombeya kirkii Mast.

Tiliaceae
  Corchorus aestuans L.
Turneraceae
  Wormskioldia longepedunculata Mast.
Urticaceae
  Laportea aestuans (L.) Chew
Verbenaceae
  Lippia javanica (Burm. f.) Spreng.

Vitaceae
  Cayratia gracilis (Guill. & Perr.) Suesseng.
  Cissus cucumerifolia Planch.
  C. grisea (Bak.) Planch.
  Cyphostemma gigantophyllum (Gilg. & Brandt) Wild & Drummond

**Taken from the checklist (Howard-Williams 1977), where specimen citations are given. Taxonomy follows Flora Zambesiaca; and Professor H. Wild, Mr. R. B. Drummond and Mrs. E Gibbs-Russell gave invaluable assistance in checking and updating nomenclature. They are not responsible for any errors which may occur in the list. Only those plants which have been recorded within the area bounded by the lowest beach terrace are presented, and plants on the islands are included. The upper shores of Nchisi Island have not been collected fully and this area is a potential source of additional material.

Part 2. Animals

This is a preliminary list of animals encountered by authors except for those groups where provisional identification is specially acknowledged below. The terrestrial fauna is certainly under-collected.

1. PLATYHELMINTHES
DIGENEA
Fasciola gigantica (Cobbold)
Schistosoma haematobium Bilharz
S. mansoni Samson
S. mattheei Vegelia & Le Roux
MONOGENEA
Cleiodiscus halli Price
CESTODA
A species in *Barbus paludinosus* Peters

2. ASCHELMINTHES
NEMATODA
Contracaecum sp. A in *Sarotherodon shiranus chilwae* Trewavas
Contracaecum sp. B in *Clarias gariepinus* Peters
Necator americanus Stiles
ROTIFERA
Monogononta
Brachionus calyciflorus Pallas
Keratella tropica (Apstein)
Filina (Tetramastix) opoliensis Zacharias
Filina spp.
Testudinella patina (Hermann)
Asplanchna brightwelli Gosse

3. ANNELIDA
OLIGOCHAETA
several species
HIRUDINEA
Gnathobellida
Hirudidae
Limnatis africana Blanchard

4. MOLLUSCA
GASTROPODA
Prosobranchia
Ampullariidae
Lanistes (Meladomus) ovum Troschel
L. sinistrosus Lea
Viviparidae
Belamya unicolor Olivier
Pulmonata
Planorbidae
Biomphalaria pfeifferi Krauss
Bulinus (Physopsis) globosus Morelet
Lymnaeidae
Lymnaea sp.
BIVALVIA– Lamellibranchiata
Mutelidae
Aspatharia (Spathopsis) wahlbergi Krs.

5. ARTHROPODA
CRUSTACEA
Cladocera
Sidoidea
Diaphanosoma excisum Sars
Daphnoidea
Daphnia barbata (Weltner)
Ceriodaphnia cornuta Sars
Moina micrura de Guerne & Richard
Alona spp.
Copepoda
Cyclopoida
Mesocyclops leukarti (Claus)
Calanoida
Tropodiaptomus kraepelini (= *cunningtoni*) Poppe & Mrazek
Ostracoda
Heterocypris giesbrechti Muller

6. ARTHROPODA
INSECTA
Ephemeroptera
Baetidae
Baetis sp.
Cloeon spp.
Polymitarcyidae
Povilla adusta Navas
*Odonata
Zygoptera
Coenagrionidae
Agriocnemis exilis Sélys
Ceriagrion glabrum (Burm.)
Ischnura senegalensis (Ramb.)
Pseudagrion cf. *helenae* Balinsky
Anisoptera
Aeshnidae
Hemianax ephippiger (Burm.)
Anax imperator Leach
Libellulidae
Orthetrum brachiale (Beauv.)
O. trinacria (Sélys)
Hemistigma albipuncta (Ramb.)
Acisoma panorpoides (Ramb.)
Diplacodes lefebvrei (Ramb.)
Brachythemis leucosticta (Burm.)
Crocothemis erythraea (Brullé)
Rhyothemis semihyalina (Desj.)
Pantala flavescens (F.)
Tramea basilaris (Beauv.)
Urothemis edwardsi (Sélys)
Corduliidae
Macromia sp.

*Compiled by Dr. M. E. Parr, Salford University, Manchester from collections 1975–77.

**Isoptera
  Kalotermitidae
    Neotermes nr. *meruensis* (Hgrn.)
    N. nr. *zuluensis* (Hgrn.)
    Cryptotermes havilandi (Sjöst.)
  Hodotermitidae
    Hodotermes mossambicus (Hag.)
  Rhinotermitidae
    Schedorhinotermes nr. *lamianus* (Sjöst.)
  Termitidae
    Ancistrotermes nr. *latinotus* (Hgrn.)
    Microtermes nr. *parvus* (Hav.)
    Odontotermes nr. *latericius* (Hav.)
    Trinervitermes nr. *bettonianus* (Sjöst.)
**From Sands, W. A. & Wilkinson, W. 1954.
Colonial Termite Research Unit. E.A.A.F.R.
Mugaga, Kenya. 144 pp.

***Orthoptera
  Acrididae
  Hemiacridinae
    Acanthoxia gladiator (Westw.)
    Leptacris monteiroi (I. Bol)
    Afroxyrrhepes procera (Burm.)
    Petamella prosternalis (Karny)
    Oxya hyla Serv.
    Tylotropidius gracilipes Brancsik
  Catantopinae
    Catantops axillaris (Thunb.)
    C. fasciatus Karny
    C. spissus adjustus (Walk.)
  Cyrtacanthacridinae
    Nomadacris septemfasciata (Serv.)
    Cyrtacanthacris tatarica (L.)
    Ornithacris cyanea (Stoll)
  Acridinae
    Acrida sulphuripennis (Gerst.)
    Duronia chloronota (Stål)
    Coryphosima stenoptera (Schaum)
    Coryphosima vicina (Dirsh)
    Paracinema tricolor (Thunb.)
    Aiolopus thalassinus (F.)
    Gastrimargus clepsydrae Sjöst.
    Locusta migratoria migratorioides (R. &
      F.)
    Morphacris fasciata (Thunb.)
    Trilophidia conturbata (Walk.)
    Acrotylus furcifer Sauss.
  Gomphocerinae
    Mesopsis laticornis (Krauss)
    Platypternodes brevipes (Stål)
    Pnorisa squalus Stål
    Dnopherula werneriana (Karny)
***Compiled from the S. African National Col-
lection of insects by Dr. H. D. Brown and from
Mr. J. A. Whellan's (1975) checklist for
Malawi.

**Hemiptera
  Heteroptera

  Pentatomidae
    Nezara viridula (L.)
  Notonectidae
    Notonecta sp.
  Naucoridae
    several species
  Belostomatidae
    Sphaerodema nepodes F.
  Corixidae
    Micronecta scutellaris Stål.
**Coleoptera
  Halipidae
    several species
  Dytiscidae
    Canthydrus notula (Erichson)
    Synchortus simplex Sharp
    Hydrovatus sp.
    Hydrocanthus sp.
  Hydrophilidae
    Berosus furcatus Boh.
    B. vitticollis Boh.
  Scarabaeidae
  Scarabinae
    Sisyphus sp.
  Coprinae
    Euoniticellus intermedius (Reiche)
    Liatongus militaris (Castel.)
    Oniticellus planatus Castel.
    Copris elphenor Klug
    C. (?) obesus Boh.
    Onitus alexis Klug
    O. uncinatus Klug
    Phalops smaragdinus Harold
    Onthophagus depressus Harold
    O. emeritus Péringuey
    O. gazella (F.)
    O. lamelliger Gerst.
    O. quadrituber D'Orb.
    O. vinctus Erich
    O. spp. (3)
  Chrysomelidae
    Trichispa sericea (Guerin)
  Curculionidae
    Sitophilus zeamais Motsch
  Meloidae
    Epicauta velata Gerst
**Diptera
  Tipulidae
    several species
  Culicidae
    Anopheles funestus Giles
    A. gambiae Giles
    Aedes aegypti L.
    Culex pipiens L.
  Chaoboridae
    species 1
  Chironomidae
    Chironomus calipterus Kieffer
    C. transvaalensis Kieffer

439

C. formosipennis Kieffer
Cryptochironomus inflexus Freeman
C. forcipatus Freeman
C. neonilicola Freeman
C. acutus Goetghebuer
C. diceras Kieffer
C. stylifer Freeman
Dicrotendipes schoutedeni Goetghebuer
D. pilosimanus 14-punctatus Goetghebuer
D. fusconotatus Kieffer
Nilodorum brevibucca Kieffer
N. brevipalpus Kieffer
Pentapedilum vittatum Freeman
Tanytarsus (Cladotanytarsus) pseudomancus Goet.
Pentaneura (Ablabesmyia) nilotica Kieffer
Tanytarsus pallidulus Freeman
T. (Rheotanytarsus) fuscus Freeman
Clinotanypus claripennis Kieffer
Clinotanypus sp.
Ceratopogonidae
  Species
Diopsidae
  Diasemopsis fasciata (Gray)
  D. hirta Lindner
  Diopsis sulcifrons Bezzi
  D. servillei Macq.
  D. gnu Hendel
  D. macrophthalma Dal.
  D. apicalis Dal.
  Sphyracephala beccarii (Rond.)
Trichoptera
  Two species
Lepidoptera
  Gelechiidae
    Brachmia sp.
    Sitotroga cerealella Oliv.
  Pyralidae
    Chilo partellus (Swinh.)
    C. diffusilineus (de Johannis)
    Maliarpha septaratella Rag.
    Nymphula depunctalis (Guen.)
    Schoenobius sp.
    Thopeutis spp.
  Hesperidae
    Borbo borbonica (Boisd.)
  Arctiidae
    Diacrisia scortilla (Wallengren)
  Noctuidae
    Grammodes geometrica F.
    Mythimna sp.
    Sesamia calamistes (Hamps.)
    Spodoptera cilium Guen.
Hymenoptera
  Trichogrammatidae
    Trichogramma kalkae SchultenFeijen
    T. pinneyi SchultenFeijen
  *Scoliidae
    Crioscolia punctum (de Sauss.)

Micromeriella hyalina insuperata forma *antennata* (Klug)
M. hyalina insuperata forma *longinerva* (Cameron)
M. hyalina nr. *meruensis* (Cameron)
M. aureola bobi Betrem
Megameris soleata soleata (Gerst.)
Cathimeris hymenaea bradleyana Betrem
C. lachesis lachesis (de Sauss.)
Campsomeriella caelebs (Sichel)
C. caelebs forma *flavata* Betrem
C. rubromarginata Betrem
C. madonensis (Buys.)
C. madonensis madonensis (Buys.)
C. madonensis zambiensis Betrem
C. madonensis transvaalensis Cameron
Trielis nyasensis Betrem

*From Schulten, G. G. M. 1975. Bull. Zool. Mus 4:59–68. Univ. Amsterdam, and 1977. 6:59–68.

  **Megachilidae
    Creightonella discolor (Smith)
    C. erythrura Pasteels
    C. rufoscopacaea (Friese)
    C. aculeata (Vachal)
    Megachile eurymera (Smith)
    M. nigrimanus Schulten
    M. seclusa Ckll.
    M. bucephala (F.)
    M. feijeni Schulten
    M. bituberculata Rits.
    M. nasalis Smith
    M. gratiosa Gerst.
    M. semiflava Ckll.
    M. rufohirtula Ckll.
    M. astridella Ckll.
    M. semivenusta Ckll.
    Chalicodoma sinuata sinuata (Friese)
    C. kigonserana (Friese)
    C. torrida pachingeri (Friese)
    C. chrysorrhea (Gerst.)
    C. bombiformis (Gerst.)
    C. cincta nigrocincta (Ritsema)
    C. felina felina (Gerst.)
  Xylocopidae
    Xylocopa anicula Vachal
**From Schulten, G. G. M. 1977 Beaufortia 26(331) 13–76.

7. CHORDATA

PISCES

Osteichthyes – Teleostei
Mormyridae
  Marcusenius macrolepidotus (Peters)   +
  M. livingstioni (Boulenger)   0
  Mormyrus longirostris Peters   −
  Petrocephalus catostoma (Gunther)   +
Characidae
  Alestes imberi Peters   +
Characidae

Hemigrammopetersius barnardi
(Herre) +
Cyprinidae
Barbus cf. *afrohamiltoni* Crass +
B. atkinsoni Bailey +
B. kerstenii Peters +
B. paludinosus Peters +
B. radiatus Peters +
B. toppini Boulenger +
B. trimaculatus Peters +
B. cf. *viviparus* Weber +
Barbus sp. A 0
Barbus sp. B +
Barbus sp. C +
Engraulicypris brevianalis
(Boulenger) −
Labeo cylindricus Peters +
Bagridae
Bagrus orientalis Boulenger −
Leptoglanis rotondiceps Hilgendorf +
Clariidae
Clarias gariepinus (Burchell) +
C. theodorae Weber +
Schilbeidae
Pareutropius longifilis (Steindachner) +
Mochokidae
Chiloglanis neumanni Boulenger +
Amphiliidae
Amphilius platychir Gunther +
Cyprinodontidae
Aplocheilichthys johnstonii (Gunther) +
Nothobranchius kirki Jubb +
Cichlidae
Haplochromis callipterus (Gunther) +
Haplochromis sp. +
Pseudocrenilabrus philander (Weber) +
Sarotherodon shiranus chilwae (Trewavas) 0
S. shiranus shiranus (Boulenger) −
Tilapia rendalli (Boulenger) +
Gobiidae
Glossogobius giuris (Hamilton-Buchanan)−

Symbols: + indicates presence of the species in both Lake Chilwa and Lake Chiuta basins; — indicates absence in Lake Chilwa basin and presence in Lake Chiuta basin; 0 indicates presence in Lake Chilwa basin and probably absence in Lake Chiuta basin. Tweddle & Willoughby, 1979a.

*AMPHIBIA
ANURA
Pipidae
Xenopus muelleri (Peters)
Bufonidae
Bufo regularis Reuss
Microhylidae
Breviceps poweri Parker
Ranidae
Afrixalus f. fornasini (Bianconi)
Chiromantis xeramphelina Peters

Hyperolius parallelus albofasciatus Hoffman
Kassina senegalensis (Dumeril and Bibron)
Leptopelis bocagei (Gunther)
Phrynobatrachus acridoides (Cope)
P. mascareniensis (Dumeril and Bibron)
P. mossambica (Peters)
Pyxicephalus adspersus Tschudi
Rana galamensis bravana (Peters)
*After Stevens 1974 and Stewart 1967

**REPTILIA
TESTUDINES
Pelomedusidae
Pelusios sinuatus (A. Smith)
P. subniger (Lacepede)
Trionychidae
Cycloderma frenatum Peters
CROCODYLIA
Crocodylidae
Crocodylus niloticus Laurenti
SQUAMATA: Sauria
Geckonidae
Hemidactylus mabonia (Jonnes)
Chamaeleonidae
Chamaeleo d. dilepis Leach
Scincidae
Riopa afer (Peters)
Lacertidae
Latastia johnstoni Boulenger
Varanidae
Varanus n. niloticus (L.)
Serpentes
Boidae
Python sebae (Gmelin)
Colubridae
Dasypeltis s. scabra (L.)
Dispholidus typus (A. Smith)
Natriciteres olivacea (Peters)
Philothemnus hoplogaster (Gunther)
Psammophis s. sibilans (L.)
Psammophylax tritaeniatus (Gunther)
Pseudaspis cane (L.)
Viperidae
Atheris superciliaris (Peters)
Bitis a. arietans (Merrem)
**After Stevens 1974
***AVES
Podicipidae
Tachybaptus ruficollis (Pallas)
Pelecanidae
Pelecanus onocrotalus L.
P. rufescens L.
Phalacrocoracidae
Phalacrocorax carbo L.
P. africanus (Gmelin)
Anhingidae
Anhinga rufa (Lacépède et Daudin)

441

Ardeidae
  Botaurus stellaris L.
  *Ixobrychus minutus L.
  *I. sturmii (Wagler)
  Nycticorax nycticorax L.
  Ardeola ralloides (Scopoli)
  A. ibis L.
  *Butorides rufiventris (Sundevall)
  Melenophoyx ardesiaca (Wagler)
  Mesophoyx intermedius (Wagler)
  Egretta garzetta L.
  E. alba L.
  Ardea cinerea L.
  A. melanocephala (Vigors & Children)
  *A. goliath (Cretzschmar)
  A. purpurea L.
Scopidae
  Scopus umbretta (Gmelin)
Ciconiidae
  *Ciconia ciconia L.
  *C. episcopus (Boddaert)
  Ephippiorhynchus senegalensis (Shaw)
  Anastomus lamelligerus (Temminck)
  *Leptoptilos crumeniferus (Lesson)
  Ibis ibis L.
Threskiornithidae
  Threskiornis aethiopica (Latham)
  Plegadis falcinellus L.
Plataleidae
  Platalea alba (Scopoli)
Phoenicopteridae
  Phoenicopterus ruber L.
  *P. minor (Geoffroy)
Anatidae
  Dendrocygna bicolor (Viellot)
  D. viduata L.
  *Alopochen aegyptiacus L.
  Plectropterus gambensis L.
  Nettapus auritus (Boddaert)
  *Anas undulata (Dubois)
  A. erythrorhynchos (Gmelin)
  A. hottentota (Eyton)
  *A. querquedula L.
  Netta erythophthalma (Weid)
  Thalassornis leuconotus (Eyton)
  Sarkidiornis melanotus (Pennant)
Sagittariidae
  *Sagittarius serpentarius (Miller)
Aquilidae
  *Gyps africanus (Salvadori)
  *Gypohierax angolensis (Gmelin)
  *Circus macrourus (Gmelin)
  C. aeruginosus ranivorus (Daudin)
  *Polyboroides radiatus (A. Smith)
  Terathopius ecaudatus (Daudin)
  Circaëtus pectoralis (A. Smith)
  *C. cinerascens (Müller)
  *Accipiter badius (Gmelin)
  *Melierax metabates (Heuglin)

  *Kaupifalco monogrammicus (Temminck)
  *Buteo buteo L.
  *Lophaëtus occipitalis (Daudin)
  *Polemaëtus bellicosus (Daudin)
  *Hieraaëtus spilogaster (Bonaparte)
  *H. pennatus (Gmelin)
  Aquila wahlbergi (Sundevall)
  Haliaëtus vocifer (Daudin)
  Milvus aegyptius (Gmelin)
  *Pernis apivorus L.
  *Elanus caeruleus (Desfontaines)
Falconidae
  Falco biarmicus (Temminck)
  *F. chicquera (Daudin)
  *F. dickinsoni (P. L. Seater)
  F. amurensis (Roddle)
  F. naumanni (Fleischer)
  *F. tinnunculus L.
Phasianidae
  *Coturnix delegorguei (Delegorgue)
Gruidae
  Balearica regulorum (Bennet)
  *Grus carunculatus (Gmelin)
Rallidae
  Rallus caerulescens (Gmelin)
  *Porzana pusilla (Pallas)
  *P. marginalis (Hartlaub)
  Limnocorax flavirostris (Swainson)
  *Sarothrura sp.
  *Gallinula angulata (Sundevall)
  G. chloropus L.
  Porphyrio porphyrio L.
  P. alleni (Thompson)
  Fulica cristata (Gmelin)
Otidae
  Lissotis melanogaster (Rüppell)
Jacanidae
  Actophilornis africanus (Gmelin)
  Microparra capensis (A. Smith)
Rostratulidae
  Rostratula benghalensis L.
Charadriidae
  Vanellus armatus (Burchell)
  Hemiparra crassirostris (Hartlaub)
  *Vanellus senegallus L.
  Charadrius hiaticula L.
  C. tricollaris (Vieillot)
  C. pecuarius (Temminck)
Scolopacidae
  *Numenius phaeopus L.
  *N. arquatus L.
  Tringa nebularia (Gunnerus)
  T. stagnatilis (Bechstein)
  T. glareola L.
  T. ochropus L.
  Actitis hypoleucos L.
  *Xenus terek (Güldenstädt)
  Gallinago media (Latham)

G. *nigripennis* (Bonaparte)
Calidris ferruginea (Pontoppidan)
C. minuta (Leisler)
*C. alba (Pallas)
Philomachus pugnax L.
Recurvirostridae
  Himantopus himantopus L.
  Recurvirostra avosetta L.
Glareolidae
  *Cursorius temminckii (Swainson)
  Glareola pratincola L.
Laridae
  Larus cirrocephalus (Vieillot)
  *L. fuscus L.
Sternidae
  *Gelochelidon nilotica (Gmelin)
  Chlidonias hybrida (Pallas)
  C. leucoptera (Temminck)
Rhynchopidae
  Rhynchops flavirostris (Vieillot)
Cerculidae
  Centropus toulou grillii Hartlaub.
Alcedinidae
  Ceryle rudis L.
  Alcedo cristata (Pallas)
Tytonidae
  Tyto capensis (A. Smith)
Bubonidae
  *Asio capensis (A. Smith)
Caprimulgidae
  Caprimulgus fossii (Hartlaub)
Meropidae
  Merops superciliosus persicus L.
Alaudidae
  Calandrella cinerea (Gmelin)
  Eremopterix leucotis (Stanley)
Turdidae
  *Oenanthe pileata L.
Sylviidae
  Bradypterus baboeculus (Vieillot)

Acrocephalus schoenobaenus L.
A. arundinaceus L.
A. baeticatus (Vieillot)
A. gracilirostris (Hartlaub)
Cisticola galactotes (Temminck)
C. juncidis (Rafinesque)
Motacillidae
  Anthus novaeseelandiae (Gmelin)
  Macronyx croceus (Viellot)
Ploceidae
  Ploceus intermedius (Rüppell)
  P. velatus (Vieillot)
  Quelea erythrops (Hartlaub)
  Q. quelea L.
  Euplectes hordeaceus L.
  E. axillaris (A. Smith)
  Estrilda astrild L.

***Taken from The Check List of Birds at Lake Chilwa (Schulten & Harrison 1975). * indicates rare record.

*MAMMALIA
Artiodactyla
  Hippopotamidae
    Hippopotamus amphibius Desmoulins
  Bovidae
    Redunca arundinum (Boddaert)
Carnivora
  Mustelidae
    Aonyx capensis Schinz
  Viverridae
    Atilax paludinosus (Cuvier)
Rodentia
  Cricetidae
    Tatera leucogaster (Peters)
  Muridae
    Mastomys natalensis (Smith)
Primates
  Cercopithecus aethiops L.
*After Sweeney 1959.

INDEX

Scientific names of genera and species described in the text and Chichewa words are printed in *italics*; organisms in the checklist (Appendix A) are not all included here. Page numbers in **bold** type refer to diagrams and photographs.

Rufiji, 178;
Rukuru, 179;
Ruo, 21, 24, 179–182;
Shire, 5, 7, 19, 24, 120, 180, **181**;
Tuchila, 21, 24, 180–**181**;
Zambesi, 5, 182
rodents, as food, 263, 347, 385
r-orientation, 410
r-strategies, 410–411
rotation of crops and grazing, 330, 336
rotifers, 75, 101, 127–128, 132, 438

saline core of lake water, 71–**72**–73, 133, 154
salinity, 14, 62–64, **65**, **67**, **68**, 71–**72**, 75, **85**, **88**, **112**, 113, 117–**118**–**119**–120, 154, 163, 177, 254
salts
  extraction of, 305:
  phases of lake, 71–**72**;
  sources, compared with East African lakes, 39, 63
Salvinia hastata, 107, 120
sand, 21, 23, 25, 66, 82, 115, 236;
  for *Sarotherodon* breeding 191
sand bar, 7, 19, 30–**31**–33;
  barrier to fish migration, 178–179;
  closing of Chilwa basin, 19, 419;
  collecting points for fisheries, 357;
  see also railway line
Sarotherodon, 203, 441
  S. esculentens, 187;
  S. grabami, 198, 200
  S. mossambicus, 203;
  S. niloticus, 187, 198;
  S. placidus, 203
  S. rovumae, 203, 205;
  S. rukwaensis, 205;
  S. shiranus chilwae
    biology of, 75, 186–187, 191–192, 194, 201, 243;
    conservation, 228, 422–423;
    DDT residues, 338;
    endemism, 178, 182;
    fisheries, 183, 194, 216–221, 225, 348, 360;
    geographical distribution, 182–183;
    juveniles and zooplankton, 128, 130, 226;
    phases of the lake, 68, 75, 193–201;
    restocking the lake, 192, 422–423;
    speciation, 202–**204**–208, 411
  S. shiranus shiranus, 178, 203–**204**–205
satellite photograph of ecosystem, 5
  exposure of lake bed, 61;
  fish refuges, 419;
  trawling area, 421
Scarabaeidae
  systematic list, 439
  see dung beetles
Scenedesmus, 96, **97**, 101, 137, 433

Schistosoma, 438
  S. haematobium, 170–171, 173, 381;
  S. mansoni, 380;
  S. matheei, 384
schools in Chilwa area, 389–**391**–392, 420, 429
Scirpus, 434
  S. littoralis, 8, 108, 110;
  S. maritimus, 109
Secchi disc, 64, 65, **73**
sediments
  lake, 66, 68–69, 74;
  swamp, 82, 85, 113, 234, 236, 241
  see also decomposition, organic matter
seeds, growth of, **119**–120
selection pressure, 15, 135, 136, 404
settlement patterns, 314–**315**–**316**–317
Shire Highlands, 8, 21, **22**, 24, 25, 28, 34, 55, 311, 313, 320, 345, 358
shorelines, ancient, 19, 28–**29**–**30**–**31**–34
silicate, 76, 95
silt, 14, 66, 76, 82, 154–156, 236, 403, 411–412, 430
sing'anga (herbalist) 372
slave trade, 7, 9, 303, 304, 313
snails, 145, **148**–**149**, 166–168, 170–173
  systematic list, 438
social organisation, 10, 313–314, 326, 328
socio-economic studies, 14, 311–341, 345–368, 371–393, 421–434
sodium
  bicarbonate, **68**, 118, 198–199;
  ions, 62, 64, **68**, 107, **118**, 199–200, 405
soils
  alkaline, 110, 330;
  conservation, 413, 430;
  cracking, 115;
  crops and 21–**22**–23, 317–**318**
  insecticides absorbed by, 339;
  pastures, 263–264, 330–332, 424;
  release of nitrate, 89, 239
solar
  energy, 427;
  radiation, **37**–38
Southern African Anticyclone, 34
spawning migrations, 188–190, 235
Spirodela polyrhiza, 108, **119**, 435
Spirogyra spp. 96, 101, 433
Spirulina major, 96, **99**, 198, 433
spash zone and epiphytes, 74
Sporobolus pyramidalis, 109, 110, **283**, 434
stability of lake,
  short term, 139, 163–164
standards of living, 320–321
stem borers, 280
stock-keeping, 328–336, 424
Stone Age, 297, 300
stratification in swamp
  horizontal, 82–**83**–**84**;
  vertical, 82–**83**

455

INDEX TO AUTHORS

458

459

461

462